Introduction to IoT

Internet of Things, as a research field, piques the interest of a growing community of academics and scholars across the world. It has witnessed massive adoption and large-scale deployment in industries and other spheres of everyday life. Considerable theoretical information about IoT as well as tutorials, courses, implementations, use-cases, etc., is available across the web. However, all this available information is so scattered that even professionals in the field have trouble obtaining concise information on IoT and connecting the dots between the various technologies relating to it. This book serves as a textbook and also as a single point of reference for readers interested in the subject. Written by leading experts in the field, this lucid and comprehensive work provides a clear understanding of the operation and scope of IoT. It discusses the basics of networking, network security, precursor technologies of IoT and the emergence of IoT. It gives an overview of various connectivity, communication, and interoperability protocols prevalent in the field. While providing a dedicated overview and scope of implementation of various analytical methods used for IoT, the book discusses numerous case studies and provides hands-on IoT exercises, enabling readers to visualize the interdisciplinary nature of IoT applications and understand how they have managed to gain a foothold in the technology sector. The book also serves curious, non-technical readers, enabling them to understand necessary concepts and terminologies associated with IoT.

Sudip Misra is a professor in the Department of Computer Science and Engineering at the Indian Institute of Technology Kharagpur. He has published over 300 research papers and 11 books in allied areas of IoT. He is Associate Editor of *IEEE TMC, TVT* and *IEEE Network*. He is a Fellow of the National Academy of Sciences (India), IET (UK), and BCS (UK), and is a distinguished lecturer of the IEEE Communications Society. His current research interests include wireless sensor networks and IoT.

Anandarup Mukherjee is the co-founder and director of Sensor Drops Networks Pvt. Ltd., an IoT startup incubated at the Indian Institute of Technology Kharagpur. He has published more than 30 research papers. He works in the domain of IoT and unmanned aerial vehicle swarms.

Arijit Roy is the co-founder and director of Sensor Drops Networks Pvt. Ltd. He is a senior researcher in the Smart Wireless Applications and Networking (SWAN) lab at the Indian Institute of Technology Kharagpur. His primary research areas include IoT, sensor networks, and sensor-cloud.

Introduction to IoT

Sudip Misra
Anandarup Mukherjee
Arijit Roy

CAMBRIDGE
UNIVERSITY PRESS

University Printing House, Cambridge CB2 8BS, United Kingdom

One Liberty Plaza, 20th Floor, New York, NY 10006, USA

477 Williamstown Road, Port Melbourne, VIC 3207, Australia

314 to 321, 3rd Floor, Plot No.3, Splendor Forum, Jasola District Centre, New Delhi 110025, India

79 Anson Road, #06–04/06, Singapore 079906

Cambridge University Press is part of the University of Cambridge.

It furthers the University's mission by disseminating knowledge in the pursuit of education, learning and research at the highest international levels of excellence.

www.cambridge.org
Information on this title: www.cambridge.org/9781108842952

First published 2021

Printed in India by Thomson Press India Ltd.

A catalogue record for this publication is available from the British Library

Library of Congress Cataloging-in-Publication Data

Names: Misra, Sudip, author. | Mukherjee, Anandarup, author. | Roy, Arijit, author.
Title: Introduction to IoT / Sudip Misra, Anandarup Mukherjee, Arijit Roy.
Description: United Kingdom ; New York : Cambridge University Press, 2020. | Includes bibliographical references and index.
Identifiers: LCCN 2020037656 (print) | LCCN 2020037657 (ebook) | ISBN 9781108842952 (hardback) | ISBN 9781108959742 (paperback) | ISBN 9781108913560 (ebook)
Subjects: LCSH: Internet of things.
Classification: LCC TK5105.8857 .I567 2020 (print) | LCC TK5105.8857 (ebook) | DDC 004.67/8–dc23
LC record available at https://lccn.loc.gov/2020037656
LC ebook record available at https://lccn.loc.gov/2020037657

ISBN 978-1-108-84295-2 Hardback
ISBN 978-1-108-95974-2 Paperback

To Our Families

Contents

PART THREE: ASSOCIATED IOT TECHNOLOGIES

Figures

Tables

Tables

Foreword

The Internet of Things (IoT) paradigm has grown by leaps and bounds in the past decade. Nowadays, IoT is a common presence in households, transportation, markets, retail, banking, industries, education, and logistics. Yet regular innovative developments in IoT continue to flood the market. IoT has given rise to many interesting applications and resulted in the development of new networking and communication technologies that are designed specifically for IoT-oriented tasks. Manually intensive, yet crucial, domains of healthcare, agriculture, and transportation now rely heavily on IoT applications. The inclusion of IoT applications in these domains has resulted in facilitating their automation, enhanced safety, and precision of operations and allowed the inclusion of scientifically optimized practices. It is popularly considered that the rapid rise of IoT resulted from the inclusion of the beneficial features from the paradigms and technologies of Internet computing, cloud computing, wireless sensor networks (WSN), cyber-physical systems (CPS), and machine-to-machine (M2M) communications. All through its development, IoT has been supported by the popular networking paradigms in distributed systems, namely of cloud computing and edge and fog computing. These have allowed the massive, yet affordable, deployment of IoT across various domains. The emergence of the recent paradigm of edge computing can be directly attributed to IoT. IoT has also motivated the development of numerous connectivity and communication protocols and technologies such as IPv6, MQTT, 6LoWPAN, and LoRA, amongst others.

Unlike other books on IoT, already available on the market, this book provides detailed and interlinked coverage of topics related to IoT networks. The authors have designed this book carefully so that it acts as a guide and a single point of reference to IoT networks for beginners, as well as those already familiar with the technologies connected to IoT. With applications in the domains of agriculture, healthcare, electronics, power sector, industries, households, consumer electronics, computing, analytics, environment, transportation, logistics, security, military, surveillance, and many others, it is no wonder that the demand for deeper insights into IoT technologies is increasing day by day. The involvement of people from diverse backgrounds makes it necessary to create a concise repository of information on this new technology. The Internet hosts much information on IoT (theory, tutorials, courses, and implementations). However, they are so scattered that even professionals in this field have trouble obtaining connected and concise information.

This book has been purposefully designed by the authors as a textbook; it gradually exposes the readers to the technical details of IoT by first providing a primer on crucial networking technologies, which will help new readers in this domain to comfortably adopt and absorb the technical details of core IoT network technologies and its associated domains. Thereafter, the book gradually shifts focus to IoT networking technologies covering the emergence of IoT, addressing strategies in IoT, sensing and actuation, processing topologies, connectivity technologies, communication protocols, and interoperability. Later, this book expands its focus to cover the popular paradigms of cloud and fog computing, along with their applications in real life. The authors also provide real-life case studies on agriculture, healthcare, and vehicular IoT, aiming to get new learners motivated in the practical applicability of IoT in real-world applications. The authors have also provided chapters on IoT hardware projects and common analytical methods/tools used in IoT.

This book is divided into five parts: 1) Introduction, 2) Internet of Things, 3) Associated IoT Technologies, 4) IoT Case Studies, and 5) IoT Hands-on. I especially found the following features quite attractive in this book:

(a)   *Preliminary and background information* on networking technologies.

(b)   *Self-descriptive illustrations* help in visualizing complicated concepts.

(c)   *Exercises at the end of chapters* test the learner's understanding of contents.

(d)   *Conceptual questions* at the end of the book test the understanding of learners.

(e)   *Real-life use cases* of IoT motivates new learns to delve deeper into IoT.

(f)   A descriptive guide to building *IoT-based hardware projects*.

The authors of this book are globally acclaimed researchers, all of whom have published a number of research papers in this domain in highly impactful journals/magazines such as the *IEEE Communications Magazine, IEEE Transactions on Communications, IEEE Transactions on Mobile Computing, IEEE Transactions on Sustainable Computing, IEEE Transactions on Vehicular Technology, IEEE Internet of Things Journal, Elsevier Computer Networks, IEEE Systems Journal*, and many more. Sudip is well known in the community for his research achievements in the broad domain of Internet of Things. He has published more than a dozen books, which are published by globally renowned publishers such as Cambridge University Press, Wiley, Springer, World Scientific, and CRC Press. Due to his significant research contributions, his work has been recognized with different fellowships and awards such as the highly prestigious Abdul Kalam Technology Innovation National Fellow (India), the Fellow of the National Academy of Sciences (India), the Fellow of IETE (India), IET (U.K.), RSPH (U.K), IEEE Communications Society Outstanding Young Researcher Award, Humboldt Fellowship (Germany), Faculty Excellence Award (IIT Kharagpur), Canadian Governor General's Academic Medal, NASI Young Scientist Award, IEI Young Engineers Award, SSI Young Systems Scientist Award and so on. He serves as the Editor of *IEEE Transactions on Vehicular Technology*, and the Associate Editor of *IEEE Transactions on Mobile Computing*, *IEEE Transactions on Sustainable Computing*, *IEEE Systems Journal*, and *IEEE Networks*. He is also the IEEE Communications Society Distinguished Lecturer for the year 2020 through 2021. Besides Sudip, the other authors, Anandarup and Ariit, are distinguished senior researchers in Sudip's Smart Wireless Applications and Networking (SWAN) Lab at IIT Kharagpur. Both of them were awarded the Gandhian Young Innovation Award (GYTI) by the President of India in 2018 for their socially relevant innovation. They also serve as co-founders and directors of their entrepreneurial venture Sensor Drops Networks Pvt. Ltd.

On a concluding note, I expect this book to be quite useful to a diverse variety of readership. I am convinced that this book serves as a guide for the readers on their journey of exploring the amazing depth of concepts, technologies, and impacts of the Internet of Things.

Professor Schahram Dustdar
IEEE Fellow & ACM Distinguished Scientist
Vienna University of Technology (TU Vienna), Austria
Co-Editor-in-Chief, *ACM Transactions on IoT*

Preface

Overview and Goals

Internet of Things (IoT) is rapidly gaining a foothold in the technology sector; it has managed to emerge as a highly sought-after field of study and research in computing sciences and electronics. The vastly interdisciplinary nature of the areas to which IoT can be applied has managed to pique the interest of the whole world. IoT finds diverse use in domains spanning industrial, military, as well as regular consumer applications. The versatility of IoT and its ability to connect anything make it one of the most demanded technologies of the modern age. The involvement of people from vastly diverse and distinct backgrounds, all point to the need for a concise repository of information on this new technology.

The Internet hosts much information on IoT, which is in the form of theory, tutorials, courses, implementations, and others. However, these discussions are so scattered that even professionals in this field have trouble obtaining integrated and concise information on IoT.

IoT is a new paradigm for connecting "things" in order to automate a system. In the context of IoT, "things" include computers, cell phones, medical devices, vehicles, wearables, and other appliances and devices for daily use. These "things" tend to be heterogeneous, which results in the development and existence of a vast number of communication solutions and protocols, which vary distinctly from each other. Consequently, communications among these "things" is a challenging issue in IoT. Another major challenge in IoT is the dynamic nature of "mobile things," which generally follow a decentralized architecture. Due to this decentralized communication and control structure, the connectivity and data transmission dynamically changes with time, in turn resulting in a new set of challenges.

Pedagogical Aids

We have included various pedagogical aids to help the reader swiftly grasp the contents and the treatment of the various topics covered in this book. We have provided a set of conceptual questions at the end of this book. For solving these questions, the reader must have completely grasped the concepts covered in this book.

Additionally, we provide visual presentations of the chapters covered in this book so that it can be used as a teaching aid in colleges and universities. Each chapter of this book has the following pedagogical components:

- *Learning outcomes*, which gives an initial glimpse into the chapter.

- *Self-descriptive illustrations*, which are easy to understand.

- *Check yourself* exercises, which encourage readers and learners to explore additional topics on their own. This will gradually enable the readers to easily find terminologies, technologies, and concepts on their own as they start understanding the more complicated parts of this book.

- *Summary*, which provides a concise outcome for each chapter.

- *Exercises*, which allow the readers of this book to brush up their knowledge at any point in the future and also enable them to test their understanding of the concepts/technologies covered in this book.

Organization and Features

This book holistically covers the significant aspects of IoT in detail, including legacy and new technologies. These technological concepts form the core of IoT, the knowledge of which is indispensable for architecting an IoT-based solution. The entire book is written in a lucid and elementary manner, which provides readers with a clear understanding of the scope and operation of each topic. This book is an excellent guide for beginners, enabling them to start architecting IoT-based solutions confidently. Keeping the need of the beginners in mind, the authors have provided an overview of the background knowledge required for working with IoT. This background information is supplied in Part I, which includes topics covering basics of networking, basics of security, and predecessor technologies of IoT.

Part II of this book describes in detail the emergence of IoT and basic IoT enabling components, including sensors, connectivity protocols, communication protocols, and others. Each chapter in Part II describes and illustrates the various components of an IoT architecture: sensors, actuators, processing, connectivity, communication, and interoperability features. Physical quantifications of environmental effects and phenomena, which are carried out by sensors, constitute one of the essential components of IoT. These sensors can be heterogeneous and are classified according to their functionalities and usage as scalar, multimedia, hybrid, and virtual sensors, which are used to sense various parameters in an IoT architecture. The planned deployment of sensor nodes in an area of interest is one of the crucial issues in IoT, which aims to minimize the energy consumption of these devices and facilitate the quick processing of sensor data. Other tasks include the geographical representation of sensor placement and connectivity establishment among them, which is also referred to as the topology. Thus, this book sheds some light on the different sensor types and the working procedure, which are commonly used in an IoT architecture.

The sensed data needs to be stored and continually updated in memory locations for further processing. As per requirement, a user has the option of utilizing the sensed information as either structured or unstructured data. In IoT, the *where* and *how* of processing is another critical issue. Therefore, the correct processing technique needs to be selected accordingly. This book covers the processing techniques required for IoT. The transmission of this processed IoT data between devices is mostly dependent on various communication and connectivity protocols. The authors illustrate and discuss these connectivity protocols in detail, which is lucid to both experts as well as novice audiences. Some of these discussed protocols have been developed, primarily due to various developments in IoT and its architectures. After successful establishment of connectivity between two or more IoT nodes, the subsequent steps involve establishing communication between these nodes. These nodes may be tasked with either sensing, actuation, or both functions with specific requirements concerning packetization, addressing, reliability, security, and other such measures. The authors discuss these needs by describing a significant number of communication protocols in a separate chapter dedicated to these communication protocols. Finally, Part II discusses the various types of interoperability and their importance concerning IoT and its technologies. As IoT involves the inclusion of legacy, present, and upcoming technologies, the primary challenge lies in the integration of hardware and software platforms, which are provided by various original equipment manufacturers (OEMs) following their proprietary solutions and technologies. Consequently, the inclusion of these devices into a single platform is a non-trivial task. Therefore, the establishment of connectivity among these diverse devices in IoT becomes a challenging issue. The paradigm of device and software interoperability is introduced to the readers, which highlights the urgency of equipment, technology, communication, and protocol standardization in the context of IoT. The authors discuss the challenges of interoperability along with its other aspects resulting in the shaping of the readers' perspective; this would enable them to come up with IoT solutions and architectures, keeping the present and future interoperability challenges in mind.

In a traditional IoT architecture, there is the option to store the sensed data from heterogeneous sensor nodes in various locations such as locally on the devices, on a remote server, on a fog, or a cloud. Part III of this book primarily covers cloud, fog, and edge computing. An IoT architecture consists of a large variety of different, multipurpose, and multifunctional devices. The authors of this book ensure that each of these storage schemes is described individually. Cloud computing, yet another crucial technology, plays a vital role in handling the massive amounts of data generated by these IoT devices. A cloud enables the features of enormous storage capabilities, processing resources, and unification of data on a single platform. This book elucidates the idea of cloud computing in a simplified manner. A few topics of cloud computing, primarily focused on its applicability in the context of IoT, are discussed. Additionally, various service models are briefly discussed as they are used with traditional cloud computing. To mitigate the issues of latency in data processing in a cloud, which can be crucial for applications such as healthcare, the authors also include brief overviews of fog and edge computing. The authors cover the different aspects of fog computing, starting from fundamental issues to issues relating to its architecture along with suitable use-cases in this book.

After discussing the technical details of IoT and its backbone technologies, the authors highlight different application domains of IoT, such as agriculture, vehicular, and healthcare. These typical application domains of IoT are discussed in Part IV of this book as case-studies. Each chapter covers the application domains one-by-one employing real-life use-cases. These use-cases provide a clear idea about the real implementation of IoT in each application domain. This part is designed to enable both beginners, as well as advanced readers, to understand the needs, implementation, technologies, solutions, and implications of sustainable IoT architectures and solutions described in Parts I–III. Finally, the last chapter of this part concludes the various theoretical and practical aspects of IoT, its components, and architectures, with a discussion on newly emerging paradigms and various enabling ones.

Finally, Part V of this book is designed to further strengthen the readers' grasp on concepts through hands-on experience with IoT applications. The authors make use of commonly available hardware, including sensors, actuators, and processor boards, to demonstrate various sample integrations. These sample integrations are designed to enable readers to envision complicated device integrations and form complex IoT network architectures. Basic knowledge of Arduino and Python programming is required for pursuing this part. However, for readers with no prior experience in these languages, the hands-on section provides sample codes and explains the functionality of the integrations and its various aspects. Another chapter focuses on data informatics and analytics popularly used in IoT. The various types of analytics and terminologies associated with them are outlined; this chapter provides new readers and knowledge-seekers a starting point into IoT analytics.

All the chapters in this book are accompanied by a set of basic questions on the topics covered in that chapter. However, the last part of this book contains a consolidated set of conceptual questions, which will require the readers to access various sources, besides this book, to be able to address them. This exercise will train the readers to selectively choose from the vast amount of information available elsewhere, and sieve the ones most crucial to their current problem.

Target Audience

This book primarily targets the undergraduate and postgraduate technical readers who are either looking to delve into the growing domain of IoT or are taking IoT targeted introductory as well as advanced courses. This book is additionally designed to address the curious, non-technical reader, as well as working professionals from non-computer science and electronics domains, enabling them to pick up the necessary concepts and terminologies associated with IoT. This multifaceted book can also be used as a quick reference introducing the concepts and challenges in IoT research, which may be of significant use to working professionals, academicians as well as researchers across various industries.

Suggested Use of This Book

This book has been designed to be used both as a textbook as well as a reference book. Undergraduate students taking an introductory course on IoT will need to first go through the introductory part, which consists of the basics of networking, security, and similar technologies preceding IoT. The solving of the exercises at the end of each chapter and the conceptual questions at the end of this book would serve as a good indicator of the undergraduate reader's progress in grasping the concepts covered in this book. Working professionals in other domains and new learners (not familiar with networking and the nuances of computing sciences) can follow the same approach.

Postgraduate and research students in electronics, computer sciences, electrical engineering, and other similar domains can use this book directly from the second part, as they would already be aware of the introductory concepts. Working professionals in the allied domains of electronics and computer sciences can also follow the same protocol for this book as the postgraduate students. This book will also serve as a ready reference for the numerous topics included in it and which are commonly encountered in designing IoT-based solutions.

Finally, the curious reader who aims to work on IoT, without having any prior knowledge of this domain and computer sciences, should first peruse the fourth part of this book, which covers various real-life case studies of IoT in various domains. This will help the reader understand the usability of IoT in the context of the unique challenges faced for each domain. The reader can also gain additional exposure to IoT through the last part, which presents a hands-on approach to building IoT-based solutions through a set of very concise and interesting experiments. The experiments are aimed at both new as well as advanced learners.

Acknowledgment

The authors would like to thank Ms. Nidhi Pathak and Mr. Pallav Deb, who are doctoral research scholars at the Indian Institute of Technology Kharagpur, India. They have been instrumental in helping with various hardware and software implementations used in this book. Additionally, the authors would like to thank the various scholars/researchers in the Smart Wireless Applications and Networking (SWAN) Lab. for taking time out to go through this work and suggest improvements. The authors sincerely acknowledge the role of various, regular as well as online, students of Professor Misra's course *Architecting Protocols for the Internet of Things* and his popular MOOC *Introduction to IoT*, which is hosted online by NPTEL (India). Finally, the authors are grateful to their families for being understanding and patient.

PART ONE
INTRODUCTION

Chapter 1

Basics of Networking

Learning Outcomes

After reading this chapter, the reader will be able to:

- Understand the basic principles of computer networking
- List the basic terminologies and technologies
- Relate new concepts of IoT with the basics of networking
- Discuss various network configurations and topologies
- Explain various OSI (open systems interconnections) and TCP/IP (transmission control protocol/Internet protocol) layers and their associated uses
- Describe basics of network addressing

1.1 Introduction

In the present era of data- and information-centric operations, everything—right from agriculture to military operations—relies heavily on information. The quality of any particular information is as good as the variety and strength of the data that generates this information. Additionally, the speed at which data is updated to all members of a team (which may be a group of individuals, an organization, or a country) dictates the advantage that the team has over others in generating useful information from the gathered data. Considering the present-day global scale of operations of various organizations or militaries of various countries, the speed and nature of germane information are crucial for maintaining an edge over others in the same area. To sum it up, today's world relies heavily on data and networking, which allows for the instant availability of information from anywhere on the earth at any moment.

Typically, networking refers to the linking of computers and communication network devices (also referred to as hosts), which interconnect through a network

(Internet or Intranet) and are separated by unique device identifiers (Internet protocol, IP addresses and media access control, MAC addresses). These hosts may be connected by a single path or through multiple paths for sending and receiving data. The data transferred between the hosts may be text, images, or videos, which are typically in the form of binary bit streams [1].

Points to ponder

- The data generated from a camera sensor tells us more about a scene compared to the data generated from, say, a proximity sensor, which only detects the presence of people in its sensing range.

- Furthermore, the simultaneous data generated from multiple cameras focusing on the same spot from various angles tell us even more about the scene than a single camera focused at that scene.

As the primary aim of this chapter is to provide the reader with an overview of networking, we have structured the text in such a manner that the general concepts are covered. Additional *Check yourself* suggestions to review various associated technologies are provided along with the topics.

We start our discussion with the different types of networks, followed by an overview of two popularly used layered network models: ISO-OSI (the open systems interconnection developed by the International Organization of Standardization) and TCP/IP (transmission control protocol/Internet protocol) suite. Subsequently, we will touch upon the various types of addressing mechanisms and set up the basic premise of how a message is transmitted between two devices/computers/hosts.

1.2 Network Types

Computer networks are classified according to various parameters: 1) Type of connection, 2) physical topology, and 3) reach of the network. These classifications are helpful in deciding the requirements of a network setup and provide insights into the appropriate selection of a network type for the setup.

1.2.1 Connection types

Depending on the way a host communicates with other hosts, computer networks are of two types—(Figure 1.1): *Point-to-point* and *Point-to-multipoint*.

(i) **Point-to-point**: Point-to-point connections are used to establish direct connections between two hosts. Day-to-day systems such as a remote control for an air conditioner or television is a point to point connection, where the connection has the whole channel dedicated to it only. These networks were designed to work over duplex links and are functional for both synchronous as well as asynchronous systems. Regarding computer networks, point to point connections find usage for specific purposes such as in optical networks.

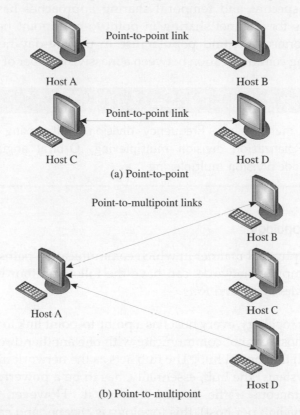

(a) Point-to-point

(b) Point-to-multipoint

Figure 1.1 Network types based on connection types

> **Point-to-point Requests for Comments (RFCs)**
>
> The following requests for comments (RFCs) are associated with point-to-point communication and its derivatives. **RFC 1332**: point-to-point (PPP) Internet protocol control protocol (IPCP); **RFC 1661**: PPP; **RFC 5072**: IP Version 6 over PPP; **RFC 2516**: PPP over Ethernet; **RFC 1963**: PPP serial data transport protocol; **RFC 1962**: PPP compression control protocol (CCP); **RFC 1990**: PPP multilink protocol (MP); **RFC 2615**: PPP over SONET/SDH (synchronous optical networking/synchronous digital hierarchy).

(ii) **Point-to-multipoint**: In a point-to-multipoint connection, more than two hosts share the same link. This type of configuration is similar to the one-to-many connection type. Point-to-multipoint connections find popular use in wireless networks and IP telephony. The channel is shared between the various hosts, either spatially or temporally. One common scheme of spatial sharing of the channel is frequency division multiple access (FDMA). Temporal sharing of channels include approaches such as time division multiple access (TDMA). Each of the spectral and temporal sharing approaches has various schemes and protocols for channel sharing in point-to-multipoint networks. Point-to-multipoint connections find popular use in present-day networks, especially while enabling communication between a massive number of connected devices.

Check yourself

Space division multiplexing, Frequency division multiplexing, Time division multiplexing, Polarization division multiplexing, Orbital angular momentum multiplexing, Code division multiplexing

---

1.2.2 Physical topology

Depending on the physical manner in which communication paths between the hosts are connected, computer networks can have the following four broad topologies—(Figure 1.2): *Star*, *Mesh*, *Bus*, and *Ring*.

(i) **Star**: In a star topology, every host has a point-to-point link to a central controller or hub. The hosts cannot communicate with one another directly; they can only do so through the central hub. The hub acts as the network traffic exchange. For large-scale systems, the hub, essentially, has to be a powerful server to handle all the simultaneous traffic flowing through it. However, as there are fewer links (only one link per host), this topology is cheaper and easier to set up. The main advantages of the star topology are easy installation and the ease of fault identification within the network. If the central hub remains uncompromised, link failures between a host and the hub do not have a big effect on the network, except for the host that is affected. However, the main disadvantage of this topology is the danger of a single point of failure. If the hub fails, the whole network fails.

(ii) **Mesh**: In a mesh topology, every host is connected to every other host using a dedicated link (in a point-to-point manner). This implies that for n hosts in a mesh, there are a total of $n(n-1)/2$ dedicated full duplex links between the hosts. This massive number of links makes the mesh topology expensive. However, it offers certain specific advantages over other topologies. The first significant advantage is the robustness and resilience of the system. Even if a link is down or broken, the network is still fully functional as there remain other pathways for the traffic to flow through. The second advantage is the security and privacy of the traffic as the data is only seen by the intended recipients and not by all members of the network. The third advantage is the reduced data load on a

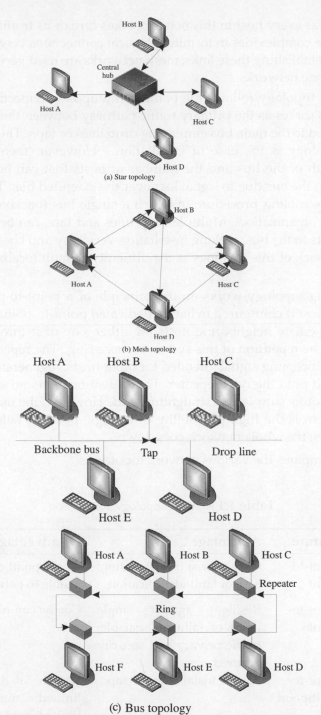

(a) Star topology

(b) Mesh topology

(c) Bus topology

Figure 1.2 Network types based on physical topologies

single host, as every host in this network takes care of its traffic load. However, owing to the complexities in forming physical connections between devices and the cost of establishing these links, mesh networks are used very selectively, such as in backbone networks.

(iii) **Bus**: A bus topology follows the point-to-multipoint connection. A backbone cable or bus serves as the primary traffic pathway between the hosts. The hosts are connected to the main bus employing drop lines or taps. The main advantage of this topology is the ease of installation. However, there is a restriction on the length of the bus and the number of hosts that can be simultaneously connected to the bus due to signal loss over the extended bus. The bus topology has a simple cabling procedure in which a single bus (backbone cable) can be used for an organization. Multiple drop lines and taps can be used to connect various hosts to the bus, making installation very easy and cheap. However, the main drawback of this topology is the difficulty in fault localization within the network.

(iv) **Ring**: A ring topology works on the principle of a point-to-point connection. Here, each host is configured to have a dedicated point-to-point connection with its two immediate neighboring hosts on either side of it through repeaters at each host. The repetition of this system forms a ring. The repeaters at each host capture the incoming signal intended for other hosts, regenerates the bit stream, and passes it onto the next repeater. Fault identification and set up of the ring topology is quite simple and straightforward. However, the main disadvantage of this system is the high probability of a single point of failure. If even one repeater fails, the whole network goes down.

Table 1.1 compares the various network topologies.

Table 1.1 Network topology comparison

Topology	Feature	Advantage	Disadvantage
Star	Point-to-point	Cheap; ease of installation; ease of fault identification	Single point of failure; traffic visible to network entities
Mesh	Point-to-point	Resilient against single point of failures; scalable; traffic privacy and security ensured	Costly; complex connections
Bus	Point-to-multipoint	Ease of installation; cheap	Length of backbone cable limited; number of hosts limited; hard to localize faults
Ring	Point-to-point	Ease of installation; cheap; ease of fault identification	Prone to single point of failure

1.2.3   Network reachability

Computer networks are divided into four broad categories based on network reachability: *personal area networks*, *local area networks*, *wide area networks*, and *metropolitan area networks*.

(i)   **Personal Area Networks (PAN)**: PANs, as the name suggests, are mostly restricted to individual usage. A good example of PANs may be connected wireless headphones, wireless speakers, laptops, smartphones, wireless keyboards, wireless mouse, and printers within a house. Generally, PANs are wireless networks, which make use of low-range and low-power technologies such as Bluetooth. The reachability of PANs lies in the range of a few centimeters to a few meters.

(ii)   **Local Area Networks (LAN)**: A LAN is a collection of hosts linked to a single network through wired or wireless connections. However, LANs are restricted to buildings, organizations, or campuses. Typically, a few leased lines connected to the Internet provide web access to the whole organization or a campus; the lines are further redistributed to multiple hosts within the LAN enabling hosts. The hosts are much more in number than the actual direct lines to the Internet to access the web from within the organization. This also allows the organization to define various access control policies for web access within its hierarchy. Typically, the present-day data access rates within the LANs range from 100 Mbps to 1000 Mbps, with very high fault-tolerance levels. Commonly used network components in a LAN are servers, hubs, routers, switches, terminals, and computers.

(iii)   **Metropolitan Area Networks (MAN)**: The reachability of a MAN lies between that of a LAN and a WAN. Typically, MANs connect various organizations or buildings within a given geographic location or city. An excellent example of a MAN is an Internet service provider (ISP) supplying Internet connectivity to various organizations within a city. As MANs are costly, they may not be owned by individuals or even single organizations. Typical networking devices/components in MANs are modems and cables. MANs tend to have moderate fault tolerance levels.

(iv)   **Wide Area Networks (WAN)**: WANs typically connect diverse geographic locations. However, they are restricted within the boundaries of a state or country. The data rate of WANs is in the order of a fraction of LAN's data rate. Typically, WANs connecting two LANs or MANs may use public switched telephone networks (PSTNs) or satellite-based links. Due to the long transmission ranges, WANs tend to have more errors and noise during transmission and are very costly to maintain. The fault tolerance of WANs are also generally low.

1.3   Layered Network Models

The intercommunication between hosts in any computer network, be it a large-scale or a small-scale one, is built upon the premise of various task-specific layers. Two of the most commonly accepted and used traditional layered network models are the open systems interconnection developed by the International Organization of Standardization (ISO-OSI) reference model and the Internet protocol suite.

1.3.1   OSI Model

The ISO-OSI model is a conceptual framework that partitions any networked communication device into seven layers of abstraction, each performing distinct tasks based on the underlying technology and internal structure of the hosts. These seven layers, from bottom-up, are as follows: 1) *Physical layer*, 2) *Data link layer*, 3) *Network layer*, 4) *Transport layer*, 5) *Session layer*, 6) *Presentation layer*, and 7) *Application layer*. The major highlights of each of these layers are explained in this section.

Points to ponder

The OSI or open system interconnect model for networked devices was standardized by the International Standards Organization (ISO). It is a conceptual framework that divides any networked communication system into seven layers, each performing specific tasks toward communicating with other systems [5], [1]. The OSI is a reference model and is maintained by the ISO under the identity of ISO/IEC 7498-1.

(i)   **Physical Layer**: This is a media layer and is also referred to as layer 1 of the OSI model. The physical layer is responsible for taking care of the electrical and mechanical operations of the host at the actual physical level. These operations include or deal with issues relating to signal generation, signal transfer, voltages, the layout of cables, physical port layout, line impedances, and signal loss. This layer is responsible for the topological layout of the network (star, mesh, bus, or ring), communication mode (simplex, duplex, full duplex), and bit rate control operations. The protocol data unit associated with this layer is referred to as a *symbol*.

(ii) **Data Link Layer**: This is a media layer and layer 2 of the OSI model. The data link layer is mainly concerned with the establishment and termination of the connection between two hosts, and the detection and correction of errors during communication between two or more connected hosts. IEEE 802 divides the OSI layer 2 further into two sub-layers [2]: Medium access control (MAC) and logical link control (LLC). MAC is responsible for access control and permissions for connecting networked devices; whereas LLC is mainly tasked with error checking, flow control, and frame synchronization. The protocol data unit associated with this layer is referred to as a *frame*.

(iii) **Network Layer**: This layer is a media layer and layer 3 of the OSI model. It provides a means of routing data to various hosts connected to different networks through logical paths called virtual circuits. These logical paths may pass through other intermediate hosts (nodes) before reaching the actual destination host. The primary tasks of this layer include addressing, sequencing of packets, congestion control, error handling, and Internetworking. The protocol data unit associated with this layer is referred to as a *packet*.

(iv) **Transport Layer**: This is layer 4 of the OSI model and is a host layer. The transport layer is tasked with end-to-end error recovery and flow control to achieve a transparent transfer of data between hosts. This layer is responsible for keeping track of acknowledgments during variable-length data transfer between hosts. In case of loss of data, or when no acknowledgment is received, the transport layer ensures that the particular erroneous data segment is re-sent to the receiving host. The protocol data unit associated with this layer is referred to as a *segment* or *datagram*.

(v) **Session Layer**: This is the OSI model's layer 5 and is a host layer. It is responsible for establishing, controlling, and terminating of communication between networked hosts. The session layer sees full utilization during operations such as remote procedure calls and remote sessions. The protocol data unit associated with this layer is referred to as *data*.

(vi) **Presentation Layer**: This layer is a host layer and layer 6 of the OSI model. It is mainly responsible for data format conversions and encryption tasks such that the syntactic compatibility of the data is maintained across the network, for which it is also referred to as the *syntax layer*. The protocol data unit associated with this layer is referred to as *data*.

(vii) **Application Layer**: This is layer 6 of the OSI model and is a host layer. It is directly accessible by an end-user through software APIs (application program interfaces) and terminals. Applications such as file transfers, FTP (file transfer protocol), e-mails, and other such operations are initiated from this layer. The application layer deals with user authentication, identification of communication hosts, quality of service, and privacy. The protocol data unit associated with this layer is referred to as *data*.

A networked communication between two hosts following the OSI model is shown in Figure 1.3. Table 1.2 summarizes the OSI layers and their features, where PDU stands for protocol data unit.

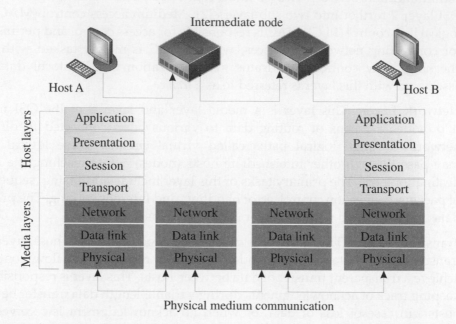

Figure 1.3 Networked communication between two hosts following the OSI model

---

Check yourself

Ethernet, FDDI, B8ZS, V.35, V.24, RJ45, PPP, FDDI, ATM, IEEE 802.5/ 802.2, IEEE 802.3/802.2, HDLC, Frame Relay, AppleTalk DDP, IP, IPX, NFS, NetBios names, RPC, SQL, ASCII, EBCDIC, TIFF, GIF, PICT, JPEG, MPEG, MIDI, NFS, SNMP, Telnet, HTTP, FTP

Table 1.2 Summary of the OSI layers and their features

Layer	Name	Location	PDU	Function	Examples
1	Physical	Media	Symbol	Communication over physical medium	Ethernet, FDDI, B8ZS, V.35, V.24, RJ45
2	Data link	Media	Frame	Reliability of communication over physical medium	IEEE 802.5 / 802.2, IEEE 802.3/802.2, PPP, HDLC, Frame Relay, ATM, FDDI
3	Network	Media	Packet	Structuring of data and routing between multiple nodes	DDP, IP, AppleTalk, IPX
4	Transport	Host	Segment	Reliability of communication over networks or between hosts	SPX, TCP, UDP
5	Session	Host	Data	Establishment, management, and termination of remote sessions	NetBios names, NFS, RPC, SQL
6	Presentation	Host	Data	Syntactic conversion of data and encryption	Encryption, ASCII, MIDI, PICT, JPEG, EBCDIC, TIFF, GIF, MPEG
7	Application	Host	Data	User identification, authentication, privacy, and quality of service	SNMP, Telnet, WWW browsers, HTTP, NFS, FTP

1.3.2   Internet protocol suite

The Internet protocol suite is yet another conceptual framework that provides levels of abstraction for ease of understanding and development of communication and networked systems on the Internet. However, the Internet protocol suite predates the OSI model and provides only four levels of abstraction: 1) Link layer, 2) Internet layer, 3) transport layer, and 4) application layer. This collection of protocols is commonly referred to as the TCP/IP protocol suite as the foundation technologies of this suite are transmission control protocol (TCP) and Internet protocol (IP) [3], [4], [6]. The TCP/IP protocol suite comprises the following four layers:

> **Points to ponder**
>
> The development of the TCP/IP protocol suite is originally attributed to DARPA, which is part of the United States Department of Defence. The Internet protocol suite or the TCP/IP protocol suite is sometimes also referred to as the Department of Defence (DoD) model.

(i)   **Link Layer**: The first and base layer of the TCP/IP protocol suite is also known as the network interface layer. This layer is synonymous with the collective physical and data link layer of the OSI model. It enables the transmission of TCP/IP packets over the physical medium. According to its design principles, the link layer is independent of the medium in use, frame format, and network access, enabling it to be used with a wide range of technologies such as the Ethernet, wireless LAN, and the asynchronous transfer mode (ATM).

(ii) **Internet Layer**: Layer 2 of the TCP/IP protocol suite is somewhat synonymous to the network layer of the OSI model. It is responsible for addressing, address translation, data packaging, data disassembly and assembly, routing, and packet delivery tracking operations. Some core protocols associated with this layer are address resolution protocol (ARP), Internet protocol (IP), Internet control message protocol (ICMP), and Internet group management protocol (IGMP). Traditionally, this layer was built upon IPv4, which is gradually shifting to IPv6, enabling the accommodation of a much more significant number of addresses and security measures.

(iii) **Transport Layer**: Layer 3 of the TCP/IP protocol suite is functionally synonymous with the transport layer of the OSI model. This layer is tasked with the functions of error control, flow control, congestion control, segmentation, and addressing in an end-to-end manner; it is also independent of the underlying network. Transmission control protocol (TCP) and user datagram protocol (UDP) are the core protocols upon which this layer is built, which in turn enables it to have the choice of providing connection-oriented or connectionless services between two or more hosts or networked devices.

(iv) **Application Layer**: The functionalities of the application layer, layer 4, of the TCP/IP protocol suite are synonymous with the collective functionalities of the OSI model's session, presentation, and application layers. This layer enables an end-user to access the services of the underlying layers and defines the protocols for the transfer of data. Hypertext transfer protocol (HTTP), file transfer protocol (FTP), simple mail transfer protocol (SMTP), domain name system (DNS), routing information protocol (RIP), and simple network management protocol (SNMP) are some of the core protocols associated with this layer.

A networked communication between two hosts following the TCP/IP model is shown in Figure 1.4

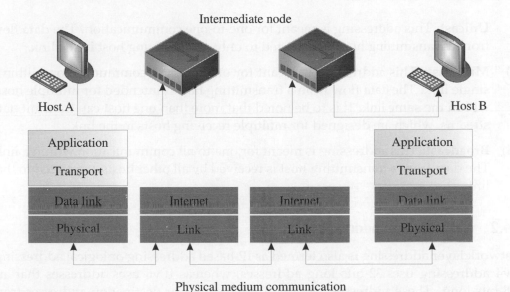

Figure 1.4 Networked communication between two hosts following the TCP/IP suite

1.4 Addressing

Addressing in networked devices plays a crucial role in ensuring the delivery of packets to the designated/intended receivers. The addressing scheme is synonymous with postal addresses used in real-life scenarios. Addressing mechanisms can be divided into two parts: one focusing on data link layer addressing, while the other focuses on network layer addressing.

1.4.1 Data link layer addressing

Data link layer addressing deals with media access control (MAC) addresses of devices, which work at the MAC sub-layer of the data link layer.

> **Points to ponder**
>
> Data link layer addressing handles the host/device network interface of physical addresses. These physical addresses are also known as MAC addresses. MAC addresses are unique 48-bit hardware addresses provided by the device manufacturers.

MAC addresses are 48-bits long; the first 24 bits are organizational identifiers, while the last 24 bits are network interface controller identifiers. These addresses are unique globally. Data link layer addressing is broadly divided into three types: *1) Unicast, 2) Multicast, and 3) Broadcast.*

(i)   **Unicast**: This addressing is meant for one-to-one communication. The data flow from a transmitting host is restricted to only one receiving host in the link.

(ii)   **Multicast**: This addressing is meant for one-to-many communication within a single link. The data flow from a transmitting host is intended for multiple hosts within the same link. It is to be noted that more than one host can transmit data streams, which are designed for multiple receiving hosts in the link.

(iii) **Broadcast**: This addressing is meant for one-to-all communication within a link. The data from a transmitting host is received by all other hosts connected to that link.

1.4.2 Network layer addressing

Network layer addressing is also termed as IP-based addressing or logical addressing. IPv4 addressing uses 32-bits long addresses, whereas IPv6 uses addresses that are 128 bits long. These addresses can identify the source or destination addresses from the address itself. The mapping of a device/host's logical address to its hardware address is done through a mechanism called address resolution protocol (ARP). During transmission of a packet from a host, the IPv4 sends an IPv4 packet, the next-hop address, and the next-hop interface to the ARP.

> **Points to ponder**
>
> Network layer addressing deals with 32-bit (in case of IPv4) or 128-bit (in case of IPv6) logical addresses assigned to networked devices. These addresses are not hard-coded or provided by the manufacturers.

Direct delivery is performed by the ARP if the destination address for delivery matches the next-hop address. In contrast, if the addresses do not match, the ARP performs an indirect delivery by forwarding the packet to a router or an intermediate node. The resolution of the mapping of a packet's next-hop address to its MAC address is made using broadcasting ARP requests. The returning ARP reply frame to the sender contains the MAC address corresponding to the packet's next-hop address.

In the context of addressing, we will discuss the structure of IPv4 and IPv6 packets, which will provide a much clearer understanding of the workings of these two protocols.

(i) **IPv4**: The IPv4 header packet shown in Figure 1.5 has 13 distinct fields, the functions of which are enumerated as follows.

- **VER**: It is 4 bits long and represents the version of IP. In the example given in Figure 1.5, it is 4 bits (binary: 0100).

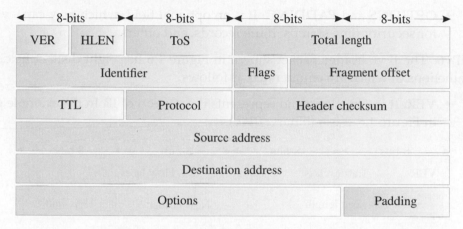

Figure 1.5 An IPv4 packet header structure

- **HLEN**: It is 4 bits long and denotes the length of the IPv4 packet header.
- **ToS**: It is 8 bits long. The first six most significant bits represent the differentiated services code point (DSCP) to be provided to this packet (by the routers). Explicit congestion notification (ECN), which gives information about the congestion witnessed in the network, is handled by the last 2 bits.
- **TOTAL LENGTH**: It is 16 bits long and identifies the length of the entire IPv4 packet, including the header and the payload.
- **IDENTIFIER**: It is 16 bits long and used to identify the original packets in case of packet fragmentation along the network.
- **FLAGS**: It is a 3-bit field with the most significant bit always set to 0. FLAGS indicates whether a packet can be fragmented or not in case the packet is too big for the network resources.
- **FRAGMENT OFFSET**: It identifies the exact offset or fragment position of the original IP packet and is 13 bits long.
- **TTL**: It is 8 bits long and prevents a packet from looping infinitely in the network. As it completes a link, its value is decremented by one.

- **PROTOCOL**: It is 8 bits long. This field identifies the protocol of the packet as user datagram protocol, UDP (17), transmission control protocol, TCP (6), or Internet control message protocol, ICMP (1). The identification is made at the network layer of the destination host.
- **HEADER CHECKSUM**: It is 16 bits long and used for identifying whether a packet is error-free or not.
- **SOURCE ADDRESS**: It indicates the origin address of the packet and is 32 bits long.
- **DESTINATION ADDRESS**: It indicates the destination address of the packet and is 32 bits long.
- **OPTIONS and PADDING**: It is an optional field, which may carry values for security, time stamps, route records, and others.

(ii) **IPv6**: The IPv6 header packet shown in Figure 1.6 has eight distinct fields, the functions of which are enumerated as follows.

- **VER**: It is 4 bits long and represents the version of IP. In the example given in Figure 1.6, it is 6 (binary: 0110).

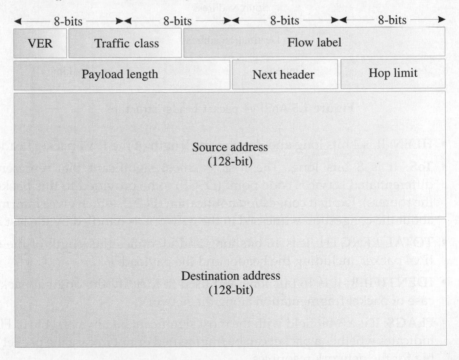

Figure 1.6 An IPv6 packet header structure

- **TRAFFIC CLASS**: It is 8 bits long. The first six most significant bits represent the type of service to be provided to this packet (by the routers); explicit congestion notification (ECN) is handled by the last 2 bits.

- **FLOW LABEL**: It is 20 bits long and designed for streaming media or real-time data. The FLOW LABEL allows for information flow ordering; it also avoids packet resequencing.

- **PAYLOAD LENGTH**: It is 16 bits long and provides a router with information about a packet's payload length or the amount of data contained in the packet's payload.

- **NEXT HEADER**: It is 8 bits long and informs the router about the type of extension header the packet is carrying. Some of the extension headers and their corresponding values are as follows: Hop-by-hop options header (0), routing header (43), fragment header (44), destination options header (60), authentication header (51), and encapsulating security payload header (50). In case an extension header is absent, it represents the upper layer protocol data units (PDUs).

- **HOP LIMIT**: It is 8 bits long and prevents a packet from looping infinitely in the network. As it completes a link, the limit's value is decremented by one.

- **SOURCE ADDRESS**: It is 128 bits long and indicates the origin address of the packet.

- **DESTINATION ADDRESS**: It is 128 bits long and indicates the destination address of the packet.

Check yourself

Classful addressing, Classless addressing, CIDR notation, Subnetting, NAT, DHCP, RARP

1.5 TCP/IP Transport layer

The transport layer is the third layer in the TCP/IP protocol suite and is an important connectivity entity as it acts as the interlocutor for the clients and the servers in a client–server paradigm. This layer forms the core of the TCP/IP protocol suite as it provides logical mechanisms for data exchanges between two or more points over the Internet. As mentioned in the previous sections, the transport layer engages in networking functionalities such as process-to-process communication, encapsulation and decapsulation of data, multiplexing and demultiplexing of virtual pathways, flow control, error control, and congestion control.

From a broader perspective, the transport layer provides two types of services: 1) connectionless and 2) connection-oriented. These service types at the transport layer determine the degree of interdependence between transmitted packets. The application layer for both service types first divides a message into smaller chunks, which are then acceptable by the transport layer for further transmission. These chunks are sequentially forwarded from the application layer to the transport layer. Upon receiving these chunks, the transport layer encapsulates these into packets for transmission. Generally, the packets in a connectionless transport layer service are independent of one another; whereas the packets in a connection-oriented service are dependent on one another. Figure 1.7 shows these two service types side-by-side for the sake of comparison of their working.

1.5.1   Connectionless service

Owing to its working, the connectionless service at the transport layer treats each incoming chunk from the application layer as independent units. After these chunks have been packetized, the packets are transmitted over the network with basic information of the source and destination addresses and ports. Even if the packets at the receiving end arrive out of order, they are submitted to the receiving host's application layer as it is, without any sequence maintenance. Additionally, no dedicated connection is established between the client and the server processes, as shown in Figure 1.7(a). The client transport layer has four message chunks M0, M1, M2, and M3, which are transmitted in that sequence. However, M2 arrives at the server at a time much later than its subsequent packet M3. As this is a connectionless service, the sequence is not maintained, and the packets are forwarded to the server's application layer as it is (out of sequence). Voice-over-IP (VoIP) is a popular usage of this service type. The most famous protocol associated with this service type is the user datagram protocol (UDP).

> **Points to ponder**
>
> Datagrams are packet-switched network transfer units that are generally used for connectionless services. There is no guarantee of the time of arrival, sequence, and surety of delivery of datagrams.

1.5.2   Connection-oriented service

The connection-oriented service, in contrast to the connectionless service, has a high dependency on the sequence of packets. Before the transmission of data to the server from the client (refer to Figure 1.7(b)), the client and server establish a connection employing handshaking using SYN and ACK frames. Once the data transmission is complete, the connection is terminated. In case another message has to be transmitted, the connection establishment process is again followed. This service type ensures that the packets arriving at the client's transport layer from its application layer are delivered in the exact sequence as in the server's application layer, in turn ensuring the quality of service (QoS) for the connection. However, ensuring QoS makes this type of service quite slow in comparison to connectionless services. Illustrating its working

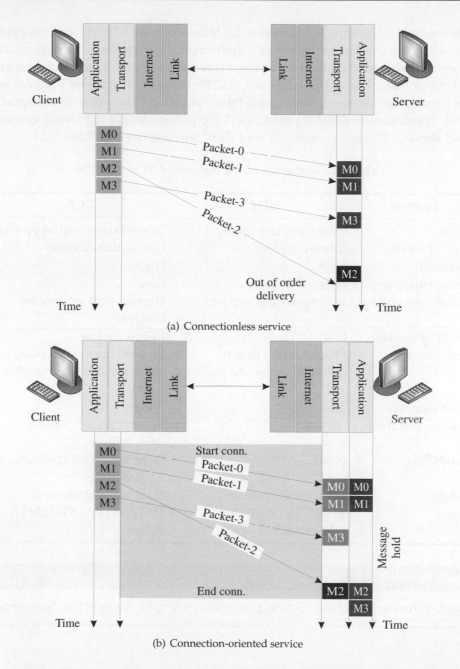

(a) Connectionless service

(b) Connection-oriented service

Figure 1.7 Transport layer service types during client–server data transfer

using Figure 1.7(b), the client transport layer has four message chunks M0, M1, M2, and M3 (from the client's application layer), which are packetized and transmitted in that sequence once a connection is established between the client and the server. Even if the M2 packet arrives out of sequence at the server's transport layer, the subsequent packets are held back until M2 is received. Upon receiving M2, M2 and the held back M3 packet are forwarded to the server's application layer in the same sequence that it was transmitted from the client's transport layer. Application layer protocols such as HTTP (hyper text transfer protocol) and HTTPS (hyper text transfer protocol secure) rely on connection-oriented services for their operation. The popular transport layer protocol, transmission control protocol (TCP), is a means of achieving connection-oriented service. The features of TCP and UDP are compared in Table 1.3.

Table 1.3 Comparison of the features of TCP and UDP

Feature	UDP	TCP
Name	User datagram protocol	Transmission control protocol
Type of service	Connectionless	Connection-oriented
Reliability	Low	High
Time-criticality	High	Low
Packet sequencing	No sequencing required	High level of sequencing involved
Speed of transfer	High	Relatively low
Error checking	Present, but it simply discards errorenous packets	Present; Errorenous packets are re-transmitted from the source
Error recovery	Absent	Present
Acknowledgment	Absent	Present; Done by means of ACK frames
Handshake	None	Done by SYN, SYN-ACK, ACK frames
Weight	Lightweight protocol	Heavyweight protocol
Usage	SNMP, TFTP, RIP, VoIP, DNS, DHCP	HTTP, HTTPs, FTP, SMTP, Telnet

Check yourself

Client–server architecture, Connection-oriented service, Connection-less service

Summary

This chapter covered the very basics of networking, which would prove handy in the following chapters covering the Internet of Things and its various associated paradigms. We discussed different network types based on connection types, topologies, and network reachability. We then outlined the two popular layered network models: the ISO-OSI model and the TCP/IP protocol suite. Subsequently, we described the various addressing terminologies and provided a brief functional outline of the IPv4 and IPv6 packet structure in this context. Finally, we concluded this chapter with a sketch of the TCP/IP transport layer and provided a general working outline on connectionless and connection-oriented services.

Exercises

(i)   Differentiate between point-to-point and point-to-multipoint connection types.

(ii)   Discuss the pros and cons of the following network topologies:

(a) Star

(b) Ring

(c) Bus

(d) Mesh

(iii)   How are PANs different from LANs?

(iv)   How are MANs different from WANs?

(v)   What is the ISO-OSI model?

(vi)   Discuss the highlights of the seven layers of the OSI stack.

(vii)   What is the Internet protocol suite?

(viii) How is the Internet protocol suite different from the ISO-OSI model?

(ix)   How is data link addressing different from network addressing?

(x)   Describe the IPv4 header format.

(xi)   Describe the IPv6 header format.

(xii)   How is IPv4 different from IPv6?

(xiii) What is meant by connectionless service?

(xiv)   How is connectionless service different from connection-oriented service?

References

[1]   Forouzan, A.B. 2007. *Data Communications and Networking* (SIE). Tata McGraw-Hill Education.

[2]   LAN/MAN Standards Committee. 2002. *IEEE Standard for Local and Metropolitan Area Networks: Overview and Architecture* (En linea). New York. USA: The Institute of Electrical and Electronics Engineers Inc.

[3]   Forouzan, B.A. and S.C. Fegan. 2006. *TCP/IP Protocol Suite* (Vol. 2). McGraw-Hill.

[4]   Wilder, F. 1998. *A Guide to the TCP/IP Protocol Suite.* Artech House, Inc.

[5]   Popescu-Zeletin, R. 1983. "Implementing the ISO-OSI Reference Model." *ACM SIGCOMM Computer Communication Review* 13(4): 56–66.

[6]   Davies, J. 2004. "Architectural Overview of the TCP/IP Protocol Suite." Microsoft Technet. https://technet.microsoft.com/en-us/library/bb726993.aspx.

Chapter 2

Basics of Network Security

Learning Outcomes

After reading this chapter, the reader will be able to:

- Understand the concepts of network security
- List the basic terminologies and technologies associated with security, privacy, and authenticity
- Explain functioning of digital signatures and key management
- Differentiate between network layer, transport layer, and application layer security
- Explain firewalls
- Relate new concepts with concepts learned before to make a smooth transition to IoT

2.1 Introduction

The range of operations dependent on computers, computer networks, and the Internet is vast. Healthcare, banking, governance, security, military, research, power, agriculture, and other fields are nowadays largely dependent on networked systems. The huge implications of the failure of one of these domains due to computer-based security lapses are undeniable. This necessitates the need for various security protocols for computer networks and computer-based systems. Typically, security in networks focuses on preventing unauthorized or forced access to a user's or organization's system or systems. The concept of security applies even to computers or systems which are not connected to a network or the Internet. The main aspects of securing a system are security, privacy, and authenticity. The security operations in computers encapsulate protection of hardware, software, data, and identity.

The various forms of network attacks are classified into two broad categories: General cyber threats, and threats to web databases [1]. Attacks such as authentication violation, non-repudiation, Trojan horses, viruses, fraud, sabotage, denial of service, and even natural disasters are categorized as general cyber threats. In contrast, attacks such as access control violations, integrity violations, confidentiality violations, privacy violations, authenticity violations, and identity thefts are categorized as threats to web databases. Most of the commonly available security tools are anti-viruses, anti-malware, anti-spyware, and firewalls. These are mostly software-based tools and used by individuals or for personal computing systems. However, costlier options such as hardware-based systems and hardware–software hybrid systems such as access control mechanisms, hardware firewalls, and proxy servers are the most opted for security measures for large organizations. These tools are designed to protect a user from a range of attacks.

> ### Points to ponder
>
> **Zero day attacks** are exploits (attacks) that make use of a previously unknown security vulnerability in a system (software or hardware) to gain access to the system or take over it. The term 'zero day' refers to the time duration of the discovery of the vulnerability and the launch of attack (the attack is launched on the day of the discovery of the vulnerability). Most zero day attacks have been attributed to faulty software development practices. The developers of software and systems are expected to provide a patch upon discovery of the vulnerability. As there is no previous signature of an attack, zero day attacks are challenging to detect.

Some basic practices on the computer or over networks can easily ward off most security threats. Examples of these practices include the following.

(i)   Choosing passwords wisely so that it is a mixture of alphabets (preferably, both uppercase and lowercase characters), numbers, and special characters; passwords need to be changed periodically.

(ii) Avoiding sharing of passwords or credentials, or storing/recording them[(iii)] in an obvious manner such as on a piece of paper, or on your desktop.

(iv) Keeping systems up to date and patched on time.

(v)   Using anti-virus, anti-spyware, and firewalls tend to reduce the scope of threats.

(vi) Avoiding download of suspicious attachments and clicking on random links or pages.

In this chapter, we will discuss the overall aims of security, privacy, and authenticity in network security. A holistic approach to Internet security is undertaken by covering the aspects of the network layer, transport layer, and application layer security. These are followed by overviews and a discussion on the importance of digital signatures, key management, and firewalls in security.

> ### Check yourself
>
> Backdoor, Denial-of-service attacks, Direct-access attacks, Eavesdropping, Spoofing, Tampering, Privilege escalation, Clickjacking, Social engineering

2.2 Security

Security in networks and computer systems work toward the following three goals: confidentiality, integrity, and availability. This is often referred to as the CIA triad.

Confidentiality pertains to protection of stored and transmitted information over the network in such a manner that the information itself is concealed and protected from unauthorized access. Attacks such as snooping and traffic analysis pose a direct threat to the confidentiality of information. A breach of confidentiality can occur if the nature of information, the information itself, or the address of the sender or receiver is revealed to an unauthorized third party.

> ### Points to ponder
>
> **Ransomware** is a type of malware that locks off a user's machine or computer by encrypting its contents and preventing other applications and software from running on the system. The system is held in this condition by the attacker until a ransom is paid. The famous *WannaCry* ransomware held many systems hostage worldwide; the systems included computers at hospitals, emergency services, personal computers, especially those running on Windows OS (operating system).

The loss of integrity of any stored or transmitted information may arise due to both intended or unintended actions. Whenever changes are made to any information in a system or network by unauthorized entities, a breach of the system or network integrity is presumed. Attacks such as repudiation, replays, modification, and masquerading may pose severe threats to the integrity of information. It is interesting to note that changes in information due to power outages or other natural causes may also be considered a breach of information integrity.

Finally, the unavailability of any information to authorized entities over a network implies infringement of availability, which is one of the three members of the CIA triad. Attacks such as denial of service or distributed denial of service directly affect the availability of information. These attacks may completely block attempts to access information or any part of it over the network by flooding the network with unscrupulous traffic. They may result in temporary or permanent loss of information from the network or the system.

2.3   Network Confidentiality

Modern-day computer systems rely on various mechanisms to address the needs of confidentiality of information being transmitted over the network. If these confidentiality measures are not present, as soon as the information leaves the relative safety of a user's computer and diffuses over the network, it can be accessed by anyone with the right tools and skill-set. Such a scenario would make modern-day operations such as online transactions, e-commerce, banking, e-mails, conversations, shared online storage, and a host of other such applications useless. Methods and schemes such as cryptography and steganography (Figure 2.1) help in achieving confidentiality over the network. Table 2.1 compares the main differences between steganography and cryptography. While cryptography focuses on encrypting the information to be transmitted, making the contents unreadable without the proper decryption credentials, steganography focuses on hiding the information in plain sight.

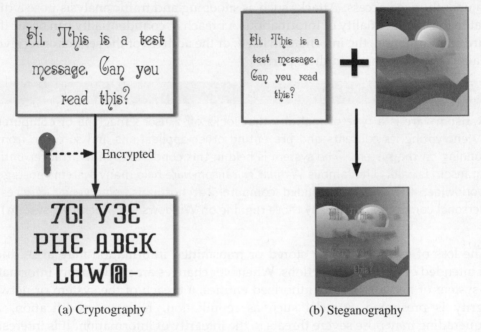

(a) Cryptography   (b) Steganography

Figure 2.1 Data confidentiality schemes

Table 2.1 An overview of the differences between cryptography and steganography

Parameters	Cryptography	Steganography
What	Method of hiding information, making it readable only by the sender and the receiver.	Method of concealing information within a non-secret medium or content.
When	Used when the message being transmitted has a high chance of being intercepted and probed.	Used when the transmitted message has a very low chance of being intercepted and probed.
Why	Used in order to obscure the transmitted content so that intercepting parties may be able to capture the message but cannot read it.	Used in order to hide the very existence of the transmitted message so that no party has any idea about the presence of the message.
Where	Messages rendered unusable by encrypting its contents, which may be text, images, sounds, or videos.	Messages hidden within lowest bit of noisy images, sound files, files in computer databases, and others.
How	Key-based encryption, hashing.	Cover synthesis, cover selection.

> **Points to ponder**
>
> **Phishing** is a type of social engineering attack wherein an unscrupulous entity poses as a trusted party to scam a target user into revealing their personal details. These details can be used by the unscrupulous entity to pretend to be the target to gain access to their social or financial accounts. Some common forms of phishing are e-mail phishing and spear phishing. Security protocols such as the two-factor authentication have proved beneficial in keeping phishing attacks under check.

These measures of incorporating and enhancing network confidentiality keep out unwanted eavesdroppers from intercepting any information transmitted over a network. However, various attacks—trojans, viruses, worms, and others—have been engineered to overcome these confidentiality measures and access information being transmitted over the network. Many non-profit, as well as commercial organizations every day, keep track of newer confidentiality-compromising attack signatures worldwide and release preventive measures as security patches or updates to various systems.

2.4 Cryptography

Cryptography, which roughly stands for *hidden writing* in Greek, is an ancient science of passing secret information by hiding its contents or making the contents obscure to the normal or unsuspecting eye. Modern-day cryptography is found in almost all forms of networked communication and transactions. The mathematically intensive and processing heavy cryptographic algorithms that form the base of information encryption over networks are theoretically breakable but without any possible means or within a possible time frame. Consider encrypted messages being transmitted between persons **A** and **C**. Person **B** (an adversary) is trying to capture and get hold of the information passing between **A** and **C**. Even if **B** gets hold of the encrypted information, decoding it would take him months, if not years, by which time either the message would have lost its significance, or the encryption key would have changed, making his attempts to extract information futile and a useless venture. Broadly speaking, cryptography is divided into two types: 1) Symmetric key, and 2) asymmetric key. Cryptography serves the following five purposes in modern-day networked systems: 1) confidentiality, 2) authentication, 3) integrity, 4) non-repudiation, and 5) key exchange [4].

Points to ponder

An attacker is trying to break-in a 40-bit encrypted message being transmitted over the network. Considering the attacker uses a brute-force approach, where he uses all the possible combinations one by one, he would have to go through 2^{40} combinations. Further, considering that each attack takes 1 ms, the attacker would take approximately 38 years to go through the combinations. It is interesting to note that present-day cryptosystems use 64 and 128 bit keys to encrypt messages.

Typically, in cryptography, the message to be transmitted is referred to as plaintext (P), which is encrypted (E) using a key (k) to generate a ciphertext (C). This ciphertext is transmitted over the network. Upon receiving the ciphertext, a receiver uses his key (k) to decrypt (D) the ciphertext message back to plaintext. The encryption process is denoted as $C = E_k(P)$, whereas the decryption is denoted as $P = D_k(C)$.

2.4.1 Symmetric key cryptography

Symmetric key cryptography is also referred to as secret key cryptography. This cryptographic technique uses a single key for both encryption and decryption. It finds primary usage in ensuring privacy and confidentiality of information. The simplest example of a symmetric key cryptosystem is a substitution additive cipher, where the message to be encoded (P) is modified by increasing the position of the alphabet by a fixed number key (k) to obtain the ciphertext ($C = E_k(P)$). The receiver must use the same key (k) to decrypt the ciphertext message into plaintext. Table 2.2 shows the process of encrypting and decrypting a message (SECRET) using a modulo five substitution additive cipher. The main drawbacks of this system of cryptography is that the shared key needs to be securely shared between the sender and the receiver. If the key falls into the hands of an adversary, the confidentiality of the encrypted message can be easily compromised. Additionally, these ciphers are prone to decryption using exhaustive key searches or brute-force attacks.

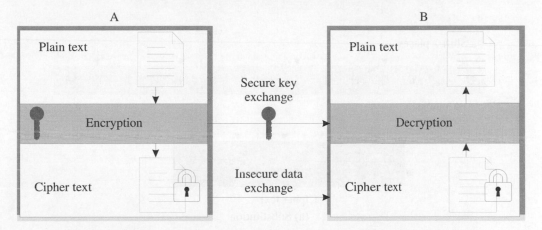

Figure 2.2 A symmetric key cryptographic mechanism

Table 2.2 Message encryption and decryption using a modulo $k(k = 5)$ substitution cipher

Plaintext	Corresponding alphabet number	Encryption (Key = 5)	Ciphertext	Decryption (Key = 5)	Plaintext
S	19	19 + 5 = 24	X	24 − 5 = 19	S
E	5	5 + 5 = 10	J	10 − 5 = 5	E
C	3	3 + 5 = 8	H	8 − 5 = 3	C
R	18	18 + 5 = 23	W	23 − 5 = 18	R
E	5	5 + 5 = 10	J	10 − 5 = 5	E
T	20	20 + 5 = 25	Y	25 − 5 = 20	T

The example of a substitution cipher demonstrated in Table 2.2 is known as the monoalphabetic cipher. These ciphers have a one-to-one mapping between plaintext and substituted ciphertext, which makes them easy to crack using brute-force attacks. To increase the complexity of the monoalphabetic system, polyalphabetic ciphers can be used, such that the plaintext message has a one-to-many relation with the ciphertext.

Modern cryptosystems use a variety of ciphers that fall under the category of symmetric key ciphers such as substitution, transposition, block, and stream ciphers (Figure 2.3). These symmetric key cryptographic algorithms are compared in Table 2.3 [2]. Data Encryption Standard (DES) is a popular modern-day symmetric key cryptographic scheme. DES falls under the category of block ciphers.

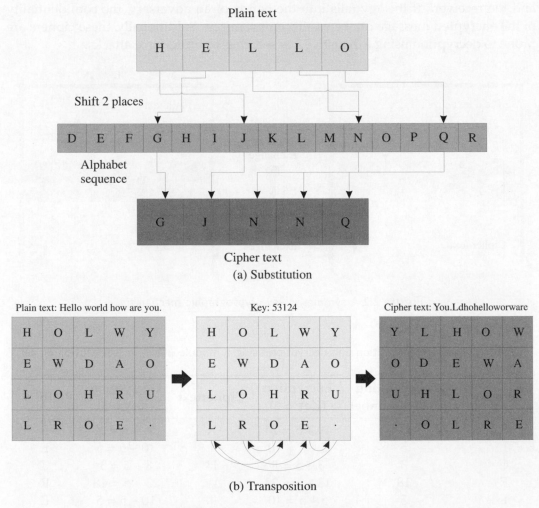

Figure 2.3 Symmetric key cryptographic primitives

Table 2.3 A comparative overview of various cipher types

Parameter	Substitution cipher	Transposition cipher	Block cipher	Stream cipher
What	Method of encryption where plaintext is replaced by a fixed system based ciphertext. The fixed system may be single letters, letter pairs, or a combination of more than two letters.	Method of encryption where bits and chunks of plaintext are transposed or shifted from its original position to generate ciphertext.	Method of encryption using a deterministic algorithm that works on blocks or a fixed length of bits using symmetric key based transposition. Substitutions and permutations are iteratively applied to blocks, allowing this method to be highly effective in providing message security.	Method of encryption where plaintext is combined with a stream of pseudo-random cipher digit. Here, each plaintext digit is encrypted with its corresponding cipher stream digit.
Advantages	Simple to formulate and use. Substitution creates confusion.	Transposition creates diffusion. Highly effective when used with other schemes such as substitution or fractionation.	Highly suitable for providing confidentiality and authenticity of messages.	Highly robust, simple and speedy during hardware implementation.
Disadvantages	Easy to guess and crack.	Regular transpositions can be easily decrypted by anagraming and genetic algorithms.	The use of invertible functions in block ciphers eventually reveals its pseudo-random nature.	Biases in choosing keystreams may allow revelation of its non-random nature.
Types	Homomorphic, polyalphabetic, polygraphic, mechanical, one-time pad.	Rail fence, route, columnar, Myszkowski.	DES, AES, Blowfish	Synchronous, self-synchronizing

2.4.2   Asymmetric key cryptography

Asymmetric key cryptography is also referred to as public key cryptography. This cryptographic technique uses two separate keys for encrypting and decrypting messages. The secret key is personal to each user and is referred to as the private key. The other key is known as the public key as it is known to all and in the public domain. The encryption and decryption of messages using this scheme require both the keys. This class of cryptography finds popular usage in establishing the tenets of authenticity and non-repudiation in communication. The essence of asymmetric key cryptosystems is the application of mathematical functions on numbers to generate new numbers. These mathematical functions are generally one-way functions, where the forward function is easy to compute, but the inverse is very complicated, if not impossible. It is interesting to mention that both the keys are unrelated to one another, and the knowledge of one does not compromise or give away any hint about the other key.

This scheme is in contrast to symmetric key cryptosystems, which mainly rely on substitution and permutation of symbols for encrypting messages. Table 2.4 compares symmetric and asymmetric key cryptographic techniques. From an operational point of view, unlike symmetric ciphers which directly use symbols, in asymmetric key cryptography, the plaintext message has to be encoded into integers before encrypting them. Similarly, for the reverse process of decryption, the decrypted message is in the form of integers, which has to be mapped back to symbols for generating the plaintext message.

For example, consider that **A** intends to send a message to **B** over a non-secure network channel. **A** encrypts the message using **B**'s public key (Figure 2.4). This encrypted message can only be decrypted using **B**'s private key. The encryption process at **A** can be denoted as $C_B = E_{B(k1)}(P)$. At **B**, the decryption process to extract the plaintext message from the ciphertext is denoted as $P = E_{B(k2)}(C_B)$ such that $k1, k2$

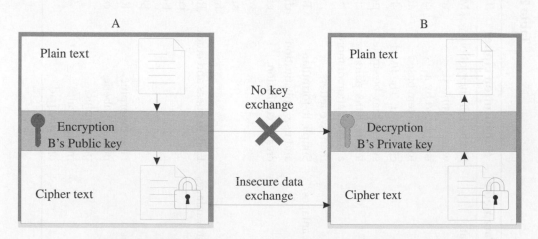

Figure 2.4 An asymmetric key cryptographic mechanism

Table 2.4 A comparative overview of symmetric and asymmetric key cryptography

Parameter	Symmetric key	Asymmetric key
Encryption/ decryption keys	Uses same key for encryption and decryption; Also called secret key cryptography.	Uses different keys for encryption and decryption; Also called public key cryptography.
Number of keys	For n users, $n(n-1)/2$ key pairs are required.	For n users, $2n$ keys required.
Key generation	Randomly generated and simple.	Complex, generally large prime numbers are used.
Security of key	Keys need to be secured.	Key security not necessary.
Example	AES-126	RSA

are respectively the public and private keys of **B**. $E_{B(k1)}$ is used only for encryption, whereas $E_{B(k2)}$ is used for decryption only.

Contrary to popular belief, the symmetric key and asymmetric key schemes complement one another and are used in conjunction with one another in a vast number of applications. For example, the shared key in symmetric key schemes is generally transmitted to the receivers using asymmetric key schemes. Rivest–Shamir–Alderman (RSA) algorithm is a popular asymmetric key cryptographic scheme.

> Check yourself
>
> AES, DES, Homomorphic substitution cipher, Columnar transposition, Cover synthesis, Hashing, RSA

2.5 Message Integrity and Authenticity

The types of information being transmitted over networks worldwide are huge. With the increase in acceptance of network-based solutions, the challenges in ensuring the quality of service and safety are increasing day by day. The massive application types give rise to various issues regarding quality of service. The concept of confidentiality is not inherent for all message types being transmitted over the networks. There are operations which focus more on the integrity of the transmitted message, rather than on its confidentiality. As a simple example, if we consider an online banking system, the account login and powers of transaction rely heavily on various security and confidentiality measures. However, in a blockchain-based banking system, the implications of message integrity far outweigh the impact of message confidentiality, so much so that all transactions are transparent in a blockchain system, yet they are immune to tampering and fraudulent manipulations [3]. Newer models of the economy are being planned upon blockchain-based systems.

The most popular scheme of ensuring message integrity over a network is hashing. Some of the more popular hashing techniques include algorithms such as SHA and MD5. A hash function applied on a message creates a digital fingerprint for that message, which is referred to as its digest (Figure 2.5). Before transmission of the message, a message digest is securely transmitted to a receiver. The message can be transmitted in any manner, and over any channel—both secure or insecure. The receiver, upon receiving this message generates its hash digest and compares it to the one received earlier. The message is considered tamper-free only if both of these digests match.

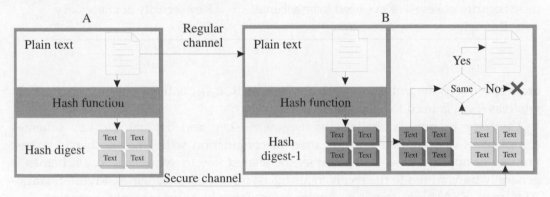

Figure 2.5 An integrity enabling hashing mechanism

The authenticity of a message (whether the sender of the message is the same person as claimed by the message) is ensured by using a pre-shared key between the sender and the receiver. The pre-shared key is applied to the hashing function to generate a message authentication code (MAC), which is transmitted along with the message. The receiver, upon receiving the message generates another MAC using the pre-shared key. If this newly generated MAC matches the one received over the network, the authenticity of the message is ensured. The main advantage of this method is that there is no need for a separate secure channel for transmitting message digests, in addition to the feature of both integrity and authenticity check of the message. However, on the flip-side, if the pre-shared key is compromised, this authentication method fails.

2.5.1 Digital signatures

A digital signature is functionally similar to paper-based signatures, which is primarily used to bind a person/user/signatory with the content of a message. In digital signatures, the binding is verifiable by the receiver of the message or even third parties, as it mainly authenticates the message. The method of digitally signing a message is akin to public key cryptography. Each signatory has a public and a private key. The private key is used for signing the message and is referred to as the signing key, whereas the public key is used for verifying the message and is known as the verifier key. Digital signatures ensure that a message complies with the features of authenticity, integrity, and non-repudiation. RSA is a commonly used algorithm for digital signatures.

Suppose, **A** is the signatory and **B** is a verifier at the two ends of a network. **A** converts the message to be sent to **B** into a hash digest by passing it through a hash function (Figure 2.6). The hash digest is then signed-on by **A**'s private/signing key. This signature is transmitted with the data to **B**. Upon receiving this signed message, the verifier applies **A**'s public key on the message to obtain the message hash digest. **B** independently hashes the data received with the signature. If the independently generated hash digest is similar to the one that has been obtained by using **A**'s public key, the signature of **A** is verified by **B**.

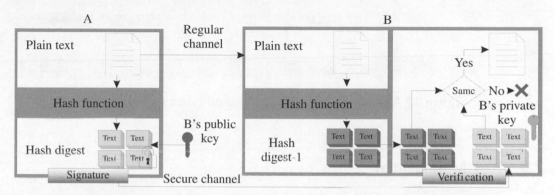

Figure 2.6 A mechanism for digital signing of electronic documents

The authentication feature of digital signatures can be grouped into two broad categories: 1) Entity authentication and 2) message authentication. Entity authentication is often referred to as peer entity authentication. It is used for binding a person to a message. An entity authentication scheme assures the receiver of a message about the sender's participation in generating the message. In contrast, message authentication, which is also known as message origin authentication, is a means to ensure receivers of a message that the message has not been tampered with during its transmission from the sender. Digital signatures are categorized into four classes: 1) Certified signatures (Figure 2.7), 2) Approval signatures, 3) Visible digital signatures, and 4) Invisible digital signatures.

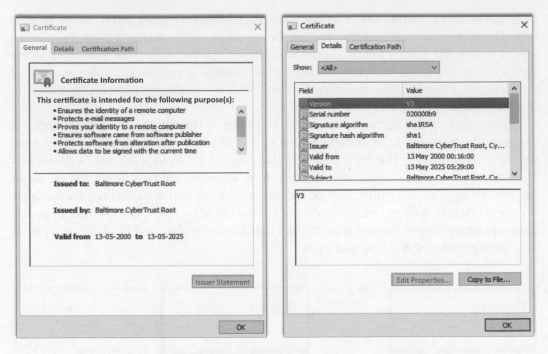

Figure 2.7 A screenshot of a third-party certificate on a host device

2.6   Key Management

Key management is one of the most crucial aspects of modern-day cryptography and deals with the administration of cryptographic keys. Generation, distribution, storage, safety, and distribution of keys are the major functionalities of a key management system. As cryptographic keys can be both symmetric as well as asymmetric, the management of these keys to provide reliable services is a very challenging, yet important task in modern-day cryptographic communications and networking.

The usability and efficiency of any modern-day cryptosystem are as good as the key management system. Supposing that despite using state-of-the-art cryptographic systems, the keys have to be somehow transmitted to the receiver of the message. At this point, the keys are the most vulnerable to hijacking or unauthorized capture. If the key itself is compromised, the layer of cryptographic encryption is automatically compromised and breached. A key management system must be robust enough to handle the challenges of scalability, security, availability, heterogeneity, and governmental policies. The overview of these challenges are outlined as follows.

(i)   **Scalability**: The key management system must be able to scale its operations on demand. Increase in the number of users must be easily managed by the system, in turn allowing for the storage and management of a large number of keys.

(ii) **Security**: The stored keys and credentials must be protected from unauthorized use or attacks, allowing the cryptosystem to function uncompromised.

(iii) **Availability**: The keys and the management servers must be accessible by authorized users at all times.

(iv) **Heterogeneity**: The use of multiple databases, variety of standards, and applications must be supported.

(v) **Governmental Policies**: The system must be robust enough to accommodate governmental or institutional regulations and policies at a very short notice. This should also enable policy-driven management of user access and privacy of users.

A modern-day key management system has the following basic components: 1) Inventory, 2) Key exchange, 3) Key use, and 4) Key storage. These components and their functionalities can be enumerated as follows.

(i) **Inventory**: It is responsible for creating and maintaining a concise list of all the crypto keys, their permissions, access rights, locations, and user mappings. The inventory is also responsible for managing certificate lists from a multitude of certifying authorities. The key inventory should be designed to take immediate measures such as replacing keys in case of breach of security.

(ii) **Key exchange**: Key exchange is a crucial part of the key management system as any slip up in the security of keys during transfer would compromise the purpose of the key management system. Symmetric cryptosystems use a separate secure channel to distribute keys, which act as an additional overhead for the whole communication system. Developments in cryptographic algorithms such as the Diffie–Hellman key exchange scheme ensure that these keys can be easily shared, even over insecure channels. Modern-day cryptosystems use techniques such as smart-card based key exchange, encrypting the key with another key, encrypting the symmetric key with an asymmetric key, and others.

(iii) **Key use**: The use of key-based encryption does not guarantee cent percent defense against attackers. The encryption, as mentioned previously, only buys the communicating parties enough time so that the message becomes obsolete, causing the key breaking exercise to become redundant. In most cases, symmetric key cryptosystems change the key after each message. This feature is highlighted by the key lifetime. If the same key is used for a very long time, the chances of an attacker gaining access to personal encrypted communication using these keys rapidly rises. The key management systems also manage the key lifetimes.

(iv) **Key storage**: The secure storage of keys ensures the success of an intrusion-free communication. The distributed storage of keys have various security mechanisms in place—user access passwords—to ensure no unauthorized access to the keys.

2.7 Internet Security

Internet security, as the name suggests, focuses more on the security of Internet accesses and devices on the Internet; it is not only restricted to simple networks. As the Internet is a huge and complex place made up of billions of devices and users, the chances of malicious or ill-intended breach of security becomes inevitable if the Internet is not secure. Attacks such as spyware, malware, Trojans, key-loggers, viruses, worms, ransomware, and other such attacks necessitates the presence of Internet security. In contrast, a regular network may not be exposed to such attacks or threats. For example, a small organizational network, which is isolated from the Internet and has networked devices within the purview of its organization only, will be faced by attacks only if someone from the inside initiates it; this type of attack is easy to track and initiate timely counter-measures.

The domain of Internet security mainly revolves around the three TCP/IP (transmission control protocol/Internet protocol) layers: 1) Network layer, 2) Transport layer, and 3) Application layer.

2.7.1 Network layer security

The network layer security encompasses security mechanisms and measures between two networked devices. The devices can be networked computers, routers, servers, and others. This layer supports security for both TCP, as well as UDP (user datagram protocol) packets arriving and going out of the networked devices.

One of the most common examples of network layer security protocol is the IP Security (IPSec), which provides packet-level security at the network level. IPSec itself consists of two modes: 1) Transport mode and 2) Tunnel mode. The transport mode is responsible for providing security only to the packet payload; it does not cover the IP header of the packet. Whereas in contrast, the IPSec tunnel mode protects both the header as well as the payload by encapsulating it in a new payload and adding a new IP header to the encapsulated packet. It is worthwhile to mention that, in the transport mode, the IPSec layer is logically positioned between the network and the transport layers of the TCP/IP stack. In contrast, the IPSec layer in the tunnel mode is positioned below the network layer and above a newly created network layer (Figure 2.8).

Additionally, IPSec defines two well-known security protocols: 1) Authentication header (AH) and 2) Encapsulating security payload (ESP). These two protocols are tasked with providing security authentication and encryption of packets at the IP level. Similar to the features of cryptographic techniques, IPSec provides for the characteristics of access control, message integrity, entity authentication, and

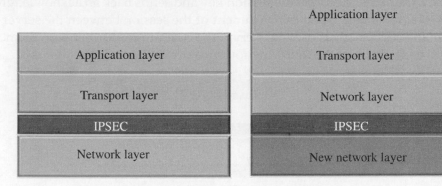

Figure 2.8 IPSec modes

confidentiality. Another essential aspect of IPSec is known as the security association (SA). SA is responsible for establishing logical connection credentials between two communication devices at the network level. It changes the connectionless nature of the IP services to a security-enabled connection-oriented service.

> Check yourself
>
> Authentication header (AH) protocol, Encapsulating security payload (ESP), Security association database, Security policy database, Internet key exchange (IKE) protocol, Virtual private networks (VPN)

2.7.2 Transport layer security

The transport layer security is based on utilizing the services of TCP to establish a connection-oriented protocol. This is why UDP based protocols are not supported by transport layer security mechanisms. The transport layer security protocols first encapsulate the packets in their packets, followed by encapsulation of TCP over the new packets. The transport layer security services logically exist between the TCP/IP stack's transport and application layers. These security protocols mainly aim to provide the features of authenticity, confidentiality, and integrity to client–server systems. Two of the most popular transport layer security protocols are as follows: 1) Secure socket layer (SSL) and 2) Transport layer security (TLS).

In SSL, a client communicating with a web server (or website) connects to it through a browser. The browser requests for the server's credentials, against which an SSL certificate with the server's public key is sent; this is verified against a list of trusted certifying authorities by the browser. Additionally, the validity and authenticity of the certificate are verified. Upon validation of the server's certificate, the browser sends back a session key (symmetric) using the server's public key.

Subsequently, the server decrypts the session key and sends back an acknowledgment for session establishment. Upon establishment of the session between the server and the browser, the session key is used to encrypt the communication between them. The SSL protocol is made up of the following four protocols (Figure 2.9).

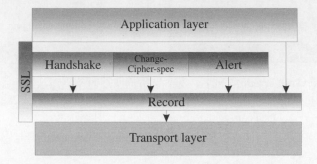

Figure 2.9 Position of the SSL protocol

(i)   **Handshake Protocol**: This protocol is responsible for establishing client–server/server–client connections, negotiating encryption algorithms, and authenticating the communicating entities.

(ii) **ChangeCipherSpec Protocol**: This protocol is used to change encryption settings which might have been set during the handshake process. A message notifies the client and server about the need for change in encryption; this is handled by the changecipherspec protocol.

(iii) **Alert Protocol**: This protocol is used for notifying the communicating systems of alerts and unusual conditions during communication between the entities.

(iv) **Record Protocol**: This protocol is responsible for breaking down the data from the upper layers into fixed sizes and compressing them. Additionally, it is responsible for encrypting messages from the upper layers coming down to the lower layers.

The TLS, which is a successor of the SSL, resides in the application layer and is functionally very similar to the SSL (Figure 2.10). It is broadly composed of two layers.

(i)   **TLS Record Protocol**: This provides connection security.

(ii) **TLS Handshake Protocol**: This allows client–server authentication and exchange of encryption algorithms and keys.

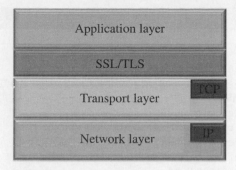

Figure 2.10 Position of the transport layer security protocol

> **Points to ponder**
>
> A **denial-of-service (DoS)** attack is a network attack that shuts down a machine or network by flooding it with unnecessary information resulting in the inability of legitimate users to access their systems. In the past, DoS attacks have been used to cripple banking organizations and financial servers. Examples of DoS attacks are buffer overflow, SYN flood, ICMP flood, and others. However, recent security protocols in place with the networks have limited the use of DoS attacks.

2.7.3 Application layer security

Application layer security protocols are designed to reside wholly within the application layer and provide security to applications such as e-mails, which is in contrast to network and transport layer security protocols. Two well-known application layer security protocols are as follows: 1) Pretty good privacy (PGP) and 2) Secure/multipurpose internet mail extension (S/MIME). Again, unlike the transport and network layer security protocols, application layer protocols must take into consideration that some communications can be one time or even unidirectional. For example, the act of sending e-mails may or may not elucidate a response mail from the receiver.

A Sample MIME e-mail Header

```
MIME-Version: 1.0
Date: Sun, 15 Jul 2018 11:03:08 +0530
Message-ID: <CAEZYB75_526bbMF-Gw9tmtyuKfBhH2O8cbcGX-OsJ7nUMrM=
    aA@mail.gmail.com>
Subject: A Book Sample Message
From: anandarup mukherjee <abc@ieee.org>
To: anandarup mukherjee <xyz@gmail.com>
Content-Type: multipart/alternative; boundary="000000000000456d3a05710309bd"
--000000000000456d3a05710309bd
Content-Type: text/plain; charset="UTF-8"
I hope it is clear now!!
--
```

2.8 Firewall

Firewalls are network security mechanisms which monitor and grant access to clean network traffic while blocking unscrupulous or malicious traffic. The firewall's access control policies are governed by a set of rules (Figure 2.11), which help them decide what to block and what to allow through the network. The mechanisms of the firewall can be hardware-based, software-based, or both. Firewalls of the modern day may have originated from network routers.

A Sample Windows Firewall Log.

```
#Version: 1.5
#Software: Microsoft Windows Firewall
#Time Format: Local
#Fields: date time action protocol src-ip dst-ip src-port dst-port size
    tcpflags tcpsyn tcpack tcpwin icmptype icmpcode info path
2018-07-15 13:50:05 DROP UDP 10.124.69.22 10.124.69.255 137 137 78 - - - - -
    - - RECEIVE
2018-07-15 13:50:08 DROP UDP 10.124.69.22 10.124.69.255 137 137 78 - - - - -
    - - RECEIVE
2018-07-15 13:50:08 ALLOW TCP 10.124.69.86 172.16.2.30 52716 8080 0 - 0 0 0
    - - - SEND
2018-07-15 13:50:09 DROP UDP 10.124.69.22 10.124.69.255 137 137 78 - - - - -
    - - RECEIVE
```

The most primitive form of firewalls worked on the Internet layer of the TCP/IP stack and were termed as packet filters [5]. These packet filters inspected incoming or outgoing packets for network addresses and ports. Depending on these parameters, the packet filters allowed or rejected the packets.

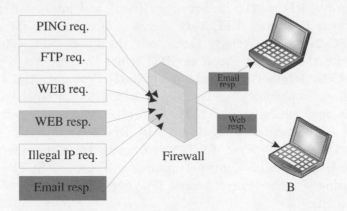

Figure 2.11 A simple rule-based firewall allowing selective traffic through it

Subsequently, the next generation of firewalls worked on the transport layer of the TCP/IP stack. These were circuit-level gateways and commonly referred to as stateful filters. The operational domain of these firewalls extended to layer four, over and beyond the realm of packet filters. The functioning of this firewall was state-based, where packets accumulated until a decision on their nature could be inferred.

Points to ponder

A **distributed DoS (DDoS)** is a type of DoS attack that makes use of a network of orchestrated or manipulated computers to launch a DoS attack on a single target machine or network. Identifying the originating source of a DDoS attack is virtually impossible due to its distributed nature. The distributed nature also allows for a much higher volume of traffic to be generated during the attack.

Finally, the current generation of firewalls mostly works in the TCP/IP stack's application layer. These firewalls can differentiate between application level processes. An understanding of application processes and their dependencies allows them to have a more robust set of rules regarding packet forwarding policies. These firewalls can detect unwanted access violations by masquerading applications on permitted ports, allowing them to have a more robust rule generating mechanism.

Check yourself

Firewalls versus Proxy servers

Security RFCs

The following are some of the RFCs (requests for comments) associated with Internet security. **RFC 6071**: IP Security (IPsec) and Internet Key Exchange (IKE) Document Roadmap, **RFC 2401**: Security Architecture for the Internet Protocol, **RFC 2411**: IP Security Document Roadmap, **RFC 4301**: Security Architecture for the Internet Protocol, **RFC 2409**: The Internet Key Exchange (IKE), **RFC 4306**: Internet Key Exchange (IKEv2) Protocol, **RFC 4718**: IKEv2 Clarifications and Implementation Guidelines, **RFC 5996**: Internet Key Exchange Protocol Version 2 (IKEv2), **RFC 1321**: The MD5 Message-Digest Algorithm, **RFC 1852**: IP Authentication using Keyed SHA, **RFC 3602**: The AES-CBC Cipher Algorithm and Its Use with IPsec, **RFC 4308**: Cryptographic Suites for IPsec, **RFC 3554**: On the Use of Stream Control Transmission Protocol (SCTP) with IPsec, **RFC 3776**: Using IPsec to Protect Mobile IPv6 Signaling Between Mobile Nodes and Home Agents.

Summary

This chapter covered the basics of network security, which would enlighten the reader about the possible gaps or vulnerabilities in their designed systems concerning various aspects of security. The various types of schemes for ensuring security and overview of basic network security mechanisms are also discussed here. Additionally, some of the exciting aspects of security measures at the network, transport, and application layers of the TCP/IP protocol stack are covered with an aim to provide insights into Internet security. Additional mechanisms such as the firewall and their types have also been outlined.

Exercises

(i)   What attacks are categorized as threats to web databases?

(ii)   What is a zero day attack?

(iii)   What is ransomware?

(iv)   Differentiate between steganography and cryptography.

(v)   What is a phishing attack?

(vi)   Describe the methodology for symmetric key cryptography.

(vii)   Differentiate between symmetric and asymmetric key cryptography.

(viii) Differentiate between block and stream ciphers.

(ix)   What are the typical methods used to ensure message integrity in IoT?

(x)   What is a digital signature?

(xi)   How is key exchange performed in the context of key management?

(xii) What are the typical methods used for network layer security?

(xiii) Discuss IPSec.

(xiv) What is SSL?

(xv)   How is SSL different from TLS?

(xvi) What are the various types of firewalls?

References

[1]   Jajodia, S., S. Noel, and B. O'berry. 2005. "Topological Analysis of Network Attack Vulnerability." In *Managing Cyber Threats* (pp. 247–266). Boston, MA: Springer.

[2]   Salomon, D., 2006. *Coding for Data and Computer Communications.* Springer Science & Business Media.

[3]   Iansiti, M. and K. R. Lakhani. 2017. "The Truth about Blockchain." *Harvard Business Review*, 95(1): 118–127.

[4]   Kessler, G. C. 1998. "An Overview of Cryptography." *The Handbook on Local Area Networks.* Auerbach.

[5]   Chapman, D. B., E. D. Zwicky, and D. Russell. 1995. *Building Internet Firewalls..* O'Reilly & Associates, Inc.

Predecessors of IoT

After reading this chapter, the reader will be able to:

- Interpret the foundation of IoT
- List the basic terminologies and technologies associated with wireless sensor networks (WSN)
- List the basic terminologies and technologies associated with machine-to-machine communications (M2M)
- List the basic terminologies and technologies associated with cyber-physical systems (CPS)
- Differentiate between WSN, M2M, and CPS
- Relate new concepts with concepts learned before to make a smooth transition to IoT

3.1 Introduction

Before delving into the details of the Internet of Things (IoT), a discussion on the base technologies, which make up the crux of IoT, is required. A majority of these technologies, before the IoT era, were used separately for sensing, decision making, and automation tasks. The range of application domains of these technologies extended from regular domains like healthcare, agriculture, home monitoring, and others to specialized domains such as military and mining. Some of these precursor technologies still being used and often re-engineered for IoT are wireless sensor networks (WSN), machine-to-machine (M2M) communications, and cyber physical systems (CPS). All of these precursor paradigms have their distinct signatures and application scopes. A basic overview of these precursor technologies is covered in the subsequent sections in this chapter.

3.2 Wireless Sensor Networks

Wireless sensor networks (WSN), as the name suggests, is a networking paradigm that makes use of spatially distributed sensors for gathering information concerning the immediate environment of the sensors and collecting the information centrally. Here, the sensors are not standalone devices but a combination of sensors, processors, and radio units—referred to as sensor nodes—sensing the environment and communicating the sensed data wirelessly to a remote location, which may or may not be connected to a backbone network. Figure 3.1 shows the block diagram of the various standard components of a typical WSN node [4]. The exact specifications of each of these blocks vary depending on the implementation requirements and the network architect's choice.

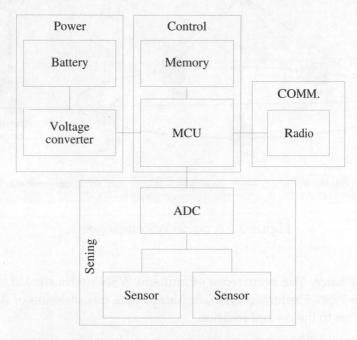

Figure 3.1 The typical constituents of a WSN node

Figure 3.2 shows a typical WSN implementation, where the master node aggregates data from multiple slave nodes, forwards it to a remote server utilizing access to the Internet through cellular connectivity. The stored data on the server can be visualized by a user or a subscriber to the system from anywhere in the world over the Internet. WSNs mainly follow a system of communication known as master–slave architecture. In a master–slave architecture, a single aggregator node, the master, is responsible for collecting data from various sensor nodes under its dominion or range of operations. The sensor nodes under the range of the master node are referred to as slave nodes. Multiple slave nodes communicate to the master node using low-power short-range wireless radios such as Zigbee, Bluetooth, and WiFi for transferring their sensed data to a remote central server. Often, in popular WSN architectures, the master node connects the WSN to the Internet and acts as the gateway for the WSN. Upon collecting data from the slave nodes, the master node pushes the aggregated data to a remotely located central server using the Internet. The master node may be linked to the Internet through cellular connections, another gateway, or directly through a backbone infrastructure. WSNs must have the following distinguishing features:

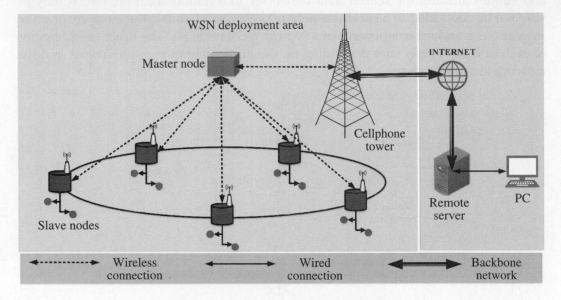

Figure 3.2 A typical WSN deployment

(i)   **Fault Tolerance**: The occurrence of faults in WSN nodes should not take down the whole WSN implementation, or hamper the transmission of data from non-faulty nodes to the central location.

(ii) **Scalability**: WSN implementations must have the feature of scalability associated with their architectures and deployments. In the event of a future increase or decrease of sensor node units, the WSN must support the scaling of the infrastructure without changing the whole implementation.

(iii) **Long lifetime**: The lifetime or the energy replenishment cycle of WSNs must be long enough to make large-scale applications feasible. WSNs have been used for monitoring remote, harsh, and hard to access environments; in these environments, it is not feasible to regularly replenish the energy source of the WSN nodes, which necessitates the need for long node lifetimes.

(iv) **Security**: The security of WSNs, if not considered, can easily compromise the security of the whole system, right back to the central server. As WSNs are used for a wide range of applications, some of which are crucial, security is one aspect that must be properly addressed to prevent intrusion and maintain the integrity of the data.

(v) **Programmability**: The programmability of WSNs is important as it ensures the robustness of these systems. WSNs deployed in one application area can be reused for other applications just as easily with a change in sensors and the backend programs associated with it. Programmability also helps in providing a means of adjusting the parameters of the system in the event of a scale-up or scale-down operation.

(vi) **Affordability**: As WSNs generally require multiple units, typically in the range of tens or hundreds of WSN nodes, the cost of the nodes and its affordability is vastly responsible for the acceptability of the system. Except for some specialized domains such as the military and the industry, where the sensing requirements are quite high and that too in harsh and challenging conditions, the majority of WSN applications are regular. To some extent, the cost of WSN deployments in these regular domains decide the acceptability of a WSN-based solution for that domain.

(vii) **Heterogeneity**: The WSNs must support a wide number and various types of sensors and solutions, thus enabling heterogeneity. In the absence of heterogeneity, the WSN will tend to become very application-specific, which in turn would require major customizations even in the event of minor changes to the network or architecture.

(viii) **Mobility**: WSNs must support the notion of mobility of nodes such that the nodes may be easily relocatable or mobile. Mobility would ensure the rapid deployability of WSN-based solutions in all environments types.

WSNs have found numerous applications in domains such as agriculture, healthcare, military, industries, mining, and others. The main reason for the popularity of WSNs is attributed to the advantages they provide in the form of enhanced monitoring times, easy installation, and multiple implementations. Implementations on a large scale are possible due to high affordability, ease of replacement or upgradation, ease of modifying system parameters, ease of additional sensor integration with the sensor nodes, and other such factors.

3.2.1 Architectural components of WSN

WSNs that are similar to regular computer network paradigms may be explained in terms of a protocol stack, which is very similar to the ISO/OSI (Open Systems Interconnection developed by the International Standards Organization) stack. However, instead of seven layers similar to the OSI model, the WSN stack is made up of five layers (Figure 3.3): 1) Physical, 2) data link, 3) network, 4) transport, and 5) application layer. In addition to these five layers, the WSN stack further comprises five cross planes concerned with management tasks such as 1) power management plane, 2) mobility management plane, 3) task management plane, 4) QoS management plane, and 5) security management plane. This section is divided into three parts: 1) Components of the WSN stack, 2) cross-layer management planes, and 3) WSN types.

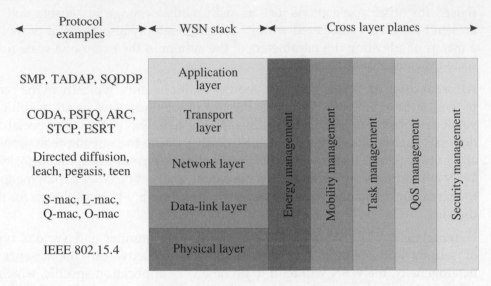

Figure 3.3 The various functional layers for a WSN communication and networking architecture

Components of the WSN Stack

We start with the physical layer, which is at the bottom of the stack and responsible for enabling transmission of signals over a physical medium between multiple WSN nodes/units. In regular computer networks, the medium can be both wired as well as wireless; however, in WSNs, the medium is strictly wireless. In WSNs, this layer is responsible for carrier frequency generation, carrier frequency selection, modulation/ demodulation, encryption/decryption, and signal detection. Typically, WSNs make use of the IEEE 802.15.4 standard for this layer because of its low cost, low energy budget, low data rate, and small form factor.

The data link layer resides above the physical layer. It is responsible for medium access control (MAC) functions such as multiplexing/ demultiplexing, framing of messages from the upper layer, frame detection, and error control. These functions help in ensuring the reliability of communication between the WSN nodes. The network layer lies on top of the data link layer. The primary function associated with this layer is the routing of packets. As routing is a demanding task and depends on many factors affecting the network elements (nodes, gateways, routers, switches, and servers), the choice of routing protocols dictates the power and memory requirements of the WSN elements such as the sensor nodes.

The transport layer sits on top of the network layer. This layer, in contrast to the network layer, plays a crucial role in ensuring reliability and congestion control of the packets arriving and leaving from each WSN node. Protocol-based mechanisms for loss recognition and loss recovery are inherent to this layer. Typically, the protocols in this layer are either packet driven or event-driven. Finally, the application layer, which is responsible for traffic management and software interfaces, sits on top of all these previous four layers. The software interfaces are responsible for the conversion of data from various application domains of WSN into an acceptable format for transfer to the layer underneath this layer.

Points to ponder

Routing: It is the selection of a path for the transport of a packet from a source to its designated destination across the same network or various networks. The networks over which routing takes place may or may not be of the same type. For example, an e-mail sent from a personal computer (PC) travels across various networks—wired, wireless, circuit-switched, and others—before reaching its destination.

Cross-layer Management Planes

The use of OSI-like stacks for outlining the functionalities of WSNs faces limitations due to specialized operations of the stacks in areas requiring prolonged deployments with constrained energy and communication infrastructure, and mobility. Unlike regular computer networks, the OSI-like WSN stack does not fully describe the functionalities of WSN-based systems. This is because its specialized nature results in a strong correlation between the five WSN stack layers. Typically, solutions addressing WSN applications and functionalities make joint use of all the five layers. It is mainly because of this reason that the cross-layer management plane structure is more popularly accepted as a means of abstraction of WSN-based systems and solutions. Table 3.1 shows the relative positioning of the WSN crosslayer management planes, in terms of the functionalities they provide.

Points to ponder

Target Tracking: Also known as object tracking, target tracking is one of the most significant areas of research in WSNs. It summarily deals with tracking of the motion of a target within a designated area of WSN coverage. This is typically achieved by the clustering of tracking nodes in the vicinity of the target. As the target moves, the clusters change, which signifies the direction and time of motion of the target.

Table 3.1 A comparison of the WSN cross-layer management planes

Sl. no.	Features	Energy management	Mobility management	Task management	QoS management	Security management
1	Functionality	Maximizing the energy of WSN nodes and overall network energy management.	Ensuring connectivity even when the WSN nodes are moving.	Distribution of tasks among WSN nodes for ensuring network lifetime.	Ensuring the quality of the service being offered by the WSNs.	Ensuring uncompromising security and integrity of the sensed and transmitted data.
2	Applications	Environment monitoring, Home monitoring, Multimedia sensors	Vehicular monitoring, Unmanned aerial vehicle networks	Environment monitoring, Agricultural monitoring, Underground and underwater sensor networks	Vehicular monitoring, Multimedia sensors	Military sensor networks, Industrial monitoring
3	Tasks	Deciding network size and deployment density.	Topology management, Sensing coverage, Communication coverage	Sensor scheduling, Processing scheduling, Data forwarding	Bandwidth allocation, Resource allocation, Error management	Security of transmission, Node security
4	Challenges	Node deployment density, Sleep scheduling, Maximizing node lifetime	Clustering, Routing	Leader election, Workforce selection	Bandwidth optimization, Jamming avoidance	Low-power security protocols

Classes of Wireless Sensor Networks

The portability and robustness of WSN solutions, in addition to the significantly enhanced operational lifetimes, made WSNs quite a popular choice for applications in various diverse areas. Some of the applications of WSNs include the following:

(i)   Military Applications: WSNs are used for the detection of enemy soldiers, vehicles, intrusion, weapon systems, and armaments.

(ii) Health Applications: WSNs in healthcare are being used to monitor patients in hospitals, ambulances, and homes. Nowadays, a new class of healthcare devices—wearable appliances—enable a user to have a miniature health sensor on them without additional discomfort.

(iii) Environmental Applications: WSNs are used for environmental monitoring of pollution, tracking of wildlife, forests, and others.

(iv) Home Applications: WSNs in the home have given rise to home automation systems and smart home connectivity systems.

(v)   Commercial Applications: WSNs are used for tracking of vehicles, packages in transport, logistics, and others.

(vi) Industrial Monitoring: WSNs in industries keep track of various industrial processes, monitor factory floors, ensure worker safety, and perform stock management.

WSNs can be organized broadly into the following domains of implementation (Figure 3.4): 1) Wireless multimedia sensor networks, 2) underwater sensor networks, 3) wireless underground sensor networks, and 4) wireless mobile sensor networks. Typically, these networks have specific challenges that need to be addressed in each of these implementation areas.

(i)   **Wireless Multimedia Sensor Networks (WMSN)**: This class of WSNs boasts of the ability to retrieve videos, audios, images, or all three in addition to regular scalar sensor readings. The sensing range and coverage area of a camera-based WMSN are defined by the field of view (FOV) of the constituent cameras. Because of their superlative capabilities, WMSNs are popularly sought after in critical domains such as surveillance and road traffic monitoring. However, due to the use of multimedia sensors, the power and processing requirements of this class of WSNs are very high as compared to the other classes of WSNs. Additionally, WMSNs typically demand high network bandwidths for convenient operation. Cell A in Figure 3.4 shows WMSNs.

(ii) **Underwater Sensor Networks (UWSN)**: This class of WSN is designed specifically to work in underwater environments. Compared to terrestrial or aerial environments, wireless underwater communications are severely restricted in terms of data rate, range, and bandwidth due to the high underwater attenuation of electromagnetic signals. The reliable use of light-based underwater communication is also limited in terms of range and noise.

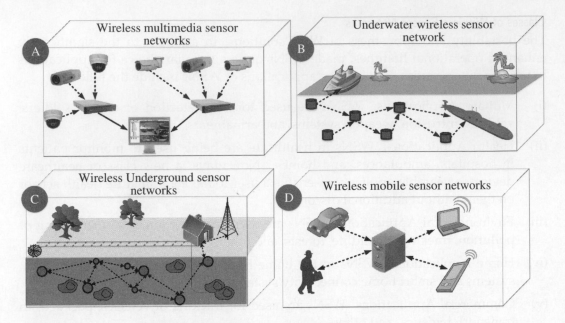

Figure 3.4 The various domains of implementation of WSNs signifying its types: A) WMSN, B) UWSN, C) WUSN, and D) MSN

The most accepted global means of underwater communication is by acoustic waves. However, the long propagation delays and uneven data rate makes it necessary to develop newer topologies and architectures, which can work under the conditions of severe limitations of the physical layer. Cell B in Figure 3.4 shows the deployment of UWSN.

(iii) **Wireless Underground Sensor Networks (WUSN)**: This class of WSNs is designed to be deployed entirely underground. The underground environment poses challenges of attenuation due to the rocks and minerals in the soil. Another significant problem associated with this class is the need for digging up of the nodes to replenish their energy sources. Typical usage scenarios of this class of WSNs are underground mines and monitoring of underground plumbing systems. WUSNs need denser deployment architectures owing to the limited range of wireless communication in underground environments. Cell C in Figure 3.4 shows the deployment of WUSN.

(iv) **Wireless Mobile Sensor Networks (MSN)**: This class of WSNs is characterized by its mobility and low power requirements. The sensor nodes are mobile, which requires them to rapidly connect to networks, disconnect from them, and then again connect to new networks until the nodes are mobile. Typical examples of MSNs include smartphone networks, wearables, vehicular networks, and others. Cell D in Figure 3.4 illustrates an MSN.

> **Points to ponder**
>
> **Scalar and Vector Sensors**: A scalar quantity is defined as a quantity that can be defined only by its magnitude. Some examples of scalar quantities include speed, volume, mass, temperature, energy, and time. A vector is a quantity that has both magnitude and direction. Some examples of vector quantities include force, velocity, acceleration, displacement, and momentum.
> A scalar sensor will give ambient values and not provide any information about the changes in these values with distance r space (e.g., a temperature sensor). A vector sensor will give precise values and their changes for every point of space it covers (e.g., a camera).

3.3 Machine-to-Machine Communications

The machine-to-machine (M2M) paradigm, as the name suggests, implies a system of communication between two or more machines/devices without human intervention. Some basic examples of M2M communication in our daily lives include ATM machines signaling banks about the need to refill them with cash, power line monitoring systems in a house alerting a generator set of possible power failures and switching to generator-based supply, vending machines updating stock of items in their inventory and alerting a remote inventory of the need to refill certain depleting items, and others [5]. Figure 3.5 outlines the significant aspects of an M2M ecosystem. The task of any sensor network system (be it WSN, M2M, CPS, or IoT) is the sensing of a physical environment and converting the acquired data into a tangible output in the form of numbers using sensors. This sensing is followed by the transfer of sensed data to a remote device or location using a network, which may be wired or wireless. The data collected at the remote device from various sensors—homogeneous or heterogeneous—is converted to usable information, which can be utilized to define the course of actions for individual scenarios. This information is processed to decide upon the most valid and optimum course of action that must be undertaken to control the sensed environment desirably or as per requirements. Finally, actuators are put to work to modify or adjust the sensed environment.

A centralized air conditioning system in a building is a good example of this paradigm. If the requirements of a particular room in that building dictate that the room must be kept at a constant temperature of $20°$ C, the sensors continuously sense the environment (temperature of the room) and direct the actuators controlling the cold air inlet to keep the room at the desired temperature.

The massive and rapid developments in the field of wireless communication have significantly helped in the widespread deployment of M2M solutions world over. The 3rd Generation Partnership Project (3GPP) [6], which is responsible for unifying and benchmarking telecommunication standards across seven different

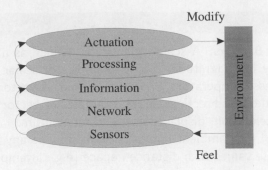

Figure 3.5 An overview of the M2M ecosystem

telecommunication standardization organizations, refers to M2M as machine type communications (MTC). 3GPP highlights the following significant characteristics of M2M: 1) heterogeneous markets, 2) low data footprint, 3) low cost, maintenance, and integration efforts, and 4) a very large number of communicating devices working without human intervention.

> **Points to ponder**
>
> **REST**: *RE*presentational *S*tate *T*ransfer is an HTTP-based web service development architectural style. The main aim of this paradigm is to enable computer/system interoperability over the Internet. The interoperability is achieved by using a predefined and seamless set of stateless operations.

The communication network in M2M serves as a transport medium for exchanging data between two or more devices. It may be wired as well as wireless. Due to the uncertain nature of deployment of a majority of the M2M systems (e.g., portable vending machines, vehicle monitoring systems, and others), preference is given to wireless cellular communication systems such as GSM-GPRS, CDMA, and EVDO. As the data between M2M devices is sporadic, their usage does not imply network congestions and frequent network bandwidth readjustments. Typically, the following features are directly identifiable with M2M; they also make M2M a very robust and interesting technology to behold:

(i)   **Negligible Mobility**: M2M devices are generally not mobile and may be considered mostly static. Even if the need for mobility arises, the displacement is generally minimal in terms of network signals and communication parameters.

(ii) **Time-restricted Transmissions**: The data transmission between M2M devices/terminals are highly time-bound and restricted. The data transmission duration is generally pre-defined.

(iii) **Delay Tolerant**: The data transmission between M2M devices is not real-time. The connected systems are designed to be delay tolerant.

(iv) **Packet Switched**: The M2M communication network is always packet-switched. The communication between the devices/terminals is in the form of packets with their respective headers, footers, and payloads.

(v) **Small Data Footprint**: The data footprint of M2M devices is tiny. However, the frequency of transmission and reception of data between the devices is typically high.

(vi) **Event Detection**: M2M devices are designed to detect and monitor events only. They do not generally react physically or actuate in the occurrence of events.

(vii) **Low-power Requirements**: Very low power requirements characterize M2M systems. This makes them easily integrable with a variety of solutions ranging from industrial and commercial systems to even household systems.

Check yourself

GPRS, GSM, CDMA, EVDO, LTE, WiMAX

The M2M paradigm finds vast avenues in applications such as environmental monitoring, civil protection and public safety, supply chain management (SCM), energy and utility distribution industry (smart grid), intelligent transport systems (ITS), healthcare, automation of buildings, military applications, agriculture, and home networks.

The standard requirements of an M2M platform include features shown in Figure 3.6 [2]. A *device management* profile should be available on the platform so that the platform can add, remove, modify, and query devices attached to the platform. The features of control and authentication of devices also fall under this category. The *user management* profile is tasked with user access restrictions, device access permissions, service access permissions, and user authentication and management tasks. The *data management* feature enables the collection of data from different devices and objects connected to the platform. It should allow the user to query the collected data, control the actuators connected to the platform, as well as allow the data to dictate the control of connected devices and actuators. The *web access* and *cloud* features are intended to provide anywhere, anytime, and anyhow access to the data and the devices connected to the M2M platform over the Internet. A provision for peer-to-peer (P2P) communication should be present to enable direct control and communication between devices over the Internet. The *P2P communication* feature facilitates the reduction of unnecessary traffic through the M2M platform by reducing unnecessary platform accesses and data storage of the platform. The *M2M area network* management profile facilitates the control of a zonal implementation of M2M, such as in sensor networks. Finally, the *connection management* allows for the interoperability between devices and communication methods by utilizing a single platform.

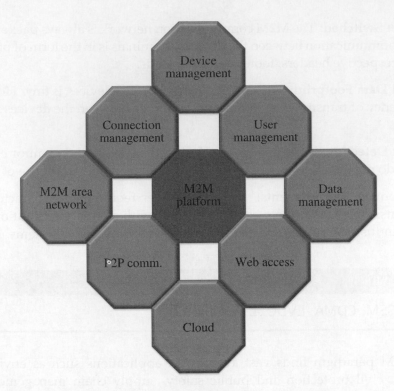

Figure 3.6 The various features desirable in an ideal M2M platform

3.3.1 Architectural components of M2M

M2M being a complex paradigm is better understood if the components are grouped under the following two categories: 1) M2M networking model and 2) M2M service ecosystem. The networking model approaches the prospective scopes and features of the M2M platform in terms of the networking components and their roles. The service ecosystem attempts to describe the M2M platform and interactions in terms of the various service providers, their roles and responsibilities.

M2M Networking Model

- **M2M Devices**: M2M devices are those entities that are capable of responding to requests for data by means of replying through networked messages almost autonomously. The devices at the end of the network, which are tasked with sensing and actuation, also fall under this category. The mode of communication of these devices may be wired or wireless. They connect to a network through a gateway, or directly using a cellular operator's network. In the latter case, the responsibility of the device's service-level agreements (SLAs) and accountability lie with the cellular network provider. For the remaining cases, accountability and SLAs are undertaken by internet service providers (ISPs). M2M devices can be broadly categorized into three types based on their interaction and features concerning an M2M platform.

(i) Low-end Devices: This device type is typically cheap and has low capabilities such as auto-configuration, power saving, and data aggregation. As devices of this type are generally static, energy-efficient, and simple, a highly dense deployment is needed to increase network lifetime and survivability. Moreover, low-end devices are resource-constrained with no IP (Internet protocol) support; they are generally used for environmental monitoring applications.

(ii) Mid-end Devices: These devices are more costly than low-end M2M devices as they may have mobility associated with them. However, these devices are less complex and energy-efficient than high-end devices. The presence of capabilities such as localization, intelligence, support for quality of service (QoS), traffic control, TCP/IP support, and power control make them appealing for applications such as home networks, SCM (supply chain management), asset management, and industrial automation.

(iii) High-end devices: These devices generally require low density of deployment. They are designed such that they can handle multimedia data (videos) with QoS requirements, even in mobile environments. The mandatory inclusion of mobility as a feature of these devices makes them costly. Generally, they are used for ITS (intelligent transportation system) and military or bio-medical applications.

- **M2M Area Network**: The M2M area network comprises multiple M2M devices, either communicating with one another or to a connected platform, which is remotely situated. The local communication between the M2M devices up to the M2M gateway can be considered as the M2M area network. It is also referred to as the *device domain*. Some examples that can be correlated to the functioning of M2M area networks include personal area networks (PANs) and local nodes in a wireless sensor network (WSN).

- **M2M Gateway**: It is responsible for enabling connectivity and communication between the M2M devices and a global communication channel such as the Internet. The gateway is responsible for distinguishing between data and control signals on the M2M platform to enable monitoring as well as maintenance of the M2M area network remotely. The gateways must additionally ensure that the M2M devices can access an outside network and that the devices themselves can be accessed from an outside network.

- **M2M Communication Network**: This is also referred to as the M2M network domain. It consists of the communication technologies and paradigms for enabling connectivity and communication between M2M gateways and various applications. Some M2M communication network enablers include WLAN, WiMAX, LTE, and others. These M2M networks can be classified as either IP-based or non-IP-based.

(i)   *IP-based Networks*: These networks are supported only by high-end M2M devices. As the other two M2M device types—low-end and mid-end—are typically resource-constrained, IP-based addressing over TCP is not supported. Both IPv4 and IPv6 are supported in IP-based M2M networks. However, IPv6-based schemes are preferred due to the provision of scalability in IPv6-based communication, which is absent in IPv4. Schemes such as IPv6 over low-power wireless personal area networks (6LoWPAN) have been used to extend IPv6 support to resource-constrained M2M devices; however, it is yet to meet consumer's expectations. Figure 3.7 shows the stack-wise communication of IP-based M2M networks. As both the M2M network and the IP network (which is global/ accessible by the Internet) follow similar stack structures, the communication between them is direct, without the need for adaptors or protocol tunnels.

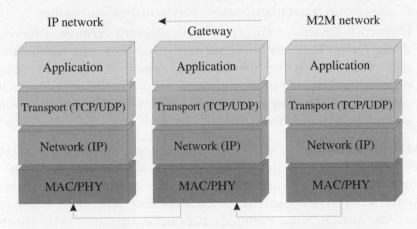

Figure 3.7 M2M communication over an IP-based network

(ii) *Non-IP-based Networks*: This scheme is generally used with low-end and mid-end M2M devices as their configurations and capabilities do not support resource-hungry protocols such as the TCP or UDP. A separate addressing scheme is designed for accommodating these resource-constrained devices and is limited to the domain within the M2M gateways. Figure 3.8 shows the non-IP-based network communication. The packets from the resource-constrained M2M devices within the M2M area network forward their packets to the gateway, which again packetizes them into IP-based packets for further transmission to an IP network.

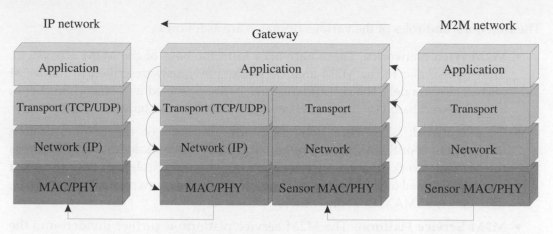

Figure 3.8 M2M communication over a non-IP-based network

M2M Service Ecosystem

The M2M service ecosystem, unlike the networking model, classifies the various components of the system based on the needs of the service offerings from the M2M platform. Figure 3.9 shows the M2M service platform's components and the ecosystem thus formed. The ecosystem can be broadly divided into four domains: 1) M2M area networks, 2) core network, 3) M2M service platform, and 4) stakeholders.

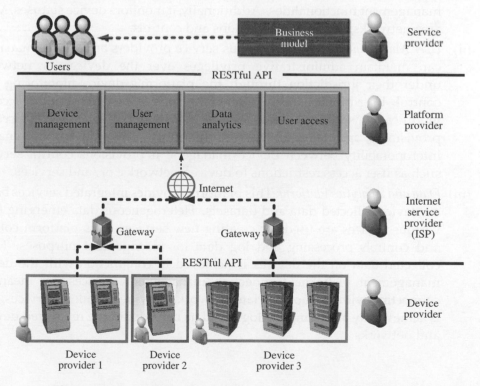

Figure 3.9 The M2M service platform components and ecosystem

The functions and roles of the various domains are as follows.

- **M2M Area Networks**: These networks form the base of the M2M ecosystem. They are similar to the M2M area networks previously described in the M2M networking model. The constituent devices are classified as low-end, mid-end, and high-end based on their functionalities and ability to handle mobility.

- **Core Network**: The core networks form the crux of the communication infrastructure of the M2M ecosystem, and carry the bulk load of traffic across the M2M network. The core network can be wired or wireless or both. Some of the conventional technologies associated with the core networks include WLAN, LTE, GSM, WiMAX, DSL, and others.

- **M2M Service Platform**: The M2M service platform is further divided into the following four parts:

  (i) *Device Management Platform*: This platform enables anytime anywhere access between Internet-connected platforms and registered objects or devices connected to the platform. During device registration, an object database is created from which information can be easily accessed by end users such as managers, users, and services. The main functions of this platform include the management of device profiles (location, device type, address, and description), authentication, authorization, and key management functionalities. Additionally, it monitors device statuses, M2M area networks, and their interactions and controls.

  (ii) *User Management Platform*: Various service providers and device managers can maintain administrative privileges over the devices or networks under their jurisdiction through the platform's device monitoring and control. User profiles and functionalities such as user registration, account modification, service charging, service inquiry, and other M2M services are provisioned and managed through this entity. The platform also enables interoperability between device managers; it provisions control services such as user access restrictions to devices, networks, or/and services.

  (iii) *Data and Analytics Platform*: This platform provides integrated services based on device-collected data and datasets. Heterogeneous data emerging from various devices are used for creating new services. This platform collects and controls processing and log data for management purposes. These collected data on the devices is achieved in conjunction with the device management platform. Connection management services, by means of connecting with the appropriate network, provide seamless services; this is achieved by analyzing the log and data behavior of the registered devices and networks.

(iv) *User Access Platform*: This platform provides a smartphone and web access environment to users. It redirects requests to service providers who have a mapping of the registered devices, users, and the services subscribed. Provisions for modifications to a device or user-specific mapping is also provided.

- **Stakeholders**: The stakeholders in an M2M service ecosystem can be divided into five different types: Device providers, Internet service providers (ISPs), Platform providers, Service providers, and Service users. The functional jurisdiction of each of these five classes of stakeholders is well defined and devised in such a manner that they do not overlap and may be considered mutually exclusive in terms of their offerings. However, at the time of functioning, all these stakeholders have to work together to ensure the smooth functioning of the M2M service ecosystem. Each of these stakeholders and their domains is outlined in Figure 3.9.

3.4 Cyber Physical Systems

Cyber physical systems (CPS) are Internet-based and networked monitoring and controlling systems that are regulated and governed by feedback-based intelligent control algorithms. These systems work in a highly interdisciplinary domain that involves expertise in lots of domains such as mechanical, electrical, computing, electronics, and many more. They are mainly designed to monitor and control physical world processes linked to businesses and industries [1]. The most interesting aspect of CPS is the involvement of the concept of human-in-the-loop, which is an integral part of many CPS-based solutions. The human-in-the loop concept simply signifies the involvement of humans in the CPS control cycle. The striking difference of CPS from paradigms such as WSN and M2M is the inclusion of a compulsory feedback system.

The typical functioning of a CPS includes the components shown in Figure 3.10. The sensing mechanism senses an environment. Various networked sensors simultaneously generate data for the environment, which is sent over the Internet to a processing cum controlling unit. Depending on the intelligent monitoring and control algorithms in the control unit, feedback is provided to the actuators controlling the state of the environment for which the sensed data was transmitted. The changes are again sensed and forwarded to the controller via the previously defined flow. The algorithms decide whether the desired state of the environment is achieved or not; they keep sending adjusted feedbacks to the actuators until the desired state is achieved.

CPS is used in a vast range of applications such as backhaul communications, smart grids, healthcare, industrial manufacturing, smart homes and buildings, military and surveillance, robotics, and even transportation. The following features generally characterize CPS.

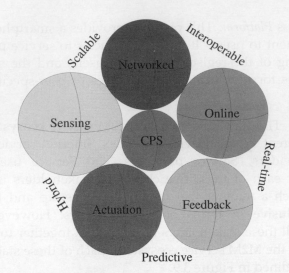

Figure 3.10 The basic overview of CPS features

(i) **Real timeliness**: CPS depends on real-time communication, processing, and feedback to effectively provide control to the environment they are deployed in. For example, in a CPS-based industrial chemical concentration monitoring system, the real-timeliness of the process from sensing to feedback to actuation is crucial for maintaining the operations of the chemical manufacturing plant and preventing disasters.

(ii) **Intelligence**: Intelligent and adaptive decision making is crucial for the maintenance of CPS-based functionalities. If random or sudden changes in the environment need to be controlled, this feature ensures effective control of the environment and effective coordination between the various dependent sub-processes and systems. For example, in case of an electrical fault in a section of a smart-grid system, the intelligence feature would enable the re-routing of the electrical supply flow through other paths instead of bringing the whole system to a complete standstill.

(iii) **Predictive**: This feature enables the prediction of outputs and events based on past behavior under similar constraints and conditions. The prediction of events enables the activation of precautionary measures to control the damage if the harmful event does occur. For example, the trend of minor line disturbances and noise in a communication channel might lead to network data loss in a backhaul network. The predictive feature would help in the timely activation of preventive countermeasures to avoid network outage in case the network does start massively dropping packets.

(iv) **Interoperable**: The vast and massive deployment zones of CPS-based systems may include software as well as hardware from a variety of manufacturers. This would lead to data, speed, and format mismatch under normal circumstances;

however, CPS-based interoperability prevents this and enables systems from various vendors to work in sync with one another as a single system. Interoperability also ensures that legacy systems already in place are not replaced but are added to the CPS infrastructure.

(v)   **Heterogeneous**: Heterogeneity in CPS-based systems may be in the types of actuators, sensors, processors, and data formats being used, besides the sensing types, software, and application types in a given CPS deployment. However, provisions are already present in CPS to accommodate these types of challenges.

(vi) **Scalable**: Scalability in CPS-based systems may be in terms of network bandwidth being required due to various sensing types (scalar or multimedia), number of sensors and actuators, size of deployment zones, and other factors. CPS systems should be able to handle such demands even after preliminary deployment. For example, a smart building wants to incorporate human presence detectors to control the central cooling for the whole building. Initially, conventional scalar sensors were deployed on all floors and corridors to monitor the approximate headcount of people in the building. However, after some years, the building management upgrades the scalar sensors by replacing them with camera sensors. Cameras generate huge volumes of data as compared to the scalar sensors previously used. The deployed CPS should be able to accommodate this upgrade without changing the whole system.

(vii) **Secure**: The security of CPS is crucial as almost all of the traffic flows through a network and eventually over the Internet. Provisions should be in place to avoid unauthenticated use of the CPS and its hijacking by unscrupulous elements or even attacks, which may reduce the response of the system or eventually bring it down all together.

Points to ponder

Digital Twin: Digital twins are behavioral and functional mathematical models of actual physical systems [7]. These are similar to simulations and are mostly used in industrial and machine health monitoring. As most industrial machinery and systems are very expensive, irreplaceable, and often cannot be isolated to run health and system diagnostics, the concept of digital twins is used to gauge the performance of these systems under various constraints and operating conditions. During the use of digital twins, which are virtual models, in the event of any failure or damage during experimentation or diagnostics, no harm comes to the actual pieces of machinery and processes. This results in huge savings in terms of productivity, costs, and time. Digital twins can also be easily put under various conditions and constraints to predict how the actual physical system would behave under similar conditions.

3.4.1 Architectural components of CPS

CPS is widely sought after by various organizations and industries due to its ability to conveniently, accurately, and timely control and automate various high-risk tasks and processes. However, the use of a wide variety of device and solution vendors in industries poses a real challenge to the implementation of CPS. Additionally, as most of the industries already have devices, plants, and systems in place, the replacement of which is not an option, the CPS-based deployment solution must incorporate new as well as legacy systems for the process of control and automation. The deployment architectures for the CPS-based solution must be robust enough to handle the basic features of CPS as previously described, as well as incorporate these legacy systems to achieve an effective and cost-efficient solution. One of the most accepted CPS architectures is termed the 5C architecture [1]. The 5Cs—connection, conversion, cyber, cognition, and configuration—aptly describes the CPS control flow and functionalities and is shown in Figure 3.11. Each of these Cs can be described as follows:

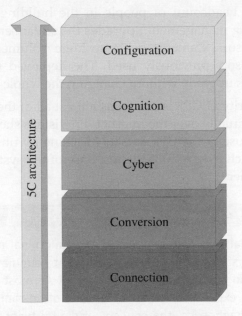

Figure 3.11 The 5C architecture for CPS

- **Connection**: The sensed data from the base of the architecture should be accurate and reliable enough to actuate effectual feedback for the whole system. The sensed data from various sensor units should be collected in a hassle-free and organized manner. The best possible solution is the use of tether-free communication systems, which should be able to support plug-and-play features of these sensing units.

- **Conversion**: The collected data should be converted to a standard unified format. Post data standardization, usable information must be extracted from the sensed data. Data from various sensor types and sources need to be correlated to generate practical information from vastly multi-dimensional data. This data can be used to predict changes to the monitored environments, machinery malfunctions, and failures.

- **Cyber**: This acts as the central nodal point of data collection and the holistic analysis of the system under control of the CPS. Data from various machine networks, environments, systems, and processes arrive at this point. Detailed and advanced analytics on the obtained data is performed to gather statistical trends. These trends can be used to predict the future behavior of machine systems and processes. The prediction can be based on digital twins of the actual systems, comparative performance of a machine with other machines, and temporal and regression results of machine health and performance.

- **Cognition**: This level is mainly responsible for the amalgamation of the collective health of the running systems and processes. The information is presented in the form of human-readable visualizations and trends. This helps in prioritizing actions and control of processes and systems under the purview of the CPS.

- **Configuration**: This stage is responsible for generating feedback for adjusting the environment being controlled. The feedback systems need to be highly adaptive, self-configuring, and resilient for effective control of the system as a whole.

Check yourself

Industry 4.0, Condition based monitoring (CBM), Prognostics and health management (PHM), Decision support system (DSS), Resilient control system (RCS)

Summary

This chapter briefly covered the various technologies considered as the immediate precursors of IoT—WSN, M2M, and CPS. Each of these technologies, their applications, their highlighting features, and architectural components are described. This overview would provide the reader with a functional knowledge of these domains, in turn enabling them to appreciate the rich history behind the emergence of IoT as it is today. The reader would be in a good position to understand the various nuances of IoT, as most of the terms and paradigms covered in this chapter would be repeatedly encountered in the subsequent chapters. The heterogeneity of application domains of each of these discussed paradigms also highlights the unifying nature of

IoT as a comprehensive and unifying solution for various application scenarios and under various constraints.

Exercises

(i)   What are WSNs?

(ii) What are the typical components of a WSN system?

(iii) What factors determine the utility of WSN?

(iv) Discuss the cross-layer management plane in WSN.

(v)   What are the various classes of WSNs?

(vi) How are wireless multimedia sensor networks different from wireless mobile sensor networks?

(vii) Describe the M2M paradigm.

(viii) How is M2M different from WSNs?

(ix) Describe the CPS paradigm.

(x)   How is CPS different from WSN?

(xi) Differentiate between CPS and M2M.

(xii) What are the components of the M2M networking model?

(xiii) What is the M2M service ecosystem?

(xiv) What are non-IP-based networks?

(xv) What are the various architectural components of CPS?

References

[1]   Lee, J., B. Bagheri, and H. A. Kao. 2015. "A Cyber-physical Systems Architecture for Industry 4.0-based Manufacturing Systems." *Manufacturing Letters* 3: 18–23.

[2]   Kim, J., J. Lee, Kim, and J. Yun. 2013. "M2M Service Platforms: Survey, Issues, and Enabling Technologies." *EEE Communications Surveys & Tutorials* 16(1): 61–76.

[3]   Anghel, A., G. Vasile, R. Boudon, G. d'Urso, A. Girard, D. Boldo, and V. Bost. 2016. "Combining Spaceborne SAR Images with 3D Point Clouds for Infrastructure Monitoring Applications." *ISPRS Journal of Photogrammetry and Remote Sensing* 111: 45–61.

[4]   Ojha, T., S. Misra, and N. S. Raghuwanshi. 2015. "Wireless Sensor Networks for Agriculture: The State-of-the-art in Practice and Future Challenges." *Computers and Electronics in Agriculture* 118: 66–84.

[5]   Verma, P. K., R. Verma, A. Prakash, K. Agrawal, A. Naik, R. Tripathi, M. Alsabaan, T. Khalifa, T. Abdelkader, and A. Abogharaf. 2016. "Machine-to-Machine (M2M) Communications: A Survey." *Journal of Network and Computer Applications* 66: 83–105.

[6] Damnjanovic, A., J. Montojo, Y. Wei, T. Ji, T. Luo, M. Vajapeyam, T. Yoo, O. Song, and D. Malladi. 2011. "A Survey on 3GPP Heterogeneous Networks." *IEEE Wireless Communications* 18(3): 10–21.

[7] Rosen, R., G. Von Wichert, G. Lo, and K. D. Bettenhausen. 2015. "About the Importance of Autonomy and Digital Twins for the Future of Manufacturing." *IFAC-PapersOnLine* 48(3): 567–572.

PART TWO
INTERNET OF THINGS

Chapter 4

Emergence of IoT

Learning Outcomes

After reading this chapter, the reader will be able to:

- Explain the chronology for the evolution of Internet of Things (IoT)
- Relate new concepts with concepts learned earlier to make a smooth transition to IoT
- List the reasons for a prevailing universal networked paradigm, which is IoT
- Compare and correlate IoT with its precursors such as WSN, M2M, and CPS
- List the various enablers of IoT
- Understand IoT networking components and various networking topologies
- Recognize the unique features of IoT which set it apart from other similar paradigms

4.1 Introduction

The modern-day advent of network-connected devices has given rise to the popular paradigm of the Internet of Things (IoT). Each second, the present-day Internet allows massively heterogeneous traffic through it. This network traffic consists of images, videos, music, speech, text, numbers, binary codes, machine status, banking messages, data from sensors and actuators, healthcare data, data from vehicles, home automation system status and control messages, military communications, and many more. This huge variety of data is generated from a massive number of connected devices, which may be directly connected to the Internet or connected through gateway devices. According to statistics from the Information Handling Services [7], the total number of connected devices globally is estimated to be around 25 billion. This figure is projected to triple within a short span of 5 years by the year 2025. Figure 4.1 shows the global trend and projection for connected devices worldwide.

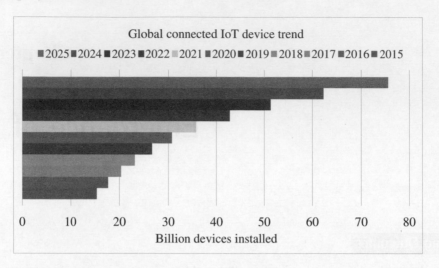

Figure 4.1 10-year global trend and projection of connected devices (statistics sourced from the Information Handling Services [7])

The traffic flowing through the Internet can be attributed to legacy systems as well as modern-day systems. The miniaturization of electronics and the cheap affordability of technology is resulting in a surge of connected devices, which in turn is leading to an explosion of traffic flowing through the Internet.

> **Points to ponder**
>
> "The Internet of Things (IoT) is the network of physical objects that contain embedded technology to communicate and sense or interact with their internal states or the external environment."
>
> —Gartner Research [5]

One of the best examples of this explosion is the evolution of smartphones. In the late 1990's, cellular technology was still expensive and which could be afforded only by a select few. Moreover, these particular devices had only the basic features of voice calling, text messaging, and sharing of low-quality multimedia. Within the next 10 years, cellular technology had become common and easily affordable. With time, the features of these devices evolved, and the dependence of various applications and services on these gadgets on packet-based Internet accesses started rapidly increasing. The present-day mobile phones (commonly referred to as smartphones) are more or less Internet-based. The range of applications on these gadgets such as messaging, video calling, e-mails, games, music streaming, video streaming, and others are solely dependent on network provider allocated Internet access or WiFi. Most of the present-day consumers of smartphone technology tend to carry more than one of these units. In line with this trend, other connected devices have rapidly increased in numbers resulting in the number of devices exceeding the number of humans on Earth by multiple times. Now imagine that as all technologies and domains are moving toward smart management of systems, the number of sensor/actuator-based systems is rapidly increasing. With time, the need for location-independent access to monitored and controlled systems keep on rising. This rise in number leads to a further rise in the number of Internet-connected devices.

The original Internet intended for sending simple messages is now connected with all sorts of "Things". These things can be legacy devices, modern-day computers, sensors, actuators, household appliances, toys, clothes, shoes, vehicles, cameras, and anything which may benefit a product by increasing its scientific value, accuracy, or even its cosmetic value.

Internet of Things

"In the 2000s, we are heading into a new era of ubiquity, where the 'users' of the Internet will be counted in billions and where humans may become the minority as generators and receivers of traffic. Instead, most of the traffic will flow between devices and all kinds of "Things", thereby creating a much wider and more complex Internet of Things."

—ITU Internet Report 2005 [6]

IoT is an anytime, anywhere, and anything (as shown in Figure 4.2) network of Internet-connected physical devices or systems capable of sensing an environment and affecting the sensed environment intelligently. This is generally achieved using low-power and low-form-factor embedded processors on-board the "things" connected to the Internet. In other words, IoT may be considered to be made up of connecting devices, machines, and tools; these things are made up of sensors/actuators and processors, which connect to the Internet through wireless technologies. Another school of thought also considers wired Internet access to be inherent to the IoT paradigm. For the sake of harmony, in this book, we will consider any technology enabling access to the Internet—be it wired or wireless—to be an IoT enabling technology. However, most of the focus on the discussion of various IoT enablers will be restricted to wireless IoT systems due to the much more severe operating constraints and challenges faced by wireless devices as compared to wired systems. Typically, IoT systems can be characterized by the following features [2]:

- Associated architectures, which are also efficient and scalable.
- No ambiguity in naming and addressing.
- Massive number of constrained devices, sleeping nodes, mobile devices, and non-IP devices.
- Intermittent and often unstable connectivity.

Figure 4.2 The three characteristic features—anytime, anywhere, and anything—highlight the robustness and dynamic nature of IoT

IoT is speculated to have achieved faster and higher technology acceptance as compared to electricity and telephony. These speculations are not ill placed as evident from the various statistics shown in Figures 4.3, 4.4, and 4.5.

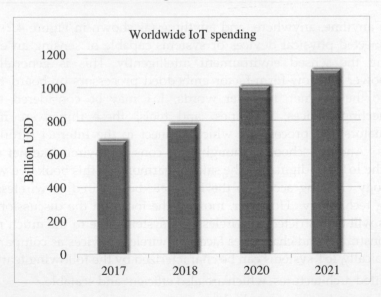

Figure 4.3 The global IoT spending across various organizations and industries and its subsequent projection until the year 2021 (sourced from International Data Corporation [1])

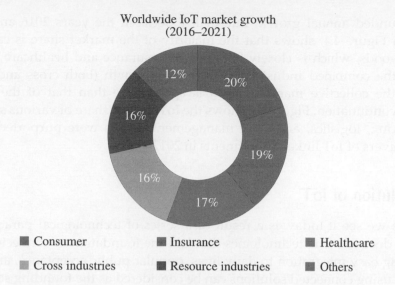

Figure 4.4 The compound annual growth rate (CAGR) of the IoT market (statistics sourced from [1])

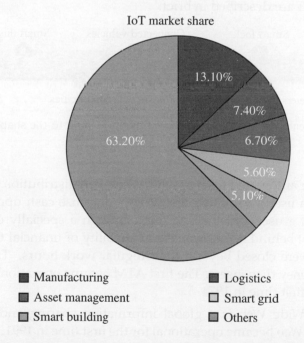

Figure 4.5 The IoT market share across various industries (statistics sourced from International Data Corporation [8])

According to an International Data Corporation (IDC) report, worldwide spending on IoT is reported to have crossed USD 700 billion. The projected spending on IoT-based technologies worldwide is estimated to be about USD 1.1 trillion [1]. Similarly, the compounded annual growth rate of IoT between the years 2016 and 2021, as depicted in Figure 4.4, shows that the majority of the market share is captured by consumer goods, which is closely followed by insurance and healthcare industries. However, the combined industrial share of IoT growth (both cross and resource) is 32% of the collective market, which is again more than that of the consumer market. In continuation, Figure 4.5 shows the IoT market share of various sectors. The manufacturing, logistics, and asset management sectors were purported to be the largest receivers of IoT-linked investments in 2017 [8].

4.2   Evolution of IoT

The IoT, as we see it today, is a result of a series of technological paradigm shifts over a few decades. The technologies that laid the foundation of connected systems by achieving easy integration to daily lives, popular public acceptance, and massive benefits by using connected solutions can be considered as the founding solutions for the development of IoT. Figure 4.6 shows the sequence of technological advancements for shaping the IoT as it is today. These sequence of technical developments toward the emergence of IoT are described in brief:

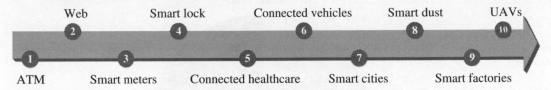

Figure 4.6 The sequence of technological developments leading to the shaping of the modern-day IoT

- ATM: ATMs or automated teller machines are cash distribution machines, which are linked to a user's bank account. ATMs dispense cash upon verification of the identity of a user and their account through a specially coded card. The central concept behind ATMs was the availability of financial transactions even when banks were closed beyond their regular work hours. These ATMs were ubiquitous money dispensers. The first ATM became operational and connected online for the first time in 1974.

- Web: World Wide Web is a global information sharing and communication platform. The Web became operational for the first time in 1991. Since then, it has been massively responsible for the many revolutions in the field of computing and communication.

- Smart Meters: The earliest smart meter was a power meter, which became operational in early 2000. These power meters were capable of communicating remotely with the power grid. They enabled remote monitoring of subscribers' power usage and eased the process of billing and power allocation from grids.

- Digital Locks: Digital locks can be considered as one of the earlier attempts at connected home-automation systems. Present-day digital locks are so robust that smartphones can be used to control them. Operations such as locking and unlocking doors, changing key codes, including new members in the access lists, can be easily performed, and that too remotely using smartphones.

- Connected Healthcare: Here, healthcare devices connect to hospitals, doctors, and relatives to alert them of medical emergencies and take preventive measures. The devices may be simple wearable appliances, monitoring just the heart rate and pulse of the wearer, as well as regular medical devices and monitors in hospitals. The connected nature of these systems makes the availability of medical records and test results much faster, cheaper, and convenient for both patients as well as hospital authorities.

- Connected Vehicles: Connected vehicles may communicate to the Internet or with other vehicles, or even with sensors and actuators contained within it. These vehicles self-diagnose themselves and alert owners about system failures.

- Smart Cities: This is a city-wide implementation of smart sensing, monitoring, and actuation systems. The city-wide infrastructure communicating amongst themselves enables unified and synchronized operations and information dissemination. Some of the facilities which may benefit are parking, transportation, and others.

- Smart Dust: These are microscopic computers. Smaller than a grain of sand each, they can be used in numerous beneficial ways, where regular computers cannot operate. For example, smart dust can be sprayed to measure chemicals in the soil or even to diagnose problems in the human body.

- Smart Factories: These factories can monitor plant processes, assembly lines, distribution lines, and manage factory floors all on their own. The reduction in mishaps due to human errors in judgment or unoptimized processes is drastically reduced.

- UAVs: UAVs or unmanned aerial vehicles have emerged as robust public-domain solutions tasked with applications ranging from agriculture, surveys, surveillance, deliveries, stock maintenance, asset management, and other tasks.

The present-day IoT spans across various domains and applications. The major highlight of this paradigm is its ability to function as a cross-domain technology enabler. Multiple domains can be supported and operated upon simultaneously over IoT-based platforms. Support for legacy technologies and standalone paradigms, along with modern developments, makes IoT quite robust and economical for commercial, industrial, as well as consumer applications. IoT is being used in vivid and diverse areas such as smart parking, smartphone detection, traffic congestion, smart lighting, waste management, smart roads, structural health, urban noise maps, river floods, water flow, silos stock calculation, water leakages, radiation levels, explosive and hazardous gases, perimeter access control, snow level monitoring, liquid presence, forest fire detection, air pollution, smart grid, tank level, photovoltaic installations, NFC (near-field communications) payments, intelligent shopping applications, landslide and avalanche prevention, early detection of earthquakes, supply chain control, smart product management, and others.

Figure 4.7 shows the various technological interdependencies of IoT with other domains and networking paradigms such as M2M, CPS, the Internet of environment (IoE), the Internet of people (IoP), and Industry 4.0. Each of these networking paradigms is a massive domain on its own, but the omnipresent nature of IoT implies that these domains act as subsets of IoT. The paradigms are briefly discussed here:

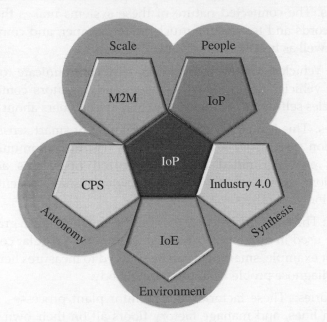

Figure 4.7 The interdependence and reach of IoT over various application domains and networking paradigms

(i)   **M2M**: The M2M or the machine-to-machine paradigm signifies a system of connected machines and devices, which can talk amongst themselves without human intervention. The communication between the machines can be for updates on machine status (stocks, health, power status, and others), collaborative task completion, overall knowledge of the systems and the environment, and others.

(ii) **CPS**: The CPS or the cyber physical system paradigm insinuates a closed control loop—from sensing, processing, and finally to actuation—using a feedback mechanism. CPS helps in maintaining the state of an environment through the feedback control loop, which ensures that until the desired state is attained, the system keeps on actuating and sensing. Humans have a simple supervisory role in CPS-based systems; most of the ground-level operations are automated.

(iii) **IoE**: The IoE paradigm is mainly concerned with minimizing and even reversing the ill-effects of the permeation of Internet-based technologies on the environment [3]. The major focus areas of this paradigm include smart and sustainable farming, sustainable and energy-efficient habitats, enhancing the energy efficiency of systems and processes, and others. In brief, we can safely assume that any aspect of IoT that concerns and affects the environment, falls under the purview of IoE.

(iv) **Industry 4.0**: Industry 4.0 is commonly referred to as the fourth industrial revolution pertaining to digitization in the manufacturing industry. The previous revolutions chronologically dealt with mechanization, mass production, and the industrial revolution, respectively. This paradigm strongly puts forward the concept of smart factories, where machines talk to one another without much human involvement based on a framework of CPS and IoT. The digitization and connectedness in Industry 4.0 translate to better resource and workforce management, optimization of production time and resources, and better upkeep and lifetimes of industrial systems.

(v) **IoP**: IoP is a new technological movement on the Internet which aims to decentralize online social interactions, payments, transactions, and other tasks while maintaining confidentiality and privacy of its user's data. A famous site for IoP states that as the introduction of the Bitcoin has severely limited the power of banks and governments, the acceptance of IoP will limit the power of corporations, governments, and their spy agencies [4].

4.2.1 IoT versus M2M

M2M or the machine-to-machine paradigm refers to communications and interactions between various machines and devices. These interactions can be enabled through a cloud computing infrastructure, a server, or simply a local network hub. M2M collects data from machinery and sensors, while also enabling device management and device interaction. Telecommunication services providers introduced the term M2M, and technically emphasized on machine interactions via one or more communication networks (e.g., 3G, 4G, 5G, satellite, public networks). M2M is part of the IoT and is considered as one of its sub-domains, as shown in Figure 4.7. M2M standards occupy a core place in the IoT landscape. However, in terms of operational and functional scope, IoT is vaster than M2M and comprises a broader range of interactions such as the interactions between devices/things, things, and people, things and applications, and people with applications; M2M enables the amalgamation of workflows comprising such interactions within IoT. Internet connectivity is central to the IoT theme but is not necessarily focused on the use of telecom networks.

4.2.2   IoT versus CPS

Cyber physical systems (CPS) encompasses sensing, control, actuation, and feedback as a complete package. In other words, a digital twin is attached to a CPS-based system. As mentioned earlier, a digital twin is a virtual system–model relation, in which the system signifies a physical system or equipment or a piece of machinery, while the model represents the mathematical model or representation of the physical system's behavior or operation. Many a time, a digital twin is used parallel to a physical system, especially in CPS as it allows for the comparison of the physical system's output, performance, and health. Based on feedback from the digital twin, a physical system can be easily given corrective directions/commands to obtain desirable outputs. In contrast, the IoT paradigm does not compulsorily need feedback or a digital twin system. IoT is more focused on networking than controls. Some of the constituent sub-systems in an IoT environment (such as those formed by CPS-based instruments and networks) may include feedback and controls too. In this light, CPS may be considered as one of the sub-domains of IoT, as shown in Figure 4.7.

4.2.3   IoT versus WoT

From a developer's perspective, the Web of Things (WoT) paradigm enables access and control over IoT resources and applications. These resources and applications are generally built using technologies such as HTML 5.0, JavaScript, Ajax, PHP, and others. REST (representational state transfer) is one of the key enablers of WoT. The use of RESTful principles and RESTful APIs (application program interface) enables both developers and deployers to benefit from the recognition, acceptance, and maturity of existing web technologies without having to redesign and redeploy solutions from scratch. Still, designing and building the WoT paradigm has various adaptability and security challenges, especially when trying to build a globally uniform WoT. As IoT is focused on creating networks comprising objects, things, people, systems, and applications, which often do not consider the unification aspect and the limitations of the Internet, the need for WoT, which aims to integrate the various focus areas of IoT into the existing Web is really invaluable. Technically, WoT can be thought of as an application layer-based hat added over the network layer. However, the scope of IoT applications is much broader; IoT also which includes non-IP-based systems that are not accessible through the web.

4.3   Enabling IoT and the Complex Interdependence of Technologies

IoT is a paradigm built upon complex interdependencies of technologies (both legacy and modern), which occur at various planes of this paradigm. Regarding Figure 4.8, we can divide the IoT paradigm into four planes: services, local connectivity, global connectivity, and processing. If we consider a bottom-up view, the services offered fall under the control and purview of service providers. The service plane is composed of two parts: 1) things or devices and 2) low-power connectivity.

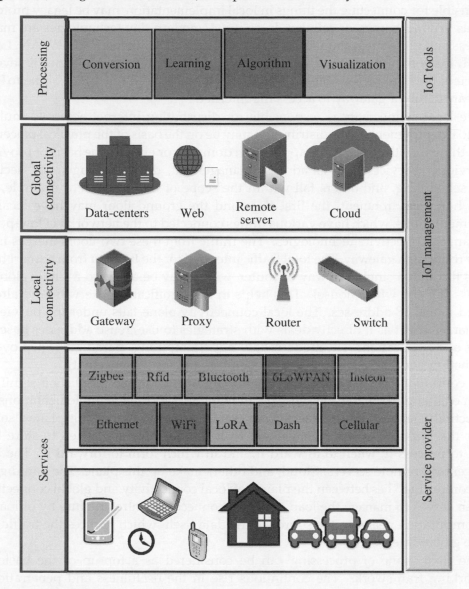

Figure 4.8 The IoT planes, various enablers of IoT, and the complex interdependencies among them

Typically, the services offered in this layer are a combination of things and low-power connectivity. For example, any IoT application requires the basic setup of sensing, followed by rudimentary processing (often), and a low-power, low-range network, which is mainly built upon the IEEE 802.15.4 protocol. The things may be wearables, computers, smartphones, household appliances, smart glasses, factory machinery, vending machines, vehicles, UAVs, robots, and other such contraptions (which may even be just a sensor). The immediate low-power connectivity, which is responsible for connecting the things in local implementation, may be legacy protocols such as WiFi, Ethernet, or cellular. In contrast, modern-day technologies are mainly wireless and often programmable such as Zigbee, RFID, Bluetooth, 6LoWPAN, LoRA, DASH, Insteon, and others. The range of these connectivity technologies is severely restricted; they are responsible for the connectivity between the things of the IoT and the nearest hub or gateway to access the Internet.

The local connectivity is responsible for distributing Internet access to multiple local IoT deployments. This distribution may be on the basis of the physical placement of the things, on the basis of the application domains, or even on the basis of providers of services. Services such as address management, device management, security, sleep scheduling, and others fall within the scope of this plane. For example, in a smart home environment, the first floor and the ground floor may have local IoT implementations, which have various things connected to the network via low-power, low-range connectivity technologies. The traffic from these two floors merges into a single router or a gateway. The total traffic intended for the Internet from a smart home leaves through a single gateway or router, which may be assigned a single global IP address (for the whole house). This helps in the significant conservation of already limited global IP addresses. The local connectivity plane falls under the purview of IoT management as it directly deals with strategies to use/reuse addresses based on things and applications. The modern-day "edge computing" paradigm is deployed in conjunction with these first two planes: services and local connectivity.

In continuation, the penultimate plane of global connectivity plays a significant role in enabling IoT in the real sense by allowing for worldwide implementations and connectivity between things, users, controllers, and applications. This plane also falls under the purview of IoT management as it decides how and when to store data, when to process it, when to forward it, and in which form to forward it. The Web, data-centers, remote servers, Cloud, and others make up this plane. The paradigm of "fog computing" lies between the planes of local connectivity and global connectivity. It often serves to manage the load of global connectivity infrastructure by offloading the computation nearer to the source of the data itself, which reduces the traffic load on the global Internet.

The final plane of processing can be considered as a top-up of the basic IoT networking framework. The continuous rise in the usefulness and penetration of IoT in various application areas such as industries, transportation, healthcare, and others is the result of this plane. The members in this plane may be termed as IoT tools, simply because they wring-out useful and human-readable information from all the raw data that flows from various IoT devices and deployments. The various sub-domains of this plane include intelligence, conversion (data and format conversion, and data cleaning), learning (making sense of temporal and spatial data patterns), cognition (recognizing patterns and mapping it to already known patterns), algorithms (various control and monitoring algorithms), visualization (rendering numbers and strings in the form of collective trends, graphs, charts, and projections), and analysis (estimating the usefulness of the generated information, making sense of the information with respect to the application and place of data generation, and estimating future trends based on past and present patterns of information obtained). Various computing paradigms such as "big data", "machine Learning", and others, fall within the scope of this domain.

4.4 IoT Networking Components

An IoT implementation is composed of several components, which may vary with their application domains. Various established works such as that by Savolainen et al. [2] generally outline five broad categories of IoT networking components. However, we outline the broad components that come into play during the establishment of any IoT network, into six types: 1) IoT node, 2) IoT router, 3) IoT LAN, 4) IoT WAN, 5) IoT gateway, and 6) IoT proxy. A typical IoT implementation from a networking perspective is shown in Figure 4.9. The individual components are briefly described here:

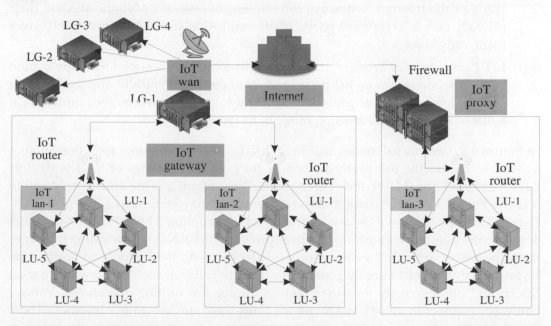

Figure 4.9 A typical IoT network ecosystem highlighting the various networking components—from IoT nodes to the Internet

(i) **IoT Node**: These are the networking devices within an IoT LAN. Each of these devices is typically made up of a sensor, a processor, and a radio, which communicates with the network infrastructure (either within the LAN or outside it). The nodes may be connected to other nodes inside a LAN directly or by means of a common gateway for that LAN. Connections outside the LAN are through gateways and proxies.

(ii) **IoT Router**: An I oT router is a piece of networking equipment that is primarily tasked with the routing of packets between various entities in the IoT network; it keeps the traffic flowing correctly within the network. A router can be repurposed as a gateway by enhancing its functionalities.

(iii) **IoT LAN**: The local area network (LAN) enables local connectivity within the purview of a single gateway. Typically, they consist of short-range connectivity technologies. IoT LANs may or may not be connected to the Internet. Generally, they are localized within a building or an organization.

(iv) **IoT WAN**: The wide area network (WAN) connects various network segments such as LANs. They are typically organizationally and geographically wide, with their operational range lying between a few kilometers to hundreds of kilometers. IoT WANs connect to the Internet and enable Internet access to the segments they are connecting.

(v) **IoT Gateway**: An IoT gateway is simply a router connecting the IoT LAN to a WAN or the Internet. Gateways can implement several LANs and WANs. Their primary task is to forward packets between LANs and WANs, and the IP layer using only layer 3.

(vi) **IoT Proxy**: Proxies actively lie on the application layer and performs application layer functions between IoT nodes and other entities. Typically, application layer proxies are a means of providing security to the network entities under it ; it helps to extend the addressing range of its network.

In Figure 4.9, various IoT nodes within an IoT LAN are configured to to one another as well as talk to the IoT router whenever they are in the range of it. The devices have locally unique (LU-*x*) device identifiers. These identifiers are unique only within a LAN. There is a high chance that these identifiers may be repeated in a new LAN. Each IoT LAN has its own unique identifier, which is denoted by IoT LAN-*x* in Figure 4.9. A router acts as a connecting link between various LANs by forwarding messages from the LANs to the IoT gateway or the IoT proxy. As the proxy is an application layer device, it is additionally possible to include features such as firewalls, packet filters, and other security measures besides the regular routing operations. Various gateways connect to an IoT WAN, which links these devices to the Internet. There may be cases where the gateway or the proxy may directly connect to the Internet. This network may be wired or wireless; however, IoT deployments heavily rely on wireless solutions. This is mainly attributed to the large number of devices that are integrated into the network; wireless technology is the only feasible and neat-enough solution to avoid the hassles of laying wires and dealing with the restricted mobility rising out of wired connections.

4.5 Addressing Strategies in IoT

Table 4.1 lists the differences in features of IPv4 and IPv6. The most interesting point to note is that as compared to IPv4, which relies more on reliable delivery of packets between source and destination, an IPv6 packet is more address-oriented. Due to the increasing rate of devices being connected to the Internet, the early developers of IPv6 felt the need for accommodating addresses as more crucial than the need for reliable transmission of packets (which was the main feature of IPv4-based routing of packets).

Table 4.1 Feature-wise difference between IPv4 and IPv6 capabilities

Feature	IPv4	IPv6
Developed	IETF 1974	IETF 1998
Address length (bits)	32	128
No. of addresses	2^{32}	2^{128}
Notation	Dotted decimal	Hexadecimal
Dynamic allocation of addresses	DHCP	DHCPv6, SLAAC
IPSec	Optional	Compulsary
Header size	Variable	Fixed
Header checksum	Yes	No
Header options	Yes	No
Broadcast addresses	Yes	No
Multicast addresses	No	Yes
Feature	Focus on reliable transmission	Focus on addressing

In the context of IoT, we will consider and center our discussions on addressing schemes primarily focused on IPv6. The IPv4 and IPv6 header packet formats are shown in Chapter 1 of this book. In continuation, Figure 4.10 shows the address format of IPv6, which is 128 bits long.

The first three blocks are designated as the global prefix, which is globally unique. The next block is designated as the subnet prefix, which identifies the subnet of an interface/gateway through which LANs may be connected to the Internet. Finally, the last four blocks (64 bits) of hexadecimal addresses are collectively known as the interface identifier (IID). IIDs may be generated based on MAC (media access control) identifiers of devices/nodes or using pseudo-random number generator algorithms [2]. The IPv6 addresses can be divided into seven separate address types, which is generally based on how these addresses are used or where they are deployed.

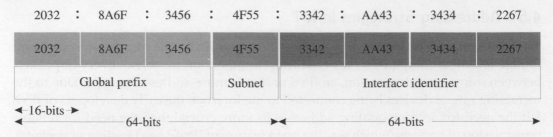

Figure 4.10 The IPv6 address format

(i)   **Global Unicast (GUA)**: These addresses are assigned to single IoT entities/interfaces; they enable the entities to transmit traffic to and from the Internet. In regular IoT deployments, these addresses are assigned to gateways, proxies, or WANs.

(ii) **Multicast**: These addresses enable transmission of messages from a single networked entity to multiple destination entities simultaneously.

(iii) **Link Local (LL)**: The operational domain of these addresses are valid only within a network segment such as LAN. These addresses may be repeated in other network segments/LANs, but are unique within that single network segment.

(iv) **Unique Local (ULA)**: Similar to LL addresses, ULA cannot be routed over the Internet. These addresses may be repeated in other network segments/LANs, but are unique within that single network segment.

(v)   **Loopback**: It is also known as the localhost address. Typically, these addresses are used by developers and network testers for diagnostics and system checks.

(vi) **Unspecified**: Here, all the bits in the IPv6 address are set to zero and the destination address is not specified.

(vii) **Solicited-node Multicast**: It is a multicast address based on the IPv6 address of an IoT node or entity.

Points to ponder

Multihoming in IoT networks: It is a network configuration in which a node/network connects to multiple networks simultaneously for improved reliability. Network proxies are used to manage multiple IP addresses and map them to LL addresses of IoT nodes in small deployments, where the allotment of address prefixes is not possible. Other approaches for multihoming include the use of gateways for assigning LL addresses to IoT nodes under the gateway's operational purview.

4.5.1 Address management classes

As discussed previously, the IoT deployment and network topology are largely dependent on where it is deployed. Unlike traditional IPv4 networked devices, the newer IoT devices largely depend on IPv6 for address allocation and management of addresses, which again is dictated by the application and the place of deployment of the IoT solution. Keeping these requirements in consideration, the addressing strategies in IoT may be broadly differentiated into seven classes, as shown in Figure 4.11. These classes are as follows:

(i) **Class 1**: The IoT nodes are not connected to any other interface or the Internet except with themselves. This class can be considered as an isolated class, where the communication between IoT nodes is restricted within a LAN only. The IoT nodes in this class are identified only by their link local (LL) addresses, as shown in Figure 4.11(a). These LL addresses may be repeated for other devices outside the purview of this network class. The communication among the nodes may be direct or through other nodes (as in a mesh configuration).

(ii) **Class 2**: The class 1 configuration is mainly utilized for enabling communication between two or more IoT LANs or WANs. The IoT nodes within the LANs cannot directly communicate to nodes in the other LANs using their LL addresses, but through their LAN gateways (which have a unique address assigned to them). Generally, ULA is used for addressing; however, in certain scenarios, GUA may also be used. Figure 4.11(b) shows a class 2 IoT network topology. L1 L5 are the LL addresses of the locally unique IoT nodes within the LAN; whereas U1 and U2 are the unique addresses of the two gateways extending communication to their LANs with the WAN. The WAN may or may not connect to the Internet.

(iii) **Class 3**: Figure 4.11(c) shows a class 3 IoT network configuration, where the IoT LAN is connected to an IoT proxy. The proxy performs a host of functions ranging from address allocation, address management to providing security to the network underneath it. In this class, the IoT proxy only uses ULA (denoted as Lx-Ux in the figure).

(iv) **Class 4**: In this class, the IoT proxy acts as a gateway between the LAN and the Internet, and provides GUA to the IoT nodes within the LAN. A globally unique prefix is allotted to this gateway, which it uses with the individual device identifiers to extend global Internet connectivity to the IoT nodes themselves. This configuration is shown in Figure 4.11(d). An important point to note in this class is that the gateway also enables local communication between the nodes without the need for the packets to be routed through the Internet. Additionally, the IoT nodes within the gateway can talk to one another directly without always involving the gateway. A proxy beyond the gateway enables global communication through the Internet.

(v) **Class 5**: This class is functionally similar to class 4. However, the main difference with class 4 is that this class follows a star topology with the gateway as the center of the star. All the communication from the IoT nodes under the gateway has to go through the gateway, as shown in Figure 4.11(e). A proxy beyond the gateway enables global communication through the Internet. The IoT nodes within a gateway's operational purview have the same GUA.

(vi) **Class 6**: The configuration of this class is again similar to class 5. However, the IoT nodes are all assigned unique global addresses (GUA), which enables a point-to-point communication network with an Internet gateway. A class 6 IoT network configuration is shown in Figure 4.11(f). Typically, this class is very selectively used for special purposes.

(vii) **Class 7**: The class 7 configuration is shown in Figure 4.11(g). Multiple gateways may be present; the configuration is such that the nodes should be reachable through any of the gateways. Typically, organizational IoT deployments follow this class of configuration. The concept of multihoming is important and inherent to this class.

Points to ponder

Tunneling: It is a networking protocol in which data from private networks can be seamlessly streamed over a public network in the form of encapsulated packets. This is mainly used for ensuring connectivity and security of data generated from various technologies and protocols that may not be supported over the public communication channel. Some of the best examples of tunneling are virtual private networks (VPNs), secure shell (SSH), and others.

4.5.2 Addressing during node mobility

Traditional networks, mainly computer networks, and even paradigms such as M2M and CPS seldom take into account the need for addressing strategies when the IoT nodes are mobile. However, in a realistic scenario, especially in modern-day IoT systems (which are low-power and have low form-factor), the need for addressing of mobile nodes is extremely crucial to avoid address clashes of addresses accommodating a large number of IoT nodes. One of the following three strategies may be to for ensure portability of addresses in the event of node mobility in IoT deployments [2] as shown in Figure 4.12:

(i) **Global Prefix Changes**: Figure 4.12(a) abstracts the addressing strategy using global prefix changes. A node from the left LAN moves to the LAN on the right. The node undergoing movement is highlighted in the figure. The nodes in the first LAN have the prefix **A**, which changes to **B** under the domain of the new gateway overseeing the operation of nodes in the new LAN. However, it may

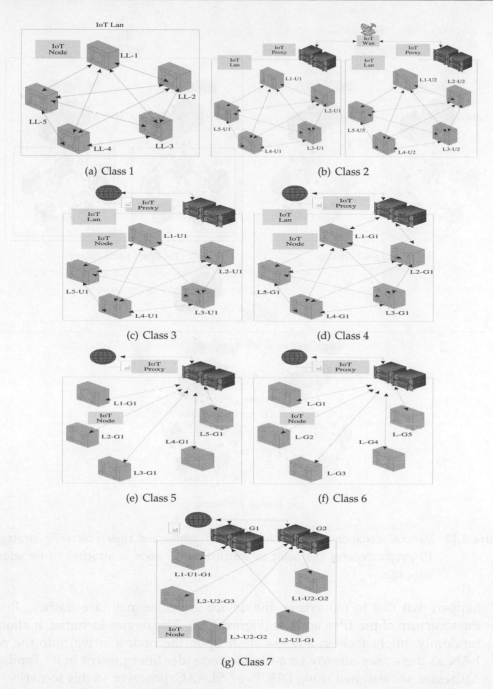

(a) Class 1

(b) Class 2

(c) Class 3

(d) Class 4

(e) Class 5

(f) Class 6

(g) Class 7

Figure 4.11 Various IoT topology configurations. LL/L denotes the link local addresses, LU denotes the locally unique link addresses (ULA), and LG denotes the globally unique link addresses (GUA)

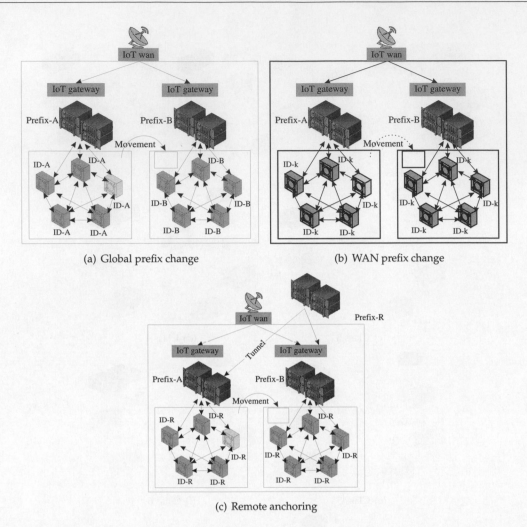

(a) Global prefix change   (b) WAN prefix change

(c) Remote anchoring

Figure 4.12 Various scenarios during mobility of IoT nodes and their addressing strategies. ID-*prefix* denotes the point to which the IoT node is attached to for address allocation

happen that due to movement, the device identifier may face clashes. Recall the structure of the IPv6 address (Figure 4.10). The device identifier, if allotted randomly, might face an address clash upon the node's arrival into the new LAN as there may already be a similar node identifier present in it. Typically, addresses are assigned using DHCPv6/ SLAAC; however, in this scenario, it is always prudent to have static node IP addresses to avoid a clash of addresses. This strategy is, in most cases, beneficial as the IoT nodes may be resource-constrained and have low-processing resources due to which it may not be able to handle protocols such as DHCPv6 or SLAAC.

(ii) **Prefix Changes within WANs**: Figure 4.12(b) abstracts the addressing strategy for prefix changes within WANs. In case the WAN changes its global prefix, the network entities underneath it must be resilient to change and function normally. The address allocation is hence delegated to entities such as gateways and proxies, which make use of ULAs to manage the network within the WAN.

(iii) **Remote Anchoring**: Figure 4.12(c) abstracts the addressing strategy using a remote anchoring point. This is applicable in certain cases which require that the IoT node's global addresses are maintained and not affected by its mobility or even the change in network prefixes. Although a bit expensive to implement, this strategy of having a remote anchoring point from which the IoT nodes obtain their global addresses through tunneling ensures that the nodes are resilient to changes and are quite stable. Even if the node's original network's (LAN) prefix changes from **A** to **B**, the node's global address remains immune to this change.

Check yourself

DHCP, DHCPv6, SLAAC, MIPv6, PMIPv6, DS-MIPv6

Summary

This chapter covered an overview of the IoT paradigm. Starting from the variations in global market trends and the rapidly expanding trend toward connected systems and devices, to the actual market capture of various IoT solutions in diverse domains, this chapter highlights the importance of IoT in the modern world. Subsequently, the emergence of IoT from its precursors, the IoT ecosystem, and thematic differences between IoT and similar technologies (M2M, CPS, WoT) are outlined. The complex technological interdependence between technologies and paradigms towards enabling IoT is described in the form of planes of functionalities. Keeping in tune with the networking theme of this book, the various networking entities in an IoT ecosystem are described, which is naturally followed by various IoT deployment topology classes and addressing schemes. This chapter concludes with a discussion on IoT node address management during node mobility.

Exercises

(i)   What is IoT?

(ii) What is smart dust?

(iii) Differentiate between IoT and M2M.

(iv) Differentiate between IoT and WoT.

(v)   What is Web of Things (WoT)?

(vi)   What are the various IoT connectivity terminologies?

(vii)   Differentiate between an IoT proxy and an IoT gateway.

(viii)   What is gateway prefix allotment?

(ix)   How are locally unique (LU) addresses different from globally unique (GU) addresses?

(x)   How is mobility handled in IoT networks?

(xi)   What is the function of a remote anchor point in IoT networks?

(xii)   What is tunneling?

(xiii)   What is multihoming in IoT networks?

References

[1]   International Data Corporation. 2017. "IDC Forecasts Worldwide Spending on the Internet of Things to Reach USD 772 Billion in 2018." https://www.idc.com/getdoc.jsp?containerId=prUS43295217.

[2]   Savolainen, T., J. Soininen, and B. Silverajan. 2013. "IPv6 Addressing Strategies for IoT." *IEEE Sensors Journal* 13(10): 3511–3519.

[3]   Malek, M. 2017. "The Development of the Internet of Environment." https://www.future-processing.com/blog/the-development-of-the-internet-of-environment/.

[4]   Brans, Cristiaan. 2018. "Internet Of People: Building A New Internet." https://iop.global/.

[5]   Gartner Research. 2016. "Internet of Things Information Handling Services." https://www.gartner.com/it-glossary/internet-of-things/.

[6]   International Telecommunication Union (ITU). 2005. *ITU Internet Reports 2005: The Internet of Things: Executive Summary.* https://www.itu.int/net/wsis/tunis/newsroom/stats/The-Internet-of-Things-2005.pdf.

[7]   IHS. "IoT Platforms: Enabling the Internet of Things." https://ihsmarkit.com/industry/telecommunications.html.

[8]   International Data Corporation. 2016. "IDC Says Worldwide Spending on the Internet of Things Forecast to Reach Nearly USD 1.4 Trillion in 2021." https://www.idc.com/getdoc.jsp?containerId=prUS42799917.

IoT Sensing and Actuation

After reading this chapter, the reader will be able to:

- List the salient features of transducers
- Differentiate between sensors and actuators
- Characterize sensors and distinguish between types of sensors
- List the multi-faceted considerations associated with sensing
- Characterize actuators and distinguish between types of actuators
- List the multi-faceted considerations associated with actuation

5.1 Introduction

A major chunk of IoT applications involves sensing in one form or the other. Almost all the applications in IoT—be it a consumer IoT, an industrial IoT, or just plain hobby-based deployments of IoT solutions—sensing forms the first step. Incidentally, actuation forms the final step in the whole operation of IoT application deployment in a majority of scenarios. The basic science of sensing and actuation is based on the process of transduction. Transduction is the process of energy conversion from one form to another. A transducer is a physical means of enabling transduction. Transducers take energy in any form (for which it is designed)—electrical, mechanical, chemical, light, sound, and others—and convert it into another, which may be electrical, mechanical, chemical, light, sound, and others. Sensors and actuators are deemed as transducers. For example, in a public announcement (PA) system, a microphone (input device) converts sound waves into electrical signals, which is amplified by an amplifier system (a process). Finally, a loudspeaker (output device) outputs this into audible sounds by converting the amplified electrical signals back into sound waves. Table 5.1 outlines the basic terminological differences between transducers, sensors, and actuators.

Table 5.1 Basic outline of the differences between transducers, sensors, and actuators

Parameters	Transducers	Sensors	Actuators
Definition	Converts energy from one form to another.	Converts various forms of energy into electrical signals.	Converts electrical signals into various forms of energy, typically mechanical energy.
Domain	Can be used to represent a sensor as well as an actuator.	It is an input transducer.	It is an output transducer.
Function	Can work as a sensor or an actuator but not simultaneously.	Used for quantifying environmental stimuli into signals.	Used for converting signals into proportional mechanical or electrical outputs.
Examples	Any sensor or actuator	Humidity sensors, Temperature sensors, Anemometers (measures flow velocity), Manometers (measures fluid pressure), Accelerometers (measures the acceleration of a body), Gas sensors (measures concentration of specific gas or gases), and others	Motors (convert electrical energy to rotary motion), Force heads (which impose a force), Pumps (which convert rotary motion of shafts into either a pressure or a fluid velocity).

5.2   Sensors

Sensors are devices that can measure, or quantify, or respond to the ambient changes in their environment or within the intended zone of their deployment. They generate responses to external stimuli or physical phenomenon through characterization of the input functions (which are these external stimuli) and their conversion into typically electrical signals. For example, heat is converted to electrical signals in a temperature sensor, or atmospheric pressure is converted to electrical signals in a barometer. A sensor is only sensitive to the measured property (e.g., a temperature sensor only senses the ambient temperature of a room). It is insensitive to any other property besides what it is designed to detect (e.g., a temperature sensor does not bother about light or pressure while sensing the temperature). Finally, a sensor does not influence the measured property (e.g., measuring the temperature does not reduce or increase the temperature). Figure 5.1 shows the simple outline of a sensing task. Here, a temperature sensor keeps on checking an environment for changes. In the event of a fire, the temperature of the environment goes up. The temperature sensor notices this change in the temperature of the room and promptly communicates this information to a remote monitor via the processor.

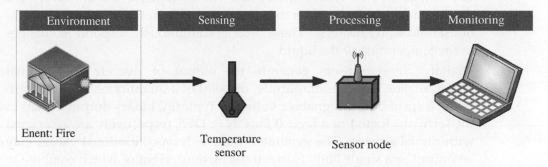

Figure 5.1 The outline of a simple sensing operation

The various sensors can be classified based on: 1) power requirements, 2) sensor output, and 3) property to be measured.

- **Power Requirements**: The way sensors operate decides the power requirements that must be provided for an IoT implementation. Some sensors need to be provided with separate power sources for them to function, whereas some sensors do not require any power sources. Depending on the requirements of power, sensors can be of two types.

  (i) Active: Active sensors do not require an external circuitry or mechanism to provide it with power. It directly responds to the external stimuli from its ambient environment and converts it into an output signal. For example, a photodiode converts light into electrical impulses.

  (ii) Passive: Passive sensors require an external mechanism to power them up. The sensed properties are modulated with the sensor's inherent characteristics to generate patterns in the output of the sensor. For example, a thermistor's resistance can be detected by applying voltage difference across it or passing a current through it.

- **Output**: The output of a sensor helps in deciding the additional components to be integrated with an IoT node or system. Typically, almost all modern-day processors are digital; digital sensors can be directly integrated to the processors.

However, the integration of analog sensors to these digital processors or IoT nodes requires additional interfacing mechanisms such as analog to digital converters (ADC), voltage level converters, and others. Sensors are broadly divided into two types, depending on the type of output generated from these sensors, as follows.

(i)   Analog: Analog sensors generate an output signal or voltage, which is proportional (linearly or non-linearly) to the quantity being measured and is continuous in time and amplitude. Physical quantities such as temperature, speed, pressure, displacement, strain, and others are all continuous and categorized as analog quantities. For example, a thermometer or a thermocouple can be used for measuring the temperature of a liquid (e.g., in household water heaters). These sensors continuously respond to changes in the temperature of the liquid.

(ii) Digital: These sensors generate the output of discrete time digital representation (time, or amplitude, or both) of a quantity being measured, in the form of output signals or voltages. Typically, binary output signals in the form of a logic **1** or a logic **0** for **ON** or **OFF**, respectively are associated with digital sensors. The generated discrete (non-continuous) values may be output as a single "bit" (serial transmission), eight of which combine to produce a single "byte" output (parallel transmission) in digital sensors.

- **Measured Property**: The property of the environment being measured by the sensors can be crucial in deciding the number of sensors in an IoT implementation. Some properties to be measured do not show high spatial variations and can be quantified only based on temporal variations in the measured property, such as ambient temperature, atmospheric pressure, and others. Whereas some properties to be measured show high spatial as well as temporal variations such as sound, image, and others. Depending on the properties to be measured, sensors can be of two types.

(i)   Scalar: Scalar sensors produce an output proportional to the magnitude of the quantity being measured. The output is in the form of a signal or voltage. Scalar physical quantities are those where only the magnitude of the signal is sufficient for describing or characterizing the phenomenon and information generation. Examples of such measurable physical quantities include color, pressure, temperature, strain, and others. A thermometer or thermocouple is an example of a scalar sensor that has the ability to detect changes in ambient or object temperatures (depending on the sensor's configuration). Factors such as changes in sensor orientation or direction do not affect these sensors (typically).

(ii) Vector: Vector sensors are affected by the magnitude as well as the direction and/or orientation of the property they are measuring. Physical quantities such as velocity and images that require additional information besides their magnitude for completely categorizing a physical phenomenon are categorized as vector quantities. Measuring such quantities are undertaken using vector sensors. For example, an electronic gyroscope, which is commonly found in all modern aircraft, is used for detecting the changes in orientation of the gyroscope with respect to the Earth's orientation along all three axes.

Points to ponder

A sensor node is made up of a combination of sensor/sensors, a processor unit, a radio unit, and a power unit. The nodes are capable of sensing the environment they are set to measure and communicate the information to other sensor nodes or a remote server. Typically, a sensor node should have low-power requirements and be wireless. This enables them to be deployed in a vast range of scenarios and environments without the constant need for changing their power sources or managing wires. The wireless nature of sensor nodes would also allow them to be freely relocatable and deployed in large numbers without bothering about managing wires. The functional outline of a typical IoT sensor node is shown in Figure 5.2.

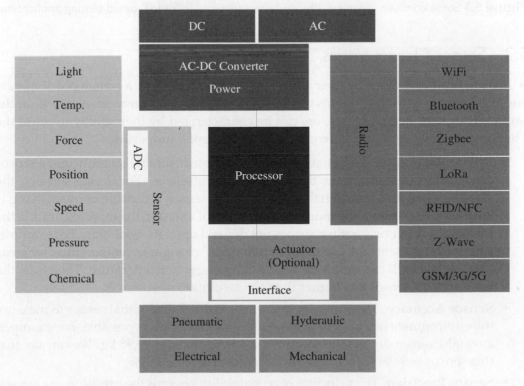

Figure 5.2 The functional blocks of a typical sensor node in IoT

Figure 5.3 shows some commercially available sensors used for sensing applications.

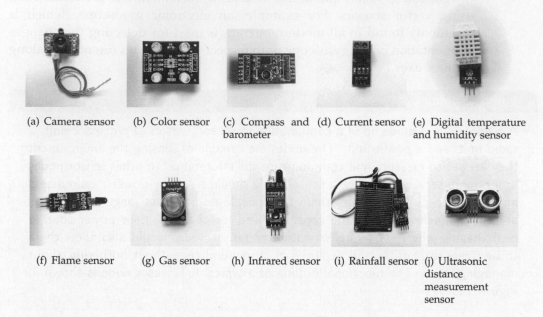

(a) Camera sensor   (b) Color sensor   (c) Compass and   (d) Current sensor   (e) Digital temperature
                                            barometer   and humidity sensor

(f) Flame sensor   (g) Gas sensor   (h) Infrared sensor (i) Rainfall sensor   (j) Ultrasonic
                                                                                        distance
                                                                                        measurement
                                                                                        sensor

Figure 5.3 Some common commercially available sensors used for IoT-based sensing applications

5.3 Sensor Characteristics

All sensors can be defined by their ability to measure or capture a certain phenomenon and report them as output signals to various other systems. However, even within the same sensor type and class, sensors can be characterized by their ability to sense the phenomenon based on the following three fundamental properties.

- **Sensor Resolution**: The smallest change in the measurable quantity that a sensor can detect is referred to as the resolution of a sensor. For digital sensors, the smallest change in the digital output that the sensor is capable of quantifying is its sensor resolution. The more the resolution of a sensor, the more accurate is the precision. A sensor's accuracy does not depend upon its resolution. For example, a temperature sensor **A** can detect up to 0.5° C changes in temperature; whereas another sensor **B** can detect up to 0.25° C changes in temperature. Therefore, the resolution of sensor **B** is higher than the resolution of sensor **A**.

- **Sensor Accuracy**: The accuracy of a sensor is the ability of that sensor to measure the environment of a system as close to its true measure as possible. For example, a weight sensor detects the weight of a 100 kg mass as 99.98 kg. We can say that this sensor is 99.98% accurate, with an error rate of ±0.02%.

- **Sensor Precision**: The principle of repeatability governs the precision of a sensor. Only if, upon multiple repetitions, the sensor is found to have the same error rate, can it be deemed as highly precise. For example, consider if the same weight sensor described earlier reports measurements of 98.28 kg, 100.34 kg, and 101.11 kg upon three repeat measurements for a mass of actual weight of 100 kg. Here, the sensor precision is not deemed high because of significant variations in the temporal measurements for the same object under the same conditions.

> **Points to ponder**
>
> The more the resolution of a sensor, the more accurate is the precision. A sensor's accuracy does not depend upon its resolution.

5.4 Sensorial Deviations

In this section, we will discuss the various sensorial deviations that are considered as errors in sensors. Most of the sensing in IoT is non-critical, where minor deviations in sensorial outputs seldom change the nature of the undertaken tasks. However, some critical applications of IoT, such as healthcare, industrial process monitoring, and others, do require sensors with high-quality measurement capabilities. As the quality of the measurement obtained from a sensor is dependent on a large number of factors, there are a few primary considerations that must be incorporated during the sensing of critical systems.

In the event of a sensor's output signal going beyond its designed maximum and minimum capacity for measurement, the sensor output is truncated to its maximum or minimum value, which is also the sensor's limits. The measurement range between a sensor's characterized minimum and maximum values is also referred to as the full-scale range of that sensor. Under real conditions, the sensitivity of a sensor may differ from the value specified for that sensor leading to *sensitivity error*. This deviation is mostly attributed to sensor fabrication errors and its calibration.

If the output of a sensor differs from the actual value to be measured by a constant, the sensor is said to have an *offset error* or *bias*. For example, while measuring an actual temperature of 0° C, a temperature sensor outputs 1.1° C every time. In this case, the sensor is said to have an offset error or bias of 1.1° C.

Similarly, some sensors have a non-linear behavior. If a sensor's transfer function (TF) deviates from a straight line transfer function, it is referred to as its non-linearity. The amount a sensor's actual output differs from the ideal TF behavior over the full range of the sensor quantifies its behavior. It is denoted as the percentage of the sensor's full range. Most sensors have linear behavior. If the output signal of a sensor changes slowly and independently of the measured property, this behavior of the sensor's output is termed as *drift*. Physical changes in the sensor or its material may result in long-term drift, which can span over months or years. Noise is a temporally varying random deviation of signals.

In contrast, if a sensor's output varies/deviates due to deviations in the sensor's previous input values, it is referred to as *hysteresis error*. The present output of the sensor depends on the past input values provided to the sensor. Typically, the phenomenon of hysteresis can be observed in analog sensors, magnetic sensors, and during heating of metal strips. One way to check for hysteresis error is to check how the sensor's output changes when we first increase, then decrease the input values to the sensor over its full range. It is generally denoted as a positive and negative percentage variation of the full-range of that sensor.

Focusing on digital sensors, if the digital output of a sensor is an approximation of the measured property, it induces *quantization error*. This error can be defined as the difference between the actual analog signal and its closest digital approximation during the sampling stage of the analog to digital conversion. Similarly, dynamic errors caused due to mishandling of sampling frequencies can give rise to *aliasing errors*. Aliasing leads to different signals of varying frequencies to be represented as a single signal in case the sampling frequency is not correctly chosen, resulting in the input signal becoming a multiple of the sampling rate.

Finally, the environment itself plays a crucial role in inducing sensorial deviations. Some sensors may be prone to external influences, which may not be directly linked to the property being measured by the sensor. This sensitivity of the sensor may lead to deviations in its output values. For example, as most sensors are semiconductor-based, they are influenced by the temperature of their environment.

5.5   Sensing Types

Sensing can be broadly divided into four different categories based on the nature of the environment being sensed and the physical sensors being used to do so (Figure 5.4): 1) scalar sensing, 2) multimedia sensing, 3) hybrid sensing, and 4) virtual sensing—[2].

5.5.1   Scalar sensing

Scalar sensing encompasses the sensing of features that can be quantified simply by measuring changes in the amplitude of the measured values with respect to time [3]. Quantities such as ambient temperature, current, atmospheric pressure, rainfall, light, humidity, flux, and others are considered as scalar values as they normally do not have a directional or spatial property assigned with them. Simply measuring the changes in their values with passing time provides enough information about these quantities. The sensors used for measuring these scalar quantities are referred to as scalar sensors, and the act is known as scalar sensing. Figures 5.3(b), 5.3(d), 5.3(e), 5.3(f), 5.3(g), 5.3(h), 5.3(i), and 5.3(j) show scalar sensors. A simple scalar temperature sensing of a fire detection event is shown in Figure 5.4(a).

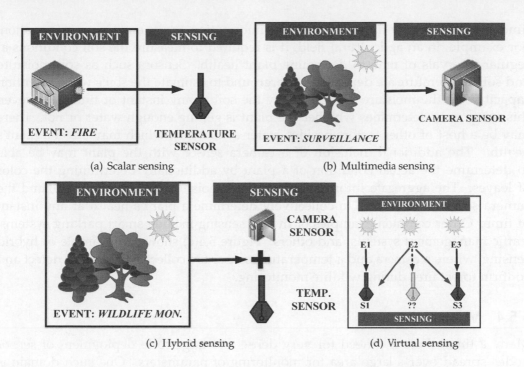

Figure 5.4(a) Scalar sensing

(b) Multimedia sensing

(c) Hybrid sensing

(d) Virtual sensing

Figure 5.4 The different sensing types commonly encountered in IoT

5.5.2 Multimedia sensing

Multimedia sensing encompasses the sensing of features that have a spatial variance property associated with the property of temporal variance [4]. Unlike scalar sensors, multimedia sensors are used for capturing the changes in amplitude of a quantifiable property concerning space (spatial) as well as time (temporal). Quantities such as images, direction, flow, speed, acceleration, sound, force, mass, energy, and momentum have both directions as well as a magnitude. Additionally, these quantities follow the vector law of addition and hence are designated as vector quantities. They might have different values in different directions for the same working condition at the same time. The sensors used for measuring these quantities are known as vector sensors. Figures 5.3(a) and 5.3(c) are vector sensors. A simple camera-based multimedia sensing using surveillance as an example is shown in Figure 5.4(b).

5.5.3 Hybrid sensing

The act of using scalar as well as multimedia sensing at the same time is referred to as hybrid sensing. Many a time, there is a need to measure certain vector as well as scalar properties of an environment at the same time. Under these conditions, a range of various sensors are employed (from the collection of scalar as well as multimedia sensors) to measure the various properties of that environment at any instant of time, and temporally map the collected information to generate new information. For example, in an agricultural field, it is required to measure the soil conditions at regular intervals of time to determine plant health. Sensors such as soil moisture and soil temperature are deployed underground to estimate the soil's water retention capacity and the moisture being held by the soil at any instant of time. However, this setup only determines whether the plant is getting enough water or not. There may be a host of other factors besides water availability, which may affect a plant's health. The additional inclusion of a camera sensor with the plant may be able to determine the actual condition of a plant by additionally determining the color of leaves. The aggregate information from soil moisture, soil temperature, and the camera sensor will be able to collectively determine a plant's health at any instant of time. Other common examples of hybrid sensing include smart parking systems, traffic management systems, and others. Figure 5.4(c) shows an example of hybrid sensing, where a camera and a temperature sensor are collectively used to detect and confirm forest fires during wildlife monitoring.

5.5.4   Virtual sensing

Many a time, there is a need for very dense and large-scale deployment of sensor nodes spread over a large area for monitoring of parameters. One such domain is agriculture [5]. Here, often, the parameters being measured, such as soil moisture, soil temperature, and water level, do not show significant spatial variations. Hence, if sensors are deployed in the fields of farmer **A**, it is highly likely that the measurements from his sensors will be able to provide almost concise measurements of his neighbor **B**'s fields; this is especially true of fields which are immediately surrounding **A**'s fields. Exploiting this property, if the data from **A**'s field is digitized using an IoT infrastructure and this system advises him regarding the appropriate watering, fertilizer, and pesticide regimen for his crops, this advisory can also be used by **B** for maintaining his crops. In short, **A** 's sensors are being used for actual measurement of parameters; whereas virtual data (which does not have actual physical sensors but uses extrapolation-based measurements) is being used for advising **B**. This is the virtual sensing paradigm. Figure 5.4(d) shows an example of virtual sensing. Two temperature sensors S1 and S3 monitor three nearby events E1, E2, and E3 (fires). The event E2 does not have a dedicated sensor for monitoring it; however, through the superposition of readings from sensors S1 and S3, the presence of fire in E2 is inferred.

5.6   Sensing Considerations

The choice of sensors in an IoT sensor node is critical and can either make or break the feasibility of an IoT deployment. The following major factors influence the choice of sensors in IoT-based sensing solutions: 1) sensing range, 2) accuracy and precision, 3) energy, and 4) device size. These factors are discussed as follows:

(i) **Sensing Range**: The sensing range of a sensor node defines the detection fidelity of that node. Typical approaches to optimize the sensing range in deployments include fixed k-coverage and dynamic k-coverage. A lifelong fixed k-coverage tends to usher in redundancy as it requires a large number of sensor nodes, the sensing range of some of which may also overlap. In contrast, dynamic k-coverage incorporates mobile sensor nodes post detection of an event, which, however, is a costly solution and may not be deployable in all operational areas and terrains [1].

Additionally, the sensing range of a sensor may also be used to signify the upper and lower bounds of a sensor's measurement range. For example, a proximity sensor has a typical sensing range of a couple of meters. In contrast, a camera has a sensing range varying between tens of meters to hundreds of meters. As the complexity of the sensor and its sensing range goes up, its cost significantly increases.

(ii) **Accuracy and Precision**: The accuracy and precision of measurements provided by a sensor are critical in deciding the operations of specific functional processes. Typically, off-the-shelf consumer sensors are low on requirements and often very cheap. However, their performance is limited to regular application domains. For example, a standard temperature sensor can be easily integrated with conventional components for hobby projects and day-to-day applications, but it is not suitable for industrial processes. Regular temperature sensors have a very low-temperature sensing range, as well as relatively low accuracy and precision. The use of these sensors in industrial applications, where a precision of up to 3–4 decimal places is required, cannot be facilitated by these sensors. Industrial sensors are typically very sophisticated, and as a result, very costly. However, these industrial sensors have very high accuracy and precision score, even under harsh operating conditions.

(iii) **Energy**: The energy consumed by a sensing solution is crucial to determine the lifetime of that solution and the estimated cost of its deployment. If the sensor or the sensor node is so energy inefficient that it requires replenishment of its energy sources quite frequently, the effort in maintaining the solution and its cost goes up; whereas its deployment feasibility goes down. Consider a scenario where sensor nodes are deployed on the top of glaciers. Once deployed, access to these nodes is not possible. If the energy requirements of the sensor nodes are too high, such a deployment will not last long, and the solution will be highly infeasible as charging or changing of the energy sources of these sensor nodes is not an option.

(iv) **Device Size**: Modern-day IoT applications have a wide penetration in all domains of life. Most of the applications of IoT require sensing solutions which are so small that they do not hinder any of the regular activities that were possible before the sensor node deployment was carried out. Larger the size of a sensor node, larger is the obstruction caused by it, higher is the cost and energy requirements, and lesser is its demand for the bulk of the IoT applications. Consider a simple human activity detector. If the detection unit is too large to be carried or too bulky to cause hindrance to regular normal movements, the demand for this solution would be low. It is because of this that the onset of wearables took off so strongly. The wearable sensors are highly energy-efficient, small in size, and almost part of the wearer's regular wardrobe.

> **Check yourself**
>
> Principle of virtualization, MEMS

5.7 Actuators

An actuator can be considered as a machine or system's component that can affect the movement or control the said mechanism or the system. Control systems affect changes to the environment or property they are controlling through actuators. The system activates the actuator through a control signal, which may be digital or analog. It elicits a response from the actuator, which is in the form of some form of mechanical motion. The control system of an actuator can be a mechanical or electronic system, a software-based system (e.g., an autonomous car control system), a human, or any other input. Figure 5.5 shows the outline of a simple actuation system. A remote user sends commands to a processor. The processor instructs a motor controlled robotic arm to perform the commanded tasks accordingly. The processor is primarily responsible for converting the human commands into sequential machine-language command sequences, which enables the robot to move. The robotic arm finally moves the designated boxes, which was its assigned task.

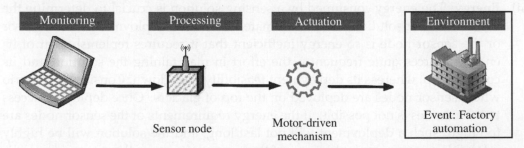

Figure 5.5 The outline of a simple actuation mechanism

5.8   Actuator Types

Broadly, actuators can be divided into seven classes: 1) Hydraulic, 2) pneumatic, 3) electrical, 4) thermal/magnetic, 5) mechanical, 6) soft, and 7) shape memory polymers. Figure 5.6 shows some of the commonly used actuators in IoT applications.

5.8.1   Hydraulic actuators

A hydraulic actuator works on the principle of compression and decompression of fluids. These actuators facilitate mechanical tasks such as lifting loads through the use of hydraulic power derived from fluids in cylinders or fluid motors. The mechanical motion applied to a hydraulic actuator is converted to either linear, rotary, or oscillatory motion. The almost incompressible property of liquids is used in hydraulic actuators for exerting significant force. These hydraulic actuators are also considered as stiff systems. The actuator's limited acceleration restricts its usage.

5.8.2   Pneumatic actuators

A pneumatic actuator works on the principle of compression and decompression of gases. These actuators use a vacuum or compressed air at high pressure and convert it into either linear or rotary motion. Pneumatic rack and pinion actuators are commonly used for valve controls of water pipes. Pneumatic actuators are considered as compliant systems. The actuators using pneumatic energy for their operation are typically characterized by the quick response to starting and stopping signals. Small pressure changes can be used for generating large forces through these actuators. Pneumatic brakes are an example of this type of actuator which is so responsive that they can convert small pressure changes applied by drives to generate the massive force required to stop or slow down a moving vehicle. Pneumatic actuators are responsible for converting pressure into force. The power source in the pneumatic actuator does not need to be stored in reserve for its operation.

5.8.3   Electric actuators

Typically, electric motors are used to power an electric actuator by generating mechanical torque. This generated torque is translated into the motion of a motor's shaft or for switching (as in relays). For example, actuating equipments such as solenoid valves control the flow of water in pipes in response to electrical signals. This class of actuators is considered one of the cheapest, cleanest and speedy actuator types available. Figures 5.6(a), 5.6(b), 5.6(c), 5.6(d), 5.6(e), 5.6(f), 5.6(i), and 5.6(j) show some of the commonly used electrical actuators.

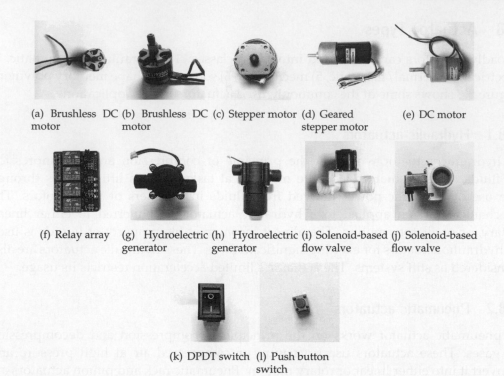

(a) Brushless DC motor (b) Brushless DC motor (c) Stepper motor (d) Geared stepper motor (e) DC motor

(f) Relay array (g) Hydroelectric generator (h) Hydroelectric generator (i) Solenoid-based flow valve (j) Solenoid-based flow valve

(k) DPDT switch (l) Push button switch

Figure 5.6 Some common commercially available actuators used for IoT-based control applications

5.8.4 Thermal or magnetic actuators

The use of thermal or magnetic energy is used for powering this class of actuators. These actuators have a very high power density and are typically compact, lightweight, and economical. One classic example of thermal actuators is shape memory materials (SMMs) such as shape memory alloys (SMAs). These actuators do not require electricity for actuation. They are not affected by vibration and can work with liquid or gases. Magnetic shape memory alloys (MSMAs) are a type of magnetic actuators.

5.8.5 Mechanical actuators

In mechanical actuation, the rotary motion of the actuator is converted into linear motion to execute some movement. The use of gears, rails, pulleys, chains, and other devices are necessary for these actuators to operate. These actuators can be easily used in conjunction with pneumatic, hydraulic, or electrical actuators. They can also work in a standalone mode. The best example of a mechanical actuator is a rack and pinion mechanism. Figures 5.6(g), 5.6(h), 5.6(k), and 5.6(l) show some of the commonly available mechanical actuators. The hydroelectric generator shown in

Figures 5.6(g) and 5.6(h) convert the water-flow induced rotary motion of a turbine into electrical energy. Similarly, the mechanical switches shown in Figures 5.6 (k) and 5.6(l) uses the mechanical motion of the switch to switch on or off an electrical circuit.

5.8.6   Soft actuators

Soft actuators (e.g., polymer-based) consists of elastomeric polymers that are used as embedded fixtures in flexible materials such as cloth, paper, fiber, particles, and others [7]. The conversion of molecular level microscopic changes into tangible macroscopic deformations is the primary working principle of this class of actuators. These actuators have a high stake in modern-day robotics. They are designed to handle fragile objects such as agricultural fruit harvesting, or performing precise operations like manipulating the internal organs during robot-assisted surgeries.

5.8.7   Shape memory polymers

Shape memory polymers (SMP) are considered as smart materials that respond to some external stimulus by changing their shape, and then revert to their original shape once the affecting stimulus is removed [6]. Features such as high strain recovery, biocompatibility, low density, and biodegradability characterize these materials. SMP-based actuators function similar to our muscles. Modern-day SMPs have been designed to respond to a wide range of stimuli such as pH changes, heat differentials, light intensity, and frequency changes, magnetic changes, and others.

Photopolymer/light-activated polymers (LAP) are a particular type of SMP, which require light as a stimulus to operate. LAP-based actuators are characterized by their rapid response times. Using only the variation of light frequency or its intensity, LAPs can be controlled remotely without any physical contact. The development of LAPs whose shape can be changed by the application of a specific frequency of light have been reported. The polymer retains its shape after removal of the activating light. In order to change the polymer back to its original shape, a light stimulus of a different frequency has to be applied to the polymer.

5.9   Actuator Characteristics

The choice or selection of actuators is crucial in an IoT deployment, where a control mechanism is required after sensing and processing of the information obtained from the sensed environment. Actuators perform the physically heavier tasks in an IoT deployment; tasks which require moving or changing the orientation of physical objects, changing the state of objects, and other such activities. The correct choice of actuators is necessary for the long-term sustenance and continuity of operations, as well as for increasing the lifetime of the actuators themselves. A set of four characteristics can define all actuators:

- **Weight**: The physical weight of actuators limits its application scope. For example, the use of heavier actuators is generally preferred for industrial applications and applications requiring no mobility of the IoT deployment. In contrast, lightweight actuators typically find common usage in portable systems in vehicles, drones, and home IoT applications. It is to be noted that this is not always true. Heavier actuators also have selective usage in mobile systems, for example, landing gears and engine motors in aircraft.

- **Power Rating**: This helps in deciding the nature of the application with which an actuator can be associated. The power rating defines the minimum and maximum operating power an actuator can safely withstand without damage to itself. Generally, it is indicated as the power-to-weight ratio for actuators. For example, smaller servo motors used in hobby projects typically have a maximum rating of 5 VDC, 500 mA, which is suitable for an operations-driven battery-based power source. Exceeding this limit might be detrimental to the performance of the actuator and may cause burnout of the motor. In contrast to this, servo motors in larger applications have a rating of 460 VAC, 2.5 A, which requires standalone power supply systems for operations. It is to be noted that actuators with still higher ratings are available and vary according to application requirements.

- **Torque to Weight Ratio**: The ratio of torque to the weight of the moving part of an instrument/device is referred to as its torque/weight ratio. This indicates the sensitivity of the actuator. Higher is the weight of the moving part; lower will be its torque to weight ratio for a given power.

- **Stiffness and Compliance**: The resistance of a material against deformation is known as its stiffness, whereas compliance of a material is the opposite of stiffness. Stiffness can be directly related to the modulus of elasticity of that material. Stiff systems are considered more accurate than compliant systems as they have a faster response to the change in load applied to it. For example, hydraulic systems are considered as stiff and non-compliant, whereas pneumatic systems are considered as compliant.

Check yourself

Operation of PLC and SCADA, Working principle of electric motors, applications of pneumatic and hydraulic actuators, Differences between pneumatic, hydraulic, electrical, and mechanical actuators

---

Summary

This chapter covered the basics of sensing and actuation in order to help the readers grasp the intricacies of designing an IoT solution keeping in mind the need to select the proper sensors and actuators. The first part of this chapter discusses sensors, sensing characteristics, considerations of various sensorial deviations, and the sensing types possible in a typical IoT-based implementation of a sensing solution. This part concludes with a discussion on the various considerations to be thought of while selecting sensors for architecting a viable IoT-based sensing solution. The second part of this chapter focuses on actuators and the broad classes of actuators available. This part concludes with a discussion on the various considerations to be thought of while selecting actuators for architecting a viable IoT-based control solution using actuators. After completing this chapter, the reader will be able to decide upon the most appropriate sensing and actuation solutions to use with their IoT-based applications.

Exercises

(i) Differentiate between sensors and actuators.

(ii) Differentiate between sensors and transducers.

(iii) How is sensor resolution different from its accuracy?

(iv) Differentiate between scalar and vector sensors.

(v) Differentiate between analog and digital sensors.

(vi) What is a an offset error?

(vii) What is a hysteresis error?

(viii) What is a quantization error?

(ix) What is aliasing error?

(x) Differentiate between hydraulic and pneumatic actuators with examples.

(xi) What are shape memory alloys (SMA)?

(xii) What are soft actuators?

(xiii) What are the main features of shape memory polymers?

(xiv) What are light activated polymers?

References

[1] Alam, Kh Mahmudul, Joarder Kamruzzaman, Gour Karmakar, and Manzur Murshed. 2014. "Dynamic Adjustment of Sensing Range for Event Coverage in Wireless Sensor Networks." *Journal of Network and Computer Applications* 46: 139–153. Elsevier.

[2] Popović, T., N. Latinović, A. Pešić, Z. Zečević, B. Krstajić, and S. Djukanović. 2017. "Architecting an IoT-enabled Platform for Precision Agriculture and Ecological Monitoring: A Case Study." *Computers and Electronics in Agriculture* 140: 255–265.

[3]   Kelly, S. D. T., N. K. Suryadevara, and S. C. Mukhopadhyay. 2013. "Towards the Implementation of IoT for Environmental Condition Monitoring in Homes." *IEEE Sensors Journal* 13(10): 3846–3853.

[4]   Rosário, D., Z. Zhao, A. Santos, T. Braun, and E. Cerqueira. 2014. "A Beaconless Opportunistic Routing based on a Cross-layer Approach for Efficient Video Dissemination in Mobile Multimedia IoT Applications." *Computer Communications* 45: 21–31.

[5]   Ojha, T., S. Misra, N. S. Raghuwanshi, and H. Poddar. 2019. "DVSP: Dynamic Virtual Sensor Provisioning in Sensor-Cloud based Internet of Things." *IEEE Internet of Things Journal* 6 (3): 5265–5272.

[6]   Lendlein, A., and O. E. Gould. 2019. "Reprogrammable Recovery and Actuation Behaviour of Shape-memory Polymers." *Nature Reviews Materials* 4(2): 116–133.

[7]   Lessing, J. A., R. V. Martinez, A. S. Tayi, J. M. Ting, and G. M. Whitesides. 2019. "Flexible and Stretchable Electronic Strain-limited Layer for Soft Actuators." U. S. Patent Application 15/972, 412.

Chapter 6

IoT Processing Topologies and Types

Learning Outcomes

After reading this chapter, the reader will be able to:

- List common data types in IoT applications
- Understand the importance of processing
- Explain the various processing topologies in IoT
- Understand the importance of processing off-loading toward achieving scalability and cost-effectiveness of IoT solutions
- Determine the importance of choosing the right processing topologies and associated considerations while designing IoT applications
- Determine the requirements that are associated with IoT-based processing of sensed and communicated data.

6.1 Data Format

The Internet is a vast space where huge quantities and varieties of data are generated regularly and flow freely. As of January 2018, there are a reported 4.021 billion Internet users worldwide. The massive volume of data generated by this huge number of users is further enhanced by the multiple devices utilized by most users. In addition to these data-generating sources, non-human data generation sources such as sensor nodes and automated monitoring systems further add to the data load on the Internet. This huge data volume is composed of a variety of data such as e-mails, text documents (Word docs, PDFs, and others), social media posts, videos, audio files, and images, as shown in Figure 6.1. However, these data can be broadly grouped into two types based on how they can be accessed and stored: 1) Structured data and 2) unstructured data.

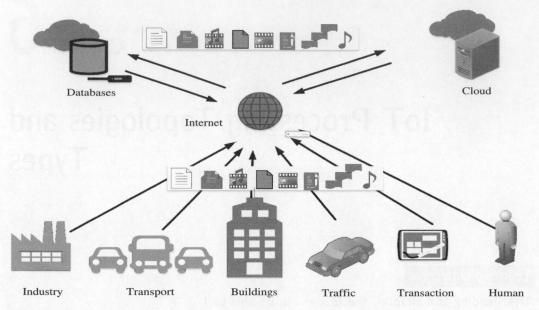

Figure 6.1 The various data generating and storage sources connected to the Internet and the plethora of data types contained within it

6.1.1 Structured data

These are typically text data that have a pre-defined structure [1]. Structured data are associated with relational database management systems (RDBMS). These are primarily created by using length-limited data fields such as phone numbers, social security numbers, and other such information. Even if the data is human or machine-generated, these data are easily searchable by querying algorithms as well as human-generated queries. Common usage of this type of data is associated with flight or train reservation systems, banking systems, inventory controls, and other similar systems. Established languages such as Structured Query Language (SQL) are used for accessing these data in RDBMS. However, in the context of IoT, structured data holds a minor share of the total generated data over the Internet.

6.1.2 Unstructured data

In simple words, all the data on the Internet, which is not structured, is categorized as unstructured. These data types have no pre-defined structure and can vary according to applications and data-generating sources. Some of the common examples of human-generated unstructured data include text, e-mails, videos, images, phone recordings, chats, and others [2]. Some common examples of machine-generated unstructured data include sensor data from traffic, buildings, industries, satellite imagery, surveillance videos, and others. As already evident from its examples, this data type does not have fixed formats associated with it, which makes it very difficult for querying algorithms to perform a look-up. Querying languages such as NoSQL are generally used for this data type.

6.2   Importance of Processing in IoT

The vast amount and types of data flowing through the Internet necessitate the need for intelligent and resourceful processing techniques. This necessity has become even more crucial with the rapid advancements in IoT, which is laying enormous pressure on the existing network infrastructure globally. Given these urgencies, it is important to decide—*when to process and what to process*? Before deciding upon the processing to pursue, we first divide the data to be processed into three types based on the urgency of processing: 1) Very time critical, 2) time critical, and 3) normal. Data from sources such as flight control systems [3], healthcare, and other such sources, which need immediate decision support, are deemed as very critical. These data have a very low threshold of processing latency, typically in the range of a few milliseconds.

Data from sources that can tolerate normal processing latency are deemed as time-critical data. These data, generally associated with sources such as vehicles, traffic, machine systems, smart home systems, surveillance systems, and others, which can tolerate a latency of a few seconds fall in this category. Finally, the last category of data, normal data,can tolerate a processing latency of a few minutes to a few hours and are typically associated with less data-sensitive domains such as agriculture, environmental monitoring, and others.

Considering the requirements of data processing, the processing requirements of data from very time-critical sources are exceptionally high. Here, the need for processing the data in place or almost nearer to the source is crucial in achieving the deployment success of such domains. Similarly, considering the requirements of processing from category 2 data sources (time-critical), the processing requirements allow for the transmission of data to be processed to remote locations/processors such as clouds or through collaborative processing. Finally, the last category of data sources (normal) typically have no particular time requirements for processing urgently and are pursued leisurely as such.

> ### Check yourself
>
> Difference between microprocessors and microcontrollers, network bandwidth, network latency

6.3   Processing Topologies

The identification and intelligent selection of processing requirement of an IoT application are one of the crucial steps in deciding the architecture of the deployment. A properly designed IoT architecture would result in massive savings in network bandwidth and conserve significant amounts of overall energy in the architecture while providing the proper and allowable processing latencies for the solutions associated with the architecture. Regarding the importance of processing in IoT as outlined in Section 6.2, we can divide the various processing solutions into two large topologies: 1) On-site and 2) Off-site. The off-site processing topology can be further divided into the following: 1) Remote processing and 2) Collaborative processing.

6.3.1   On-site processing

As evident from the name, the on-site processing topology signifies that the data is processed at the source itself. This is crucial in applications that have a very low tolerance for latencies. These latencies may result from the processing hardware or the network (during transmission of the data for processing away from the processor). Applications such as those associated with healthcare and flight control systems (real-time systems) have a breakneck data generation rate. These additionally show rapid temporal changes that can be missed (leading to catastrophic damages) unless the processing infrastructure is fast and robust enough to handle such data. Figure 6.2 shows the on-site processing topology, where an event (here, fire) is detected utilizing a temperature sensor connected to a sensor node. The sensor node processes the information from the sensed event and generates an alert. The node additionally has the option of forwarding the data to a remote infrastructure for further analysis and storage.

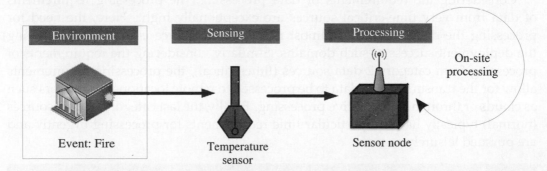

Figure 6.2 Event detection using an on-site processing topology

6.3.2 Off-site processing

The off-site processing paradigm, as opposed to the on-site processing paradigms, allows for latencies (due to processing or network latencies); it is significantly cheaper than on-site processing topologies. This difference in cost is mainly due to the low demands and requirements of processing at the source itself. Often, the sensor nodes are not required to process data on an urgent basis, so having a dedicated and expensive on-site processing infrastructure is not sustainable for large-scale deployments typical of IoT deployments. In the off-site processing topology, the sensor node is responsible for the collection and framing of data that is eventually to be transmitted to another location for processing. Unlike the on-site processing topology, the off-site topology has a few dedicated high-processing enabled devices, which can be borrowed by multiple simpler sensor nodes to accomplish their tasks. At the same time, this arrangement keeps the costs of large-scale deployments extremely manageable [5]. In the off-site topology, the data from these sensor nodes (data generating sources) is transmitted either to a remote location (which can either be a server or a cloud) or to multiple processing nodes. Multiple nodes can come together to share their processing power in order to collaboratively process the data (which is important in case a feasible communication pathway or connection to a remote location cannot be established by a single node).

Remote processing

This is one of the most common processing topologies prevalent in present-day IoT solutions. It encompasses sensing of data by various sensor nodes; the data is then forwarded to a remote server or a cloud-based infrastructure for further processing and analytics. The processing of data from hundreds and thousands of sensor nodes can be simultaneously offloaded to a single, powerful computing platform; this results in massive cost and energy savings by enabling the reuse and reallocation of the same processing resource while also enabling the deployment of smaller and simpler processing nodes at the site of deployment [4]. This setup also ensures massive scalability of solutions, without significantly affecting the cost of the deployment. Figure 6.3 shows the outline of one such paradigm, where the sensing of an event is performed locally, and the decision making is outsourced to a remote processor (here, cloud). However, this paradigm tends to use up a lot of network bandwidth and relies heavily on the presence of network connectivity between the sensor nodes and the remote processing infrastructure.

Collaborative processing

This processing topology typically finds use in scenarios with limited or no network connectivity, especially systems lacking a backbone network. Additionally, this topology can be quite economical for large-scale deployments spread over vast areas, where providing networked access to a remote infrastructure is not viable. In such scenarios, the simplest solution is to club together the processing power of nearby

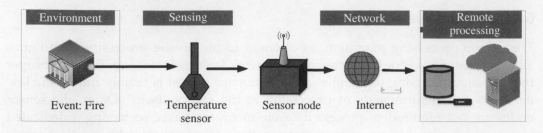

Figure 6.3 Event detection using an off-site remote processing topology

processing nodes and collaboratively process the data in the vicinity of the data source itself. This approach also reduces latencies due to the transfer of data over the network. Additionally, it conserves bandwidth of the network, especially ones connecting to the Internet. Figure 6.4 shows the collaborative processing topology for collaboratively processing data locally. This topology can be quite beneficial for applications such as agriculture, where an intense and temporally high frequency of data processing is not required as agricultural data is generally logged after significantly long intervals (in the range of hours). One important point to mention about this topology is the preference of mesh networks for easy implementation of this topology.

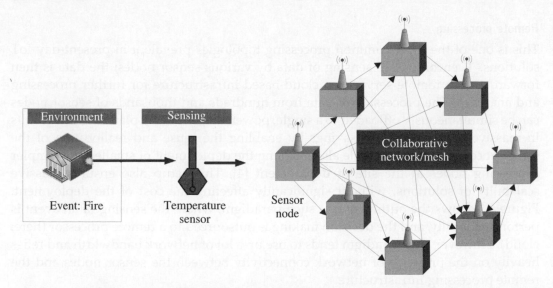

Figure 6.4 Event detection using a collaborative processing topology

6.4   IoT Device Design and Selection Considerations

The main consideration of minutely defining an IoT solution is the selection of the processor for developing the sensing solution (i.e., the sensor node). This selection is governed by many parameters that affect the usability, design, and affordability of the designed IoT sensing and processing solution. In this chapter, we mainly focus on the deciding factors for selecting a processor for the design of a sensor node. The main factor governing the IoT device design and selection for various applications is the processor. However, the other important considerations are as follows.

- **Size**: This is one of the crucial factors for deciding the form factor and the energy consumption of a sensor node. It has been observed that larger the form factor, larger is the energy consumption of the hardware. Additionally, large form factors are not suitable for a significant bulk of IoT applications, which rely on minimal form factor solutions (e.g., wearables).

- **Energy**: The energy requirements of a processor is the most important deciding factor in designing IoT-based sensing solutions. Higher the energy requirements, higher is the energy source (battery) replacement frequency. This principle automatically lowers the long-term sustainability of sensing hardware, especially for IoT-based applications.

- **Cost**: The cost of a processor, besides the cost of sensors, is the driving force in deciding the density of deployment of sensor nodes for IoT-based solutions. Cheaper cost of the hardware enables a much higher density of hardware deployment by users of an IoT solution. For example, cheaper gas and fire detection solutions would enable users to include much more sensing hardware for a lesser cost.

- **Memory**: The memory requirements (both volatile and non-volatile memory) of IoT devices determines the capabilities the device can be armed with. Features such as local data processing, data storage, data filtering, data formatting, and a host of other features rely heavily on the memory capabilities of devices. However, devices with higher memory tend to be costlier for obvious reasons.

- **Processing power**: As covered in earlier sections, processing power is vital (comparable to memory) in deciding what type of sensors can be accommodated with the IoT device/node, and what processing features can integrate on-site with the IoT device. The processing power also decides the type of applications the device can be associated with. Typically, applications that handle video and image data require IoT devices with higher processing power as compared to applications requiring simple sensing of the environment.

- **I/O rating**: The input–output (I/O) rating of IoT device, primarily the processor, is the deciding factor in determining the circuit complexity, energy usage, and requirements for support of various sensing solutions and sensor types. Newer processors have a meager I/O voltage rating of 3.3 V, as compared to 5 V for the somewhat older processors. This translates to requiring additional voltage and logic conversion circuitry to interface legacy technologies and sensors with the newer processors. Despite low power consumption due to reduced I/O voltage levels, this additional voltage and circuitry not only affects the complexity of the circuits but also affects the costs.

- **Add-ons**: The support of various add-ons a processor or for that matter, an IoT device provides, such as analog to digital conversion (ADC) units, in-built clock circuits, connections to USB and ethernet, inbuilt wireless access capabilities, and others helps in defining the robustness and usability of a processor or IoT device in various application scenarios. Additionally, the provision for these add-ons also decides how fast a solution can be developed, especially the hardware part of the whole IoT application. As interfacing and integration of systems at the circuit level can be daunting to the uninitiated, the prior presence of these options with the processor makes the processor or device highly lucrative to the users/developers.

> **Check yourself**
>
> RISC versus CISC processors, volative versus non-volatile memory

6.5   Processing Offloading

The processing offloading paradigm is important for the development of densely deployable, energy-conserving, miniaturized, and cheap IoT-based solutions for sensing tasks. Building upon the basics of the off-site processing topology covered in the previous sections in this chapter, we delve a bit further into the various nuances of processing offloading in IoT.

Figure 6.5 shows the typical outline of an IoT deployment with the various layers of processing that are encountered spanning vastly different application domains—from as near as sensing the environment to as far as cloud-based infrastructure. Starting from the primary layer of sensing, we can have multiple sensing types tasked with detecting an environment (fire, surveillance, and others). The sensors enabling these sensing types are integrated with a processor using wired or wireless connections (mostly, wired). In the event that certain applications require immediate processing of the sensed data, an on-site processing topology is followed, similar to the one in Figure 6.2. However, for the majority of IoT applications, the bulk of the processing is carried out remotely in order to keep the on-site devices simple, small, and economical.

Typically, for off-site processing, data from the sensing layer can be forwarded to the fog or cloud or can be contained within the edge layer [6]. The edge layer makes use of devices within the local network to process data that which is similar to the collaborative processing topology shown in Figure 6.4. The devices within the local network, till the fog, generally communicate using short-range wireless connections. In case the data needs to be sent further up the chain to the cloud, long-range wireless connection enabling access to a backbone network is essential. Fog-based processing is still considered local because the fog nodes are typically localized within a geographic area and serve the IoT nodes within a much smaller coverage area as compared to the cloud. Fog nodes, which are at the level of gateways, may or may not be accessed by the IoT devices through the Internet.

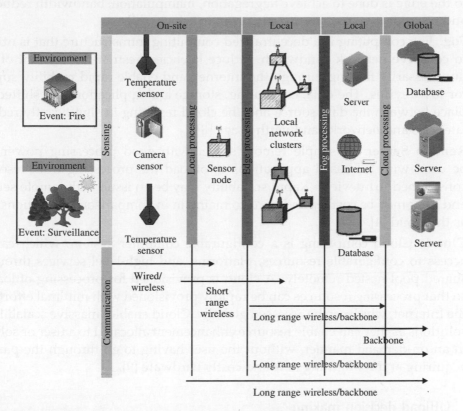

Figure 6.5 The various data generating and storage sources connected to the Internet and the plethora of data types contained within it

Finally, the approach of forwarding data to a cloud or a remote server, as shown in the topology in Figure 6.3, requires the devices to be connected to the Internet through long-range wireless/wired networks, which eventually connect to a backbone network. This approach is generally costly concerning network bandwidth, latency, as well as the complexity of the devices and the network infrastructure involved.

This section on data offloading is divided into three parts: 1) offload location (which outlines where all the processing can be offloaded in the IoT architecture), 2) offload decision making (how to choose where to offload the processing to and by how much), and finally 3) offloading considerations (deciding when to offload).

6.5.1 Offload location

The choice of offload location decides the applicability, cost, and sustainability of the IoT application and deployment. We distinguish the offload location into four types:

- **Edge**: Offloading processing to the edge implies that the data processing is facilitated to a location at or near the source of data generation itself. Offloading to the edge is done to achieve aggregation, manipulation, bandwidth reduction, and other data operations directly on an IoT device [7].

- **Fog**: Fog computing is a decentralized computing infrastructure that is utilized to conserve network bandwidth, reduce latencies, restrict the amount of data unnecessarily flowing through the Internet, and enable rapid mobility support for IoT devices. The data, computing, storage and applications are shifted to a place between the data source and the cloud resulting in significantly reduced latencies and network bandwidth usage [8].

- **Remote Server**: A simple remote server with good processing power may be used with IoT-based applications to offload the processing from resource-constrained IoT devices. Rapid scalability may be an issue with remote servers, and they may be costlier and hard to maintain in comparison to solutions such as the cloud [4].

- **Cloud**: Cloud computing is a configurable computer system, which can get access to configurable resources, platforms, and high-level services through a shared pool hosted remotely. A cloud is provisioned for processing offloading so that processing resources can be rapidly provisioned with minimal effort over the Internet, which can be accessed globally. Cloud enables massive scalability of solutions as they can enable resource enhancement allocated to a user or solution in an on-demand manner, without the user having to go through the pains of acquiring and configuring new and costly hardware [9].

6.5.2   Offload decision making

The choice of where to offload and how much to offload is one of the major deciding factors in the deployment of an offsite-processing topology-based IoT deployment architecture. The decision making is generally addressed considering data generation rate, network bandwidth, the criticality of applications, processing resource available at the offload site, and other factors. Some of these approaches are as follows.

- **Naive Approach**: This approach is typically a hard approach, without too much decision making. It can be considered as a rule-based approach in which the data from IoT devices are offloaded to the nearest location based on the achievement of certain offload criteria. Although easy to implement, this approach is never recommended, especially for dense deployments, or deployments where the data generation rate is high or the data being offloaded in complex to handle (multimedia or hybrid data types). Generally, statistical measures are consulted for generating the rules for offload decision making.

- **Bargaining based approach**: This approach, although a bit processing-intensive during the decision making stages, enables the alleviation of network traffic congestion, enhances service QoS (quality of service) parameters such as bandwidth, latencies, and others. At times, while trying to maximize multiple parameters for the whole IoT implementation, in order to provide the most optimal solution or QoS, not all parameters can be treated with equal importance. Bargaining based solutions try to maximize the QoS by trying to reach a point where the qualities of certain parameters are reduced, while the others are enhanced. This measure is undertaken so that the achieved QoS is collaboratively better for the full implementation rather than a select few devices enjoying very high QoS. Game theory is a common example of the bargaining based approach. This approach does not need to depend on historical data for decision making purposes.

- **Learning based approach**: Unlike the bargaining based approaches, the learning based approaches generally rely on past behavior and trends of data flow through the IoT architecture. The optimization of QoS parameters is pursued by learning from historical trends and trying to optimize previous solutions further and enhance the collective behavior of the IoT implementation. The memory requirements and processing requirements are high during the decision making stages. The most common example of a learning based approach is machine learning.

6.5.3 Offloading considerations

There are a few offloading parameters which need to be considered while deciding upon the offloading type to choose. These considerations typically arise from the nature of the IoT application and the hardware being used to interact with the application. Some of these parameters are as follows.

- **Bandwidth**: The maximum amount of data that can be simultaneously transmitted over the network between two points is the bandwidth of that network. The bandwidth of a wired or wireless network is also considered to be its data-carrying capacity and often used to describe the data rate of that network.

- **Latency**: It is the time delay incurred between the start and completion of an operation. In the present context, latency can be due to the network (network latency) or the processor (processing latency). In either case, latency arises due to the physical limitations of the infrastructure, which is associated with an operation. The operation can be data transfer over a network or processing of a data at a processor.

- **Criticality**: It defines the importance of a task being pursued by an IoT application. The more critical a task is, the lesser latency is expected from the IoT solution. For example, detection of fires using an IoT solution has higher criticality than detection of agricultural field parameters. The former requires a response time in the tune of milliseconds, whereas the latter can be addressed within hours or even days.

- **Resources**: It signifies the actual capabilities of an offload location. These capabilities may be the processing power, the suite of analytical algorithms, and others. For example, it is futile and wasteful to allocate processing resources reserved for real-time multimedia processing (which are highly energy-intensive and can process and analyze huge volumes of data in a short duration) to scalar data (which can be addressed using nominal resources without wasting much energy).

- **Data volume**: The amount of data generated by a source or sources that can be simultaneously handled by the offload location is referred to as its data volume handling capacity. Typically, for large and dense IoT deployments, the offload location should be robust enough to address the processing issues related to massive data volumes.

Summary

This chapter started with an overview of the various data formats available on the Internet and to which various IoT solutions are exposed. The complexities in handling the numerous data formats available present a significant challenge to the design of IoT-based solutions. In order to address these challenges, the importance of processing in IoT is discussed. This discussion is followed by an introduction to various processing topologies, which can be chosen to address the challenges of IoT processing. These topologies are broadly made up of two categories: 1) On-site processing and 2) Off-site processing. The off-site processing is typically composed of approaches to offload data to locations which are not the same as the one from which the data was generated. A discussion on processing offloading follows these topologies. Various offload location types, means of deciding offload location and quantity are explained. Finally, the various parameters to be considered for offloading are discussed to enable the reader to grasp the nuances of processing in IoT.

Exercises

(i)   What are the different data formats found in IoT network traffic streams?

(ii)   Depending on the urgency of data processing, how are IoT data classified?

(iii) Highlight the pros and cons of on-site and off-site processing.

(iv)   Differentiate between structured and unstructured data.

(v)   How is collaborative processing different from remote processing?

(vi)   What are the critical factors to be considered during the design of IoT devices?

(vii) What are the typical data offload locations available in the context of IoT?

(viii) What are the various decision making approaches chosen for offloading data in IoT?

(ix) What factors are to be considered while deciding on the data offload location?

References

[1] Misra, S., A. Mukherjee, and A. Roy. 2018. "Knowledge Discovery for Enabling Smart Internet of Things: A Survey." *Wiley Interdisciplinary Reviews: Data Mining and Knowledge Discovery* 8(6): 1276.

[2] Jiang, L., L. Da Xu, H. Cai, Z. Jiang, F. Bu, and B. Xu. 2014. "An IoT-oriented Data Storage Framework in Cloud Computing Platform." *IEEE Transactions on Industrial Informatics* 10(2): 1443–1451.

[3] Mukherjee, A., S. Misra, V. S. P. Chandra, and M. S. Obaidat. 2018. "Resource-Optimized Multi-Armed Bandit Based Offload Path Selection in Edge UAV Swarms." *IEEE Internet of Things Journal* 6(3): 4889–4896.

[4] Mukherjee, A., S. Misra, N. S. Raghuwanshi, and S. Mitra. 2018. "Blind Entity Identification for Agricultural IoT Deployments." *IEEE Internet of Things Journal* 6(2): 3156–3163.

[5] Mukherjee, A., N. Pathak, S. Misra, and S. Mitra. 2018. "Predictive Intra-Edge Packet-Source Mapping in Agricultural Internet of Things." In 2018 *IEEE Globecom Workshops (GC Wkshps)* (pp. 1–6). IEEE. doi: 10.1109/GLOCOMW.2018.8644296

[6] Cheng, N., F. Lyu, W. Quan, C. Zhou, H. He, W. Shi, and X. Shen. 2019. "Space/Aerial-Assisted Computing Offloading for IoT Applications: A Learning-Based Approach." *IEEE Journal on Selected Areas in Communications* 37(5): 1117–1129.

[7] Huang, L., X. Feng, C. Zhang, L. Qian, and Y. Wu. 2019. "Deep Reinforcement Learning-based Joint Task Offloading and Bandwidth Allocation for Multi-user Mobile Edge Computing." *Digital Communications and Networks* 5(1): 10–17.

[8] Adhikari, M., M. Mukherjee, and S. N. Srirama. 2019. "DPTO: A Deadline and Priority-aware Task Offloading in Fog Computing Framework Leveraging Multi-level Feedback Queueing." *IEEE Internet of Things Journal*. doi: 10.1109/JIOT.2019.2946426.

[9] Mahmoodi, S. E., K. Subbalakshmi, and R. N. Uma. 2019. "Classification of Mobile Cloud Offloading." In *Spectrum-Aware Mobile Computing* (pp. 7–11). Springer, Cham.

Chapter 7

IoT Connectivity Technologies

After reading this chapter, the reader will be able to:

- List common connectivity protocols in IoT
- Identify the salient features and application scope of each connectivity protocol
- Understand the terminologies and technologies associated with IoT connectivity
- Determine the requirements associated with each of these connectivity protocols in real-world solutions
- Determine the most appropriate connectivity protocol for each segment of their IoT implementation

7.1 Introduction

This chapter outlines the main features of fifteen identified commonly used and upcoming IoT connectivity enablers. These connectivity technologies can be integrated with existing sensing, actuation, and processing solutions for extending connectivity to them. Some of these solutions necessarily require integration with a minimal form of processing infrastructure, such as Wi-Fi. In contrast, others, such as Zigbee, can work in a standalone mode altogether, without the need for external processing and hardware support. These solutions are outlined in the subsequent sections in this chapter.

7.2 IEEE 802.15.4

The IEEE 802.15.4 standard represents the most popular standard for low data rate wireless personal area networks (WPAN) [1]. This standard was developed to enable monitoring and control applications with lower data rate and extend the operational life for uses with low-power consumption. This standard uses only the first two layers—physical and data link—for operation along with two new layers above it: 1) logical link control (LLC) and 2) service-specific convergence sublayer (SSCS). The additional layers help in the communication of the lower layers with the upper layers. Figure 7.1 shows the IEEE 802.15.4 operational layers. The IEEE 802.15.4 standard was curated to operate in the ISM (industrial, scientific, and medical) band.

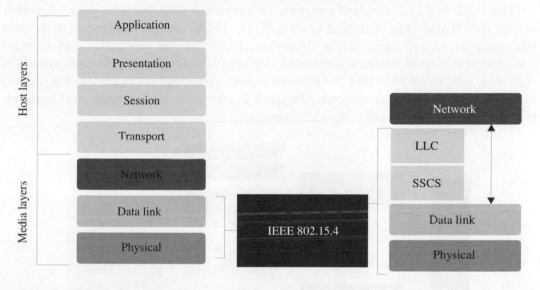

Figure 7.1 The operational part of IEEE 802.15.4's protocol stack in comparison to the OSI stack

The direct sequence spread spectrum (DSSS) modulation technique is used in IEEE 802.15.4 for communication purposes, enabling a wider bandwidth of operation with enhanced security by the modulating pseudo-random noise signal. This standard exhibits high tolerance to noise and interference and offers better measures for improving link reliability. Typically, the low-speed versions of the IEEE 802.15.4 standard use binary phase shift keying (BPSK), whereas the versions with high data rate implement offset quadrature phase shift keying (O-QPSK) for encoding the message to be communicated. Carrier sense multiple access with collision avoidance (CSMA-CA) is the channel access method used for maintaining the sequence of transmitted signals and preventing deadlocks due to multiple sources trying to access the same channel. Temporal multiplexing enables access to the same channel by multiple users or nodes at different times in a maximally interference-free manner.

The IEEE 802.15.4 standard [2] utilizes infrequently occurring and very short packet transmissions with a low duty cycle (typically, < 1%) to minimize the power consumption. The minimum power level defined is –3 dBm or 0.5 mW for the radios utilizing this standard. The transmission, for most cases, is line of sight (LOS), with the standard transmission range varying between 10 m to 75 m. The best-case transmission range achieved outdoors can be up to 1000 m.

This standard typically defines two networking topologies: 1) Star and 2) mesh. There are seven variants identified with in IEEE 802.15.4—A, B. C, D, E, F, and G. Variants A/B are the base versions, C is assigned for China, and D for Japan. Variants E, F, and G are assigned respectively for industrial applications, active RFID (radio frequency identification) uses, and smart utility systems.

The IEEE 802.15.4 standard supports two types of devices: 1) reduced function device (RFD) and 2) full function devices (FFD). FFDs can talk to all types of devices and support full protocol stacks. However, these devices are costly and energy-consuming due to increased requirements for support of full stacks. In contrast, RFDs can only talk to an FFD and have lower power consumption requirements due to minimal CPU/RAM requirements. Figure 7.2 shows the device types and network types supported by the IEEE 802.15.4 standard.

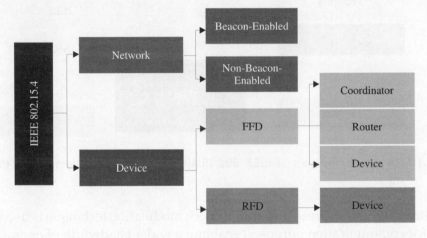

Figure 7.2 The various device and network types supported in the IEEE 802.15.4 standard

The IEEE 802.15.4 standard supports two network types: 1) Beacon-enabled networks and 2) non-beacon-enabled networks. The periodic transmission of beacon messages characterizes beacon-enabled networks. Here, the data frames sent via slotted CSMA/CA with a superframe structure managed by a personal area network (PAN) coordinator. These beacons are used for synchronization and association of other nodes with the coordinator. The scope of operation of this network type spans the whole network.

In contrast, for non-beacon-enabled networks, unslotted CSMA/CA (contention-based) is used for transmission of data frames, and beacons are used only for link layer discovery. This network typically requires both source and destination IDs of the communicating nodes. As the IEEE 802.15.4 is primarily a mesh protocol, all protocol addressing must adhere to mesh configurations such that there is a decentralized communication amongst nodes.

Figure 7.3 shows the frame types associated with the IEEE 802.15.4 standard. Beacon frames are used for signaling and synchronization; data transmission is done through the data frames; and message reception is confirmed using the acknowledgment frames. MAC and command frames are used for association requests/responses, dissociation requests, data requests, beacon requests, coordinator realignment, and orphan notifications.

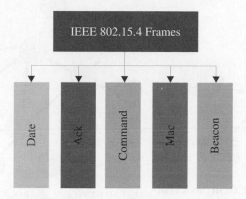

Figure 7.3 Various frame types supported in the IEEE 802.15.4 standard

Check yourself

CSMA/CD versus CSMS/CA, transmission power, DSSS, BPSK, OQPSK

7.3 Zigbee

The Zigbee radio communication is designed for enabling wireless personal area networks (WPANs). It uses the IEEE 802.15.4 standard for defining its physical and medium access control (layers 1 and 2 of the OSI stack). Zigbee finds common usage in sensor and control networks [4]. It was designed for low-powered mesh networks at low cost, which can be broadly implemented for controlling and monitoring applications, typically in the range of 10–100 meters [3]. The PHY and MAC layers in this communication are designed to handle multiple low data rate operating devices. The frequencies of 2.4 GHz, 902–928 MHz or 868 MHz are commonly associated with Zigbee WPAN operations. The Zigbee commonly uses 250 kbps data rate which is optimal for both periodic and intermittent full-duplex data transmission between two Zigbee entities.

Zigbee supports various network configurations such as master-to-master communication or master-to-slave communication. Several network topologies are supported in Zigbee, namely the star (Figure 7.4(a)), mesh (Figure 7.4(b)), and cluster tree (Figure 7.4(c)). Any of the supported topologies may consist of a single or multiple coordinators. In star topology, a coordinator initiates and manages the other devices in the Zigbee network. The other devices which communicate with the coordinator are called end devices. As the star topology is easy to maintain and deploy, it finds widespread usage in applications where a single central controller manages multiple devices.

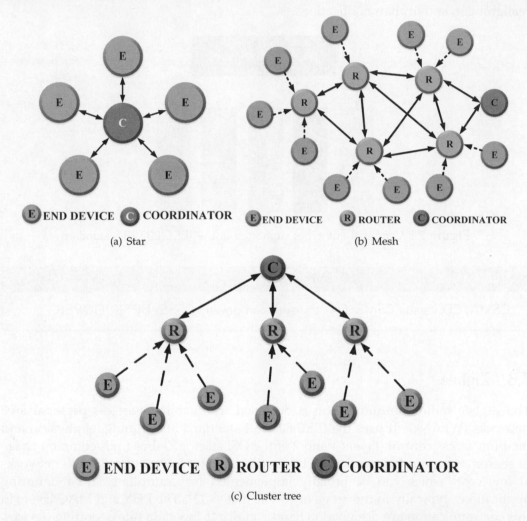

Figure 7.4 Various communication topologies in Zigbee

A network can be significantly extended in the Zigbee mesh and tree topologies by using multiple routers where the root of the topology is the coordinator. These configurations allow any Zigbee device or node to communicate with any other adjacent node. In case of the failure of one or more nodes, the information is automatically forwarded to other devices through other functional devices. In a Zigbee cluster tree network, a coordinator is placed in the leaf node position of the cluster, which is, in turn, connected to a parent coordinator who initiates the entire network.

A typical Zigbee network structure can consist of three different device types, namely the Zigbee coordinator, router, and end device, as shown in Figure 7.4. Every Zigbee network has a minimum of one coordinator device type who acts as the root; it also functions as the network bridge. The coordinator performs data handling and storing operations. The Zigbee routers play the role of intermediate nodes that connect two or more Zigbee devices, which may be of the same or different types. Finally, the end devices have restricted functionality; communication is limited to the parent nodes. This reduced functionality enables them to have a lower power consumption requirement, allowing them to operate for an extended duration. There are provisions to operate Zigbee in different modes to save power and prolong the deployed network lifetime.

The PHY and MAC layers of the IEEE 802.15.4 standard are used to build the protocol for Zigbee architecture; the protocol is then accentuated by network and application layers designed especially for Zigbee. Figure 7.5 shows the Zigbee protocol stack. The various layer of the Zigbee stack are as follows.

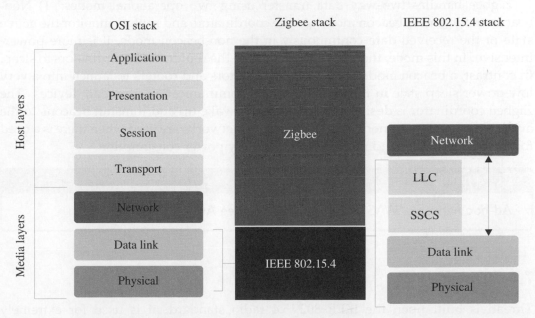

Figure 7.5 The Zigbee protocol stack in comparison to the OSI stack

- **Physical Layer**: This layer is tasked with transmitting and receiving signals, and performing modulation and demodulation operations on them, respectively. The

Zigbee physical layer consists of 3 bands made up of 27 channels: the 2.4 GHz band has 16 channels at 250 kbps the 868.3 MHz has one channel at 20 kbps; and the 902-928 MHz has ten channels at 40 kbps.

- **MAC Layer**: This layer ensures channel access and reliability of data transmission. CSMA-CA is used for channel access and intra-channel interference avoidance. This layer handles communication synchronization using beacon frames.

- **Network Layer**: This layer handles operations such as setting up the network, connecting and disconnecting the devices, configuring the devices, and routing.

- **Application Support Sub-Layer**: This layer handles the interfacing services, control services, bridge between network and other layers, and enables the necessary services to interface with the lower layers for Zigbee device object (ZDO) and Zigbee application objects (ZAO). This layer is primarily tasked with data management services and is responsible for service-based device matching.

- **Application Framework**: Two types of data services are provided by the application framework: provision of a key-value pair and generation of generic messages. A key-value pair is used for getting attributes within the application objects, whereas a generic message is a developer-defined structure.

Zigbee handles two-way data transfer using two operational modes: 1) Non-beacon mode and 2) beacon mode. As the coordinators and routers monitor the active state of the received data continuously in the non-beacon mode, it is more power-intensive. In this mode, there is no provision for the routers and coordinators to sleep. In contrast, a beacon mode allows the coordinators and routers to launch into a very low-power sleep state in the absence of data communication from end devices. The Zigbee coordinator is designed to periodically wake up and transmit beacons to the available routers in the network. These beacon networks are used when there is a need for lower duty cycles and more extended battery power consumption.

> Check yourself
>
> Ad-hoc networks, WASN, Zigbee ZDO, Zigbee APS

7.4   Thread

Thread is built upon the IEEE 802.15.4 radio standard; it is used for extremely low power consumption and low latency deployments [5]. Unlike Zigbee, Thread can extend direct Internet connectivity to the devices it is connected with. Thread removes the need for a mobile phone or a proprietary gateway to be in the range of devices for accessing the Internet. It is specially designed for IoT with the need for interoperability, security, power, and architecture addressed in a single radio platform.

Figure 7.6 shows the comparison of the Thread stack against the standard ISO-OSI stack. Thread is built on open standards to achieve a low-power wireless mesh

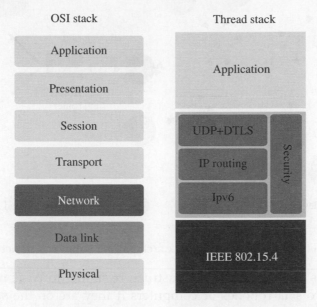

Figure 7.6 The functional protocol stack of Thread in comparison to the OSI stack

networking protocol with universal Internet Protocol (IP) support. The standard is easy to set up and simple to use; it can reliably connect thousands of devices to the Internet or a cloud with no single point of failure. It has the distinctive feature of self-healing and reconfiguration in the event of the addition or removal of a device. Figure 7.7 shows the Thread network architecture.

Thread enables IoT interoperability by utilizing a certification application that validates a device's conformance to the specification as well as its interoperability against multiple certified stacks. This feature ensures the resilience of connectivity, even with diverse networks, in turn enabling its users to have consistent operational experience.

Empowering low-power wireless devices with IP connectivity enables Thread to seamlessly accommodate itself with larger IP-based networks and be a robust option for most IoT applications such as smart homes/buildings, connected vehicles, and others. This feature of Thread devices removes the need for Internet-enabled proprietary gateways and cross-stack translators for connection between other technologies. The additional benefits of this feature include better resilience to single point of failures, highly economical deployments, less complex infrastructure, and enhanced IoT end-to-end device security on the Internet.

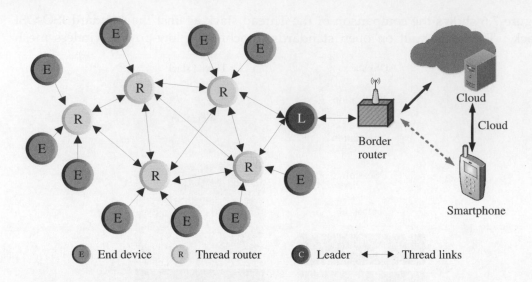

Figure 7.7 Outline of the Thread network architecture (from end devices to the cloud)

Thread devices can use common infrastructure similar to Wi-Fi networks and can connect directly to smartphones or computers if they are on the same IP network, without any additional setup for Thread.

> Check yourself
>
> IP-networks versus non-IP-networks, types of interoperability

7.5   ISA100.11A

The ISA100.11A is a very low power communication standard and has been developed and managed by ISA (International Society of Automation) [7]. Similar to the previous protocols, it uses the IEEE 802.15.4 standard as a base for building its protocol. The standard was mainly proposed for industrial plant automation systems. The ISA100.11A is characterized by an IoT compliant protocol stack, which can also be integrated with wired networks using Ethernet, support for open access protocols and device-level interoperability; it boasts of a 128-bit AES (Advanced Encryption Standard) encryption securing all communications. The security in ISA100.11A is in two layers: Transport layer and data link layer. ISA100.11A provides extensive support for IPv6 and UDP and uses TDMA (time-division multiple access)-based resource sharing with CSMA-CA. Both IPv6 and UDP as well as star topologies are supported by this standard. The utilization of IPv6 provides certain distinct benefits to ISA100.11A, such as increased address sizes, enhanced IPSec-based security measures, savings in network bandwidth by virtue of multicasting and auto address configuration.

An ISA100.11A wireless network utilizes the 2.4 GHz frequency band for communication, similar to Wi-Fi and Bluetooth. To avoid interference over wireless channels in the same band, it uses frequency hopping spread spectrum (FHSS) over a total of 16 channels. A definitive feature of this protocol is channel blacklisting, which blacklists the channels already in use by other protocols. This enables the protocol to perform even better by further achieving immunity from interference. Figure 7.8

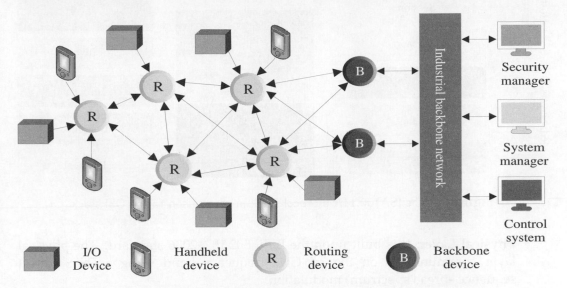

Figure 7.8 A typical ISA100.11A network architecture

shows the ISA100.11A network architecture. The ISA100.11A architecture consists of the following: 1) field devices and 2) backbone devices. Field devices may be non-routing I/O devices, handheld devices, routing devices, which may or may not be fixed or mobile. For industrial usage, the inclusion of portable and mobile devices is highly desirable as it allows floor supervisors and workers to keep checking various parts of the plant without the need for dedicated devices for each part. In contrast, backbone devices include backbone routers, gateways, the system manager, and the security manager, which are kept fixed and not portable. The ISA100.11A architecture provides support for mesh, star, and star–mesh topologies. The connected devices in ISA100.11A are collectively referred to as the downLink (DL) subnet. A wireless industrial sensor network (WISN) gateway connects the ISA100.11A network to the plant network.

The average ISA100.11A protocol stack consists of five different layers: 1) Application layer, 2) transport layer, 3) network layer, 4) data link layer, and 5) physical layer. Figure 7.9 compares the ISA100.11A stack with the standard ISO-OSI stack. A central system manager handles network routing by scheduling communication. The functionalities of the ISA100.11A protocol stack can be outlined as follows:

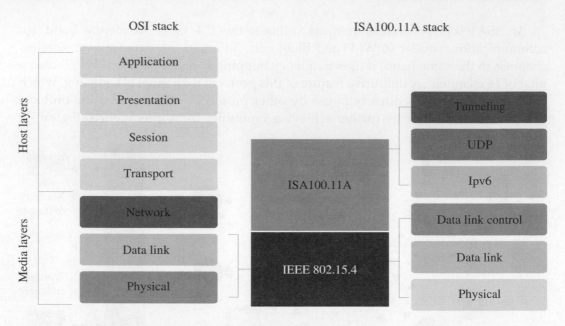

Figure 7.9 The ISA100.11A protocol stack in comparison to the OSI stack

- **Physical Layer**: It is built upon the IEEE 802.15.4-2006 standard. The physical layer communicates on the 2.4 GHz frequency band using a DSSS (direct sequence spread spectrum) modulation.

- **Data Link Layer**: It handles the creation, maintenance, and forwarding packet functionalities in addition to typical MAC functionalities. Additionally, it is responsible for operations dealing with the structure of the data packet, formation of the frame, detecting the error, and bus arbitration. A data link control (DLC) layer in ISA100.11A, which uses a graph-based routing, is responsible for specific distinctive functions such as adaptive channel hopping, detection and recovery of message loss, and clock synchronization.

- **Network Layer**: The ISA100.11A network layer is 6LoWPAN-compliant and uses IPv6 addressing for an end-to-end routing. Protocol conversion from IPv6 to 6LoWPAN and 6LoWPAN to IPv6 is executed at this layer by a router.

- **Transport Layer**: The ISA100.11A transport layer supports UDP-based connectionless services.

- **Application Layer**: The ISA100.11A stack only specifies system management application in this layer.

> **Check yourself**
>
> Wireless Industrial Sensor Networks (WISN), 6LoWPAN, field devices, routing and non-routing devices

7.6 WirelessHART

WirelessHART can be considered as the wireless evolution of the highway addressable remote transducer (HART) protocol [7]. It is a license-free protocol, which was developed for networking smart field devices in industrial environments. The lack of wires makes the adaptability of this protocol significantly advantageous over its predecessor, HART, in industrial settings. By virtue of its highly encrypted communication, wireless HART is very secure and has several advantages over traditional communication protocols. Similar to Zigbee, wirelessHART uses the IEEE 802.15.4 standard for its protocols designing.

Figure 7.10 shows the WirelessHART network architecture. WirelessHART can communicate with a central control system in any of the two ways: 1) Direct and 2) indirect. Direct communication is achieved when the devices transmit data directly to the gateway in a clear LOS (typically 250 m). Indirect communication is achieved between devices in a mesh and a gateway when messages jump from device to device until it reaches the gateway. WirelessHART communication is 99.999% reliable due to the maintenance of a tight schedule between message transmissions. All wirelessHART devices are back-compatible and allow for the integration of legacy devices as well as new ones.

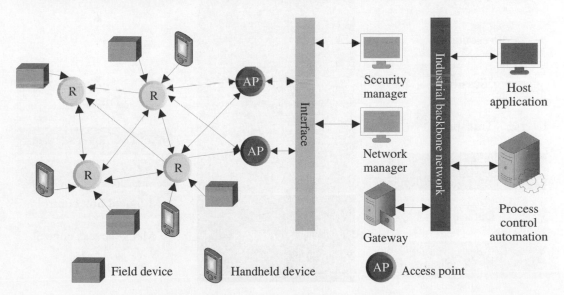

Figure 7.10 The WirelessHART network architecture

The HART encompasses the most number of field devices incorporated in any field network. WirelessHART makes device placements more accessible and cheaper, such as the top of a reaction tank, inside a pipe, or widely separated warehouses. The wired and unwired versions differ mainly in the network, data link, and physical layer. The wired HART lacks a network layer. HART ensures congestion control in the 2.4 Ghz ISM band by eliminating channel 26 because of its restricted usage in certain areas. The use of interference-prone channels is avoided by using channel switching after every transmission. The transmissions are synchronized using 10 ms time-slots. During each time-slot, all available channels can be utilized by the various nodes in the network, allowing for the simultaneous propagation of 15 packets through the network, which also minimizes the risk of collisions between channels.

A network manager supervises each node in the network and guides them on when and where to send packets. This network manager allows for collision-free and timely delivery of packets between a source and the destination. It updates information regarding neighbors, signal strength, and information needing a delivery receipt. This network manager also decides which nodes transmit, which nodes listen, and the frequency to be utilized in each time-slot. It also handles code-based network security and prevents unauthorized nodes from joining the network.

Figure 7.11 shows the comparison of the wirelessHART protocol stack against the standard ISO-OSI stack. The various layers of the wirelessHART stack are outlined as follows:

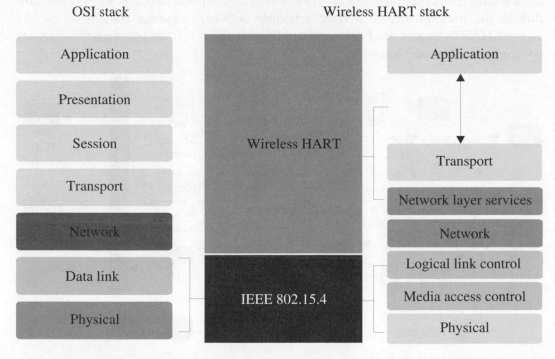

Figure 7.11 The WirelessHART protocol stack in comparison to the OSI stack

- **Physical Layer**: The IEEE 802.15.4 standard specification is used for designing the physical layer of this protocol. Its operation is limited to the use of the 2.4 GHz frequency band. The channel reliability is significantly increased by utilizing only 15 channels of the 2.4 GHz band.

- **Data Link Layer**: The data link layer avoids collisions by the use of TDMA. The communication is also made deterministic by the use of superframes.

WirelessHART superframes consist of 10 ms wide time-slots that are grouped together. The use of superframes ensures better controllability of the transmission timing, collision avoidance, and communication reliability. This layer incorporates channel hopping and channel blacklisting to increase reliability and security. A characteristic feature of the wirelessHART is channel blacklisting. This feature identifies channels consistently affected by interference and removes them from use.

- **Network and Transport Layers**: The network and the transport layer work in tandem to address issues of network traffic, security, session initiation/termination, and routing. WirelessHART is primarily a mesh-based network, where each node can accept data from other nodes in range and forward them to the next node. All the devices in its network have an updated network graph, which defines the routing paths to be taken. Functionally, the OSI stack's network, transport, and session layers constitute the WirelessHART's network layer.

- **Application Layer**: The application layer connects gateways and devices through various command and response messages. This layer enables back-compatibility with legacy HART devices as it does not differentiate between the wired and wireless versions of HART.

Check yourself

Graph-based routing, superframes, co-channel interference

7.7 RFID

RFID stands for radio frequency identification. This technology uses tags and readers for communication. RFID tags have data encoded onto them digitally [8]. The RFID readers can read the values encoded in these tags without physically touching them. RFIDs are functionally similar to barcodes as the data read from tags are stored in a database. However, RFID does not have to rely on line of sight operation, unlike barcodes.

The automatic identification and data capture (AIDC) technology can be considered as the precursor of RFID. Similar to AIDC techniques, RFID systems are capable of automatically categorizing objects. Categorization tasks such as identifying tags, reading data, and feeding the read data directly into computer systems through radio waves outline the operation of RFID systems. Typically, RFID systems are made up of three components: 1) RFID tag or smart label, 2) RFID reader, and 3) an antenna. Figure 7.12 shows the various RFID components.

In RFID, the tags consist of an integrated circuit and an antenna, enclosed in a protective casing to protect from wear and tear and environmental effects. These

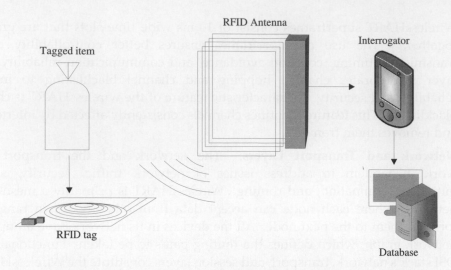

Figure 7.12 An outline of the RFID operation and communication

tags can be either active or passive. Passive tags find common usage in a variety of applications due to its low cost; however, it has to be powered using an RFID reader before data transmission. Active tags have their own power sources and do not need external activation by readers. Tags are used for transmitting the data to an RFID interrogator or an RFID reader. The radio waves are then converted to a more usable form of data by this reader. A host computer system accesses the collected data on the reader by a communication technology such as Wi-Fi or Ethernet. The data on the host system is finally updated onto a database. RFID applications span across domains such as inventory management, asset tracking, personnel tracking, and supply chain management.

Check yourself

How does RFID tackle various services, such as asset tracking and inventory management?

7.8 NFC

Near field communication (NFC) was jointly developed by Philips and Sony as a short-range wireless connectivity standard, enabling peer-to-peer (P2P) data exchange network. Communication between NFC devices is achieved by the principle of magnetic induction, whenever the devices are brought close to one another [9]. NFC can also be used with other wireless technologies such as Wi-Fi after establishing and configuring the P2P network. The communication between compatible devices requires a pair of transmitting and receiving devices. The typical NFC operating frequency for data is 13.56 MHz, which supports data rates of 106, 212, or 424 kbps. NFC devices can be grouped into two types: 1) passive NFC and 2) active NFC. Figure 7.13 shows the various NFC types, components, and its usage.

A small electric current is emitted by the NFC reader, which creates a magnetic field that acts as a bridge in the physical space between two NFC devices. The generated EM (electromagnetic) field is converted back into electrical impulses through another coil on the client device. Data such as identifiers, messages, currency, status, and others can be transmitted using NFCs. NFC communication and pairing are speedy due to the use of inductive coupling and the absence of manual pairing.

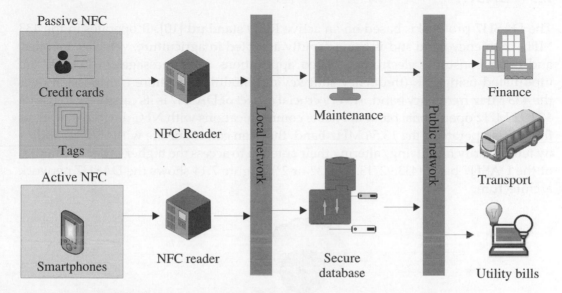

Figure 7.13 An outline of the NFC operation and communication

Passive NFC devices do not need a power source for communicating with the NFC reader. Tags and other small transmitters can act as passive NFC devices. However, passive devices cannot process information; they simply store information, which is read by an NFC reader. In contrast, active NFC devices can communicate with active as well as passive NFC devices. Active devices are capable of reading as well as writing data to other NFC terminals or devices. Some of the most commonly used NFC platforms are smartphones, public transport card readers, and commercial touch payment terminals.

NFC currently supports three information exchange modes: 1) peer-to-peer, 2) read/write, and 3) card emulation. The peer-to-peer mode is commonly used in NFC modes; it enables two NFC devices to exchange information. In the peer-to-peer mode of information exchange, the transmitting device goes active while the receiving device becomes passive. During the reverse transfer, both devices change roles. The read/write mode of information exchange allows only one-way data transmission. An active NFC device connects to a passive device to read information from it. Finally, the card emulation mode enables an NFC device (generally, smartphones) to act as a contactless credit card and make payments using just a simple tap on an NFC reader.

> Check yourself
>
> Magnetic induction, inductive coupling, peer-to-peer data exchange

7.9 DASH7

The DASH7 protocol is based on an active RFID standard [10]. It operates in the 433 MHz frequency band and is being rapidly accepted in agriculture, vehicles, mobiles, and other consumer electronics-related applications. The messages in DASH7 are modulated using FSK (frequency shift keying) modulation before transmission over the 433 MHz frequency band. A very crucial aspect of DASH7 is its capability to use its 433.92 MHz operational band to enable communications with NFC devices. Recall, as the NFCs operate in the 13.56 MHz band, they can communicate with DASH7 radios by temporarily modifying/altering their antenna to access the higher-order harmonics of the DASH7 band ($433.92/13.56 = 32$ or 2^5). Figure 7.14 shows the DASH7 network architecture.

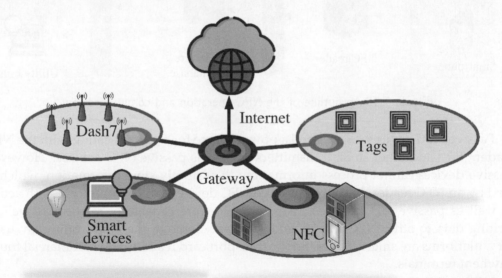

Figure 7.14 The DASH7 communication architecture

Compared to the IEEE 802.15.4 and its dependent technologies, the DASH7 protocol has a fully defined and complete OSI stack. This enables the DASH7 stack to be made adaptable to the physical layers of technologies such as Sigfox or LoRa. The

DASH7 stack includes support for cheap processing systems by virtue of its integrated file system. Figure 7.15 shows the protocol stack of DASH 7 in comparison to the ISO-OSI stack. DASH7 gateways can query devices in proximity to it without waiting for pre-defined time-slots to listen to end-device beacons.

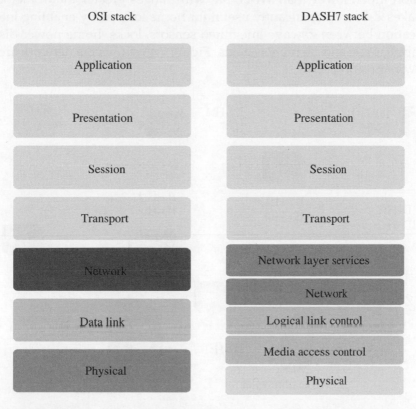

Figure 7.15 The DASH7 protocol stack in comparison to the OSI stack

DASH7 is capable of very dense deployments, has a low memory footprint, consumes minuscule power, and considered by many as a bridge between NFC and IoT communication systems. It can also be used to enable tag-to-tag communication without needing the tags to pass their information through a base station or a tag reader. This feature of DASH7 is quite synonymous with the multinode hopping mesh networks found in Zigbee and Z-wave. The reported range of DASH7 is between 1 to 10 km and a typical querying latency of 1 to 10 seconds.

Check yourself

File-system, node hopping mesh network, frequency harmonics

7.10   Z-Wave

Z-Wave is an economical and less complicated alternative to Zigbee. It was developed by Zensys, mainly for home automation solutions [11]. It boasts of a power consumption much lower than Wi-Fi, but with ranges greater than Bluetooth. This feature makes Z-Wave significantly useful for home IoT use by enabling inter-device communication between Z-wave integrated sensors, locks, home power distribution systems, appliances, and heating systems. Figure 7.16 shows the network architecture of the Z-Wave protocol.

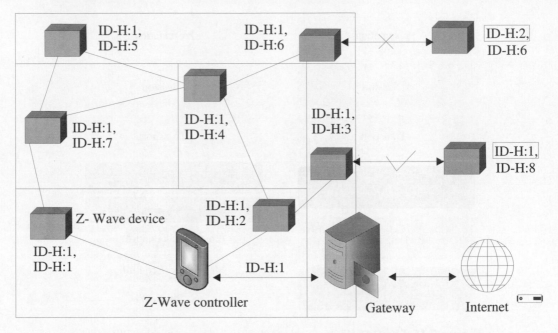

Figure 7.16 A typical Z-Wave deployment and communication architecture

Figure 7.17 shows the stack for this protocol. The Z-Wave operational frequency is in the range of 800–900 MHz, which makes it mostly immune to the interference effects of Wi-Fi and other radios utilizing the 2.4 GHz frequency band. Z-wave utilizes gaussian frequency shift keying (GFSK) modulation, where the baseband pulses are passed through a Gaussian filter before modulation. The filtering operation smoothens the pulses consisting of streams of −1 and 1 (known as pulse shaping), which limits the modulated spectrum's width. A Manchester channel encoding is applied for preparing the data for transmission over the channel.

Z-Wave stack

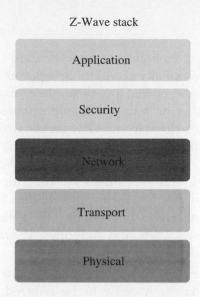

Figure 7.17 The Z-Wave protocol stack

Z-wave devices are mostly configured to connect to home-based routers and access points. These routers and access points are responsible for forwarding Z-wave messages to a central hub. Z-wave devices can also be configured to connect to the central hub directly if they are in range. Z-wave routing within the home follows a source-routed mesh network topology. When the Z-wave devices are not in range, messages are routed through different nodes to bypass obstructions created by household appliances or layouts. This process of avoiding radio dead-spots is done using a message called healing. Healing messages are a characteristic of Z-wave.

A central network controller device sets up and manages a Z-wave network (Figure 7.16), where each logical Z-wave network has one home (network) ID and multiple node IDs for the devices in it. Each network ID is 4 bytes long, whereas the node ID length is 1 byte. Z-Wave nodes with different home IDs cannot communicate with one another. The central hub is designed to be connected to the Internet, but their quantities are limited to one hub per home. Each home can have multiple devices, which can talk to the hub using Z-Wave. However, the devices themselves cannot connect to the Internet. The Z-wave can support 232 devices in a single home deployment (a single hub). This technology has been designed to be backward compatible. As Z-wave uses a source-routed static network, mobile devices are excluded from the network; only static devices are considered.

Check yourself

GFSK, Manchester encoding, pulse shaping principle

7.11   Weightless

Weightless is yet another emerging open standard for enabling networked communication in IoT; it is especially useful for low-power wide area networks [12]. It was designed for useful for low-power, low-throughput, and moderate to high latency applications supporting either or both public and private networks. The operating frequency of Weightless is restricted to sub-GHz bands, which are also exempted from the requirements of licensing such as 138 MHz, 433 MHz, 470 MHz, 780 MHz, 868 MHz, 915 MHz, and 923 MHz. Initially, three standards were released for Weightless: Weightless P, Weightless N, and Weightless W. Weightless P is the only currently accepted and used standard as it has features for bi-directional communication over both licensed as well as unlicensed ISM bands. Weightless N was designed as an LWPAN uplink-only technology, whereas Weightless W was designed to make use of the TV whitespace frequencies for communication.

   As Weightless P was the most commonly adopted and accepted standard among the three Weightless standards, it came to be referred to merely as Weightless. Weightless provides a true bi-directional and reliable means of communication, where each message transaction is validated using an acknowledgment message. As it was designed initially for dense deployments of low-complexity IoT end devices, its payload size was limited to less than 48 bytes. Weightless networks can be optimized to attain ultra-low-power consumption status compared to cellular networks. However, this is at the cost of network latency and throughput with data rates in the range of 0.625 kbps to 100 kbps. Weightless has been identified with three architectural components: end devices, base stations, and base station network (Figure 7.18). The end devices (ED) form the leaf nodes in the Weightless network. These devices are typically low complexity and low cost. The duty cycle of EDs is also low, with a nominal transmiting power of 14 dBm (which can be increased up to 30 dBm).

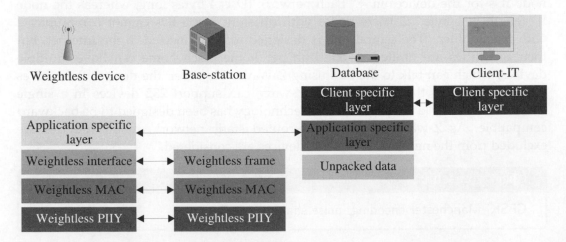

Figure 7.18 Typical components of the Weightless standard and its protocols

The base stations (BS) act as the central coordinating node in each cell. A star topology is deployed to connect the EDs to the BS. The transmit powers of a typical BS lie in the range of 27 dBm to 30 dBm. Finally, the base station network (BSN) is responsible for connecting all the BS of a single network. This enables the BSN to manage the allocation and scheduling of radio resources across the network. Additional tasks of the BSN include addressing authentication, roaming, and scheduling responsibilities.

> **Check yourself**
>
> Cellular architecture, base stations, whitespace bands

7.12 Sigfox

Sigfox is a low-power connectivity solution, which was developed for various businesses such as building automation and security, smart metering, agriculture, and others. It uses ultra-narrowband technology (192 kHz wide) for accessing and communicating through the radio spectrum [13]. The typical data rates achieved in Sigfox is in the range 100–600 bits per second. A binary phase shift keying (BPSK) is used for encoding the message transmission by changing the phase of the carrier waves, where each message is 100 Hz wide. Sigfox in Europe utilizes the 868 and 868.2 MHz spectrum, whereas it uses 902 and 928 MHz elsewhere. As the Sigfox receiver has to access only a very tiny part of the spectrum for receiving messages, the effects of noise are significantly reduced. It can even communicate in the presence of jamming signals, making this standard quite resilient.

Figure 7.19 shows the network architecture of Sigfox. Sigfox has an exciting message forwarding principle called random access, which ensures the high quality of services in this standard. Each Sigfox device emits a message at an arbitrary frequency;

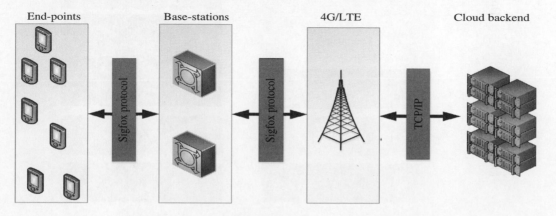

Figure 7.19 The Sigfox communication architecture

it simultaneously sends two replicas of the same message at different frequencies; it time using a principle known as time frequency diversity. Although the Sigfox devices are relatively less complicated, the base stations are very complicated as they monitor the whole 192 kHz spectrum looking for UNB (ultra narrow band) transmissions for demodulation. The base stations in Sigfox follow a cooperative reception principle. The messages in Sigfox are not attached to any base station, and any base station in the vicinity of the device can receive messages from it. This is called the principle of spatial diversity in Sigfox. The time and frequency diversity, along with the spatial diversity, ensures excellent quality of service for Sigfox.

Figure 7.20 shows the comparison of the Sigfox stack with the standard ISO-OSI stack. The Sigfox communication is bi-directional and asynchronous with a significant difference between the uplink and downlink speeds. As the devices are less complex than the base stations, the uplink budget (device to base station) is high compared to the downlink budget (base station to device). It is mainly due to this reason that the Sigfox was designed to have small message lengths ranging from 0 to 12 bytes. This 12-byte payload supports the simultaneous transfer of sensor data, the status of an event/alerts, GPS coordinates, and even application data. Sigfox boasts of excellent security features with support for authentication, integrity, and anti-replay on messages transmitted through the network. AES is supported by this standard. All

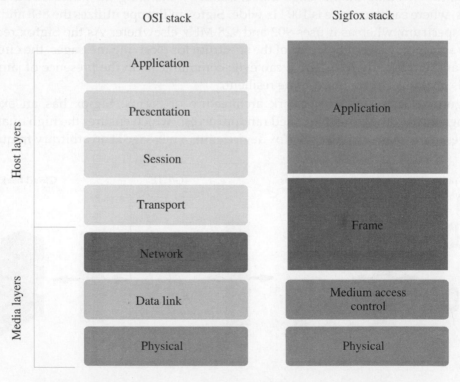

Figure 7.20 The Sigfox protocol stack in comparison to the OSI stack

these collective features of Sigfox enables it to be a low-power and resilient standard. However, due to the low data rates and asynchronous links, it is better utilized in applications requiring infrequent communication with small bursts of data. The Sigfox architecture and range supports wide and dense deployments depending on topologies and is better suited for indoor use; however, mobility is not an aspect associated with it.

> **Check yourself**
>
> AES, asynchronous versus synchronous communication

7.13 LoRa

LoRa or long range is a patented wireless technology for communication developed by Cycleo of Grenoble, France for cellular-type communications aimed at providing connectivity to M2M and IoT solutions [14]. It is a sub-GHz wireless technology that operationally uses the 169 MHz, 433 MHz, 868 MHz, and 915 MHz frequency bands for communication. LoRa uses bi-directional communication links symmetrically and a spread spectrum with a 125 kHz wideband for operating. Applications such as electric grid monitoring are typically suited for utilizing LoRa for communications. Typical communication of LoRa devices ranges from 15 to 20 km, with support for millions of devices. Figure 7.21 shows the LoRa network architecture.

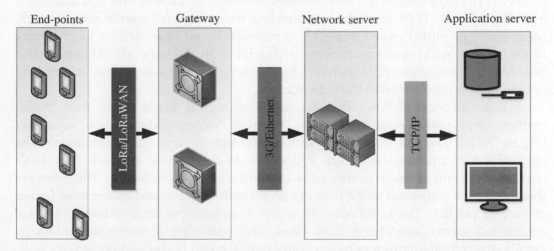

Figure 7.21 A typical LoRa deployment and communication architecture

It is a spread spectrum technology with a broader band (usually 125 kHz or more). LoRa achieves high receiver sensitivity by utilizing frequency-modulated chirp coding gain. LoRa devices provide excellent support for mobility, which makes them very useful for applications such as asset tracking and asset management. In comparison with similar technologies such as NB-IoT, LoRa devices have significantly higher battery lives, but these devices have low data rates (27 to 50 kbps) and longer latency times. Figure 7.22 shows the LoRa protocol stack.

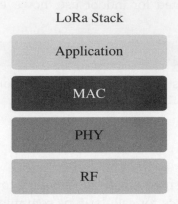

Figure 7.22 The LoRa protocol stack

LoRa devices make use of a network referred to as LoRaWAN, which enables the routing of messages between end nodes and the destination via a LoRaWAN gateway. Unlike Sigfox, LoRaWAN has a broader spectrum resulting in interference, which is solved using coding gains of the chirp signals. Additionally, unlike Sigfox, the LoRaWAN end nodes and the base stations are quite inexpensive. The LoRaWAN protocol is designed for WAN communications and is an architecture that makes use of LoRa, whereas LoRa is used as an enabling technology for a wide area network. Messages transmitted over LoRaWAN is received by all base stations in proximity to the device, which induces message redundancy in the network. However, this enhances the resilience of the network by ensuring more messages are successfully delivered between entities in the network.

A LoRa network follows the star topology and is made up of four crucial entities: end points/nodes, gateways, network server, and a remote computer (Figure 7.21). The end nodes deal with all the sensing and control solutions. The gateways forward messages from end nodes to a backhaul network. The LoRa network can comprise both or either of wired and wireless technologies. The gateways themselves are connected to the network server utilizing IP-based connections (either private or public). The LoRa network server is responsible for scheduling message acknowledgments, modifying data rates, and removing message redundancies. Finally, the remote computers have control over the end nodes and act as data sinks for data originating from these nodes.

The LoRa network security is achieved through various mechanisms such as unique network key, which ensures security on the network level, unique application key, which ensures an end-to-end security on the application level and device specific key.

> **Check yourself**
>
> Features of a chirp signal, coding gains, spread spectrum technology

7.14 NB-IoT

NB-IoT or narrowband IoT is an initiative by the Third Generation Partnership Project (3GPP) to develop a cellular standard, which can coexist with cellular systems (2G/3G/4G), be highly interoperable and that too using minimum power [15]. It is reported that a major portion of the NB-IoT applications can support a battery life of up to ten years. NB-IoT also boasts of significant improvements in reliability, spectrum efficiencies, and system capacities. NB-IoT uses orthogonal frequency division multiplexing (OFDM) modulation, which enhances the system capacity and increases spectrum efficiency (Figure 7.23). However, device complexities are quite high. NB-IoT also provides support for security features such as confidentiality, authentication, and integrity. Figure 7.24 shows the protocol stacks of the various components of NB-IoT.

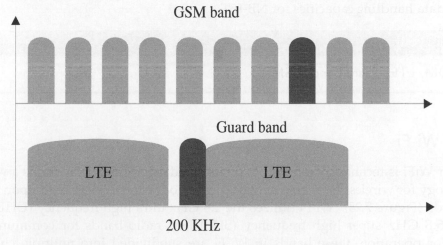

Figure 7.23 A location of NB-IoT band within the LTE spectrum

The coverage of NB-IoT supports deployments in indoor environments as well as in dense urban areas. When compared with technologies such as LoRa, NB-IoT ensures a higher quality of service as well as reduced latencies. Because of its design principles, the transfer of large messages is not efficient. NB-IoT is better suited for static deployments such as energy metering, fixed sensors, and others. Mobility support is not provided in this standard. NB-IoT communication can either make use of the available 200-kHz GSM (global system for mobile communications) bands or be allocated resource blocks on the guard bands by LTE base stations. This ensures

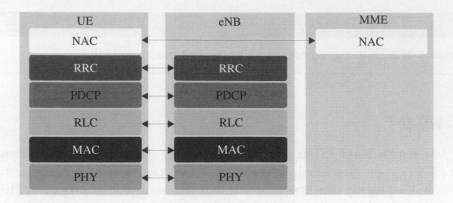

Figure 7.24 The NB-IoT protocol stack with respect to its entities

that the NB-IoT can achieve more extensive coverage while coexisting with cellular systems.

NB-IoT was developed for non-IP based applications requiring quite small volumes of daily data transactions, typically in the range of a few tens to a hundred bytes of data per device daily. Unlike technologies such as Sigfox and LoRa, the use of OFDM (orthogonal frequency division multiplexing's) faster modulation rates ensures higher data handling capacities for NB-IoT.

Check yourself

OFDM, LTE, guard band, GSM

7.15   Wi-Fi

Wi-Fi or WiFi is technically referred to by its standard, IEEE 802.11, and is a wireless technology for wireless local area networking of nodes and devices built upon similar standards (Figure 7.25). Wi-Fi utilizes the 2.4 GHz ultra high frequency (UHF) band or the 5.8 GHz super high frequency (SHF) ISM radio bands for communication [16]. For operation, these bands in Wi-Fi are subdivided into multiple channels. The communication over each of these channels is achieved by multiple devices simultaneously using time-sharing based TDMA multiplexing. It uses CSMA/CA for channel access.

Various versions of IEEE 802.11 have been popularly adapted, such as a/b/g/n. The IEEE 802.11a achieves a data rate of 54 Mbps and works on the 5 GHz band using OFDM for communication. IEEE 802.11b achieves a data rate of 11 Mbps and operates on the 2.4 GHz band. Similarly, IEEE 802.11g also works on the 2.4 GHz band but achieves higher data rates of 54 Mbps using OFDM. Finally, the newest version, IEEE 802.11n, can transmit data at a rate of 140 Mbps on the 5 GHz band.

Wi-Fi Stack

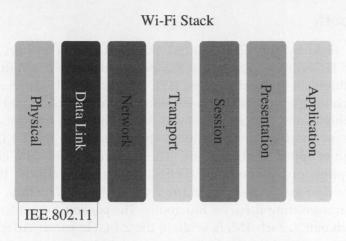

IEE.802.11

Figure 7.25 The IEEE 802.11 Wi-Fi stack

Wi-Fi devices can network using a technology referred to as wireless LAN (WLAN), as shown in Figure 7.26. A Wi-Fi enabled device has to connect to a wireless access point, which connects the device to the WLAN. WLAN is then responsible for forwarding the messages from the devices to and fro between the devices and the Internet.

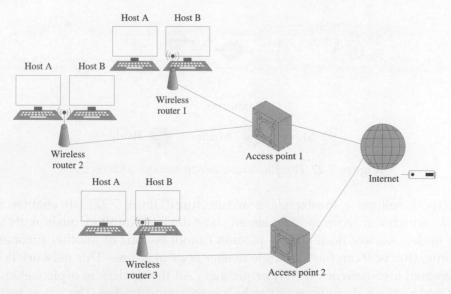

Figure 7.26 The Wi-Fi deployment architecture

Check yourself
TDMA

7.16 Bluetooth

Bluetooth is defined by the IEEE 802.15.1 standard and is a short-range wireless communication technology operating at low power to enable communication among two or more Bluetooth-enabled devices [17]. It was initially developed as a cable replacement technology for data communication between two or more mobile devices such as smartphones and laptops. This standard allows the transmission of data as well as voice-over short distances. Bluetooth functions on the 2.4 GHz ISM band and has a range of approximately 10 m. The transmission of data is done through frequency hopping spread spectrum (FHSS), which also reduces the interference caused by other devices functioning in the 2.4 GHz band. The data is divided into packets before transmitting them by Bluetooth. The packets are transmitted over the 79 designated channels, each 1MHz wide in the 2.4 GHz band. Adaptive frequency hopping (AFH) enables this standard to perform 800 hops per second over these channels. Initial versions of this standard followed Gaussian frequency shift keying (GFSK) modulation, which was known as the basic rate (BR) mode, and was capable of data rates of up to 1 Mbps. However, with the development of newer variants, modulation schemes such as $\pi/4$ DQPSK (differential quadrature phase shift keying) and 8-DPSK (differential phase shift keying) were adopted, which enabled data rates of 2 Mbps and 3 Mbps respectively.

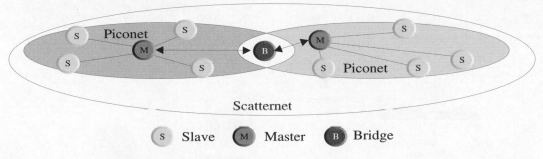

Figure 7.27 The Bluetooth device network architecture

Bluetooth follows a master–slave architecture (Figure 7.27). It enables a small network, which can accommodate seven slave devices simultaneously with a single master node. A slave node in one piconet cannot be part of another piconet at the same time, that is, it can have a single master node at a time. This network is known as a personal area network (PAN) or piconet. All the devices in a piconet share the master node's clock. Two piconets can be joined using a bridge. The whole network is also referred to as a scatternet.

Bluetooth Low Energy (BLE), the advanced variant of Bluetooth has 2 MHz wide bands, which can accommodate 40 channels. Its features include low energy consumption, low cost, multivendor interoperability, and an enhanced range of operations.

Bluetooth connections are encrypted and prevent eavesdropping of communications between devices. The inclusion of service-level security adds an additional layer of security by restricting the usage and device features and activities.

The Bluetooth standard consists of four parts: 1) core protocols, 2) cable replacement protocols, 3) telephony control protocols, and 4) adopted protocols. Figure 7.28 shows the Bluetooth protocol stack. Link Manager Protocol (LMP), Logical Link Control and Adaptation Protocol (L2CAP), Host Controller Interface (HCI), Radio Frequency Communications (RFCOMM), and Service Discovery Protocol (SDP) are some of the well-known protocols associated with Bluetooth. These protocols can be enumerated as follows:

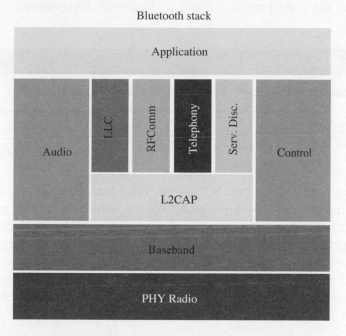

Figure 7.28 The Bluetooth protocol stack

(i) **Link Manager Protocol:** It manages the establishment, authentication, and links configuration. LMPs consist of some protocol data units (PDU), between which transmission occurs for availing services such as name requests, link address requests, connection establishment, connection authentication, mode negotiation, and data transfer.

(ii) **Host Controller Interface:** It enables access to hardware status and control registers and connects the controller with the link manager. The automatic discovery of Bluetooth devices in its proximity is one of the essential tasks of HCI.

(iii) **L2CAP:** It multiplexes logical connections between two devices. It is also tasked with data segmentation, flow control, and data integrity checks.

(iv) **Service Discovery Protocol:** It is tasked with the discovery of services provided by other Bluetooth devices.

(v) **Radio Frequency Communications:** It is a cable replacement protocol, which generates a virtual stream of serial data. This protocol supports many telephony related profiles as AT commands and Object Exchange Protocol (OBEX) over Bluetooth.

(vi) **Telephony Control Protocol – Binary (TCS BIN):** It is a bit-oriented protocol to control call signaling prior to initiation of voice or data communications between devices.

> Check yourself
>
> Bluetooth paging and inquiry, $\pi/4$ DQPSK, 8-DPSK, AFH

Summary

This chapter covered the various IoT connectivity technologies and their functional requirements and focuses on those technologies, which primarily rely on wireless media for communications. The standards covered in this chapter are a mix of the ones used for general consumer electronics, household devices, as well as speciality applications such as industries. The connectivity technologies covered here range from near-field ones to long-range ones. After going through this chapter, readers will be able to select various connectivity technologies, which will be suitable for an IoT application or architecture under consideration.

Exercises

(i) What is a piconet?

(ii) What is a scatternet? Explain the working of a scatternet with a brief description of its various members.

(iii) Describe the protocol stack of Bluetooth.

(iv) What is BLE?

(v) Differentiate between class 1, 2, and 3 Bluetooth devices.

(vi) What are the various modes of operation of Bluetooth?

(vii) Describe the L2CAP layer in Bluetooth.

(viii) Describe the RFCOMM layer in Bluetooth.

(ix) What is service discovery protocol (SDP) in Bluetooth?

(x) Describe the Bluetooth baseband.

(xi) How does Bluetooth avoid collisions between simultaneously transmitting nodes?

(xii) Explain the protocol stack of Zigbee.

(xiii) What is ZDO? How is it different from APS?

(xiv) Elaborate on the various network topologies of Zigbee.

(xv) What are the various Zigbee device types?

(xvi) Describe the Zigbee network layer.

(xvii) What is AODV? Explain with an example.

(xviii) How is Zigbee different from Bluetooth?

(xix) How is Zigbee different from 6LoWPAN?

(xx) Explain the protocol stack of IEEE 802.15.4

(xxi) How is LWPAN different from PANs?

(xxii) Explain the terms:

    (a) DSSS

    (b) BPSK

    (c) QPSK

    (d) O-QPSK

(xxiii) Differentiate between CSMA/CA and CSMA/CD.

(xxiv) Differentiate between star and mesh network topologies.

(xxv) What are the various IEEE 802.15.4 network types?

(xxvi) Differentiate between RFD and FFD.

(xxvii) Differentiate between a PAN coordinator, router, and a device in IEEE 802.15.4.

(xxviii) What are the various IEEE 802.15.4 frame types?

(xxix) What is beaconing?

(xxx) How are beacon-enabled networks different from non-Beacon enabled networks?

(xxxi) What is HART? How is it different from wirelessHART?

(xxxii) Describe the protocol stack of HART.

(xxxiii) Describe the HART physical layer.

(xxxiv) Describe the HART data link layer.

(xxxv) Describe the HART network and transport layers.

(xxxvi) What is TDMA? Describe with an example.

(xxxvii) What is channel blacklisting?

(xxxviii) What are superframes?

(xxxix) Describe the HART congestion control mechanism.

(xl) Describe the working of the wirelessHART network manager.

(xli) How is wirelessHART different from Zigbee?

(xlii) What is RFID? Explain its working.

(xliii) How is RFID different from QR codes?

(xliv) Differentiate between active and passive RFID.

(xlv) List some of the typical applications of RFID.

(xlvi) What is NFC? Describe its working.

(xlvii) How is NFC different from RFID?

(xlviii) What are the different types of NFC? Explain in detail.

(xlix) Describe the various modes of operation of NFC.

(l) List some of the popular applications of NFC.

(li) What is ISA 100.11a?

(lii) Describe the various transport services in ISA100.11a.

(liii) What are the various networks permitted in ISA100.11a?

(liv) What network topologies are allowed in ISA100.11a?

(lv) What are the various device types in ISA100.11a?

(lvi) List the salient features of ISA100.11a.

(lvii) What are the security features of ISA100.11a?

(lviii) Differentiate between an NRD and backbone device in ISA100.11a.

(lix) Differentiate between an RD and an NRD in ISA100.11a.

(lx) What are the typical usage classes in ISA100.11a?

(lxi) What is Z-Wave?

(lxii) Describe the working of a Z-Wave implementation.

(lxiii) Describe GFSK.

(lxiv) What is Manchester encoding?

(lxv) What is healing in the context of Z-Wave?

(lxvi) Differentiate between Z-Wave and Zigbee.

(lxvii) What are the different variants of Weightless? Enumerate the highlighting features of each.

(lxviii) How does Weightless provide true bi-directional communication?

(lxix) In Weightless, what topology is deployed to connect the EDs to the BS?

(lxx) What is the typical payload size restriction of Weightless?

(lxxi) What are the typical application domains of Sigfox?

(lxxii) What are the general data rates associated with Sigfox?

(lxxiii) Which encoding is used in Sigfox for transmitting messages?

(lxxiv) How does Sigfox communicate even in the presence of jamming signals?

(lxxv) What is the principle of spatial diversity in Sigfox?

(lxxvi) Why is the Sigfox uplink budget different from its downlink budget?

(lxxvii) What frequency bands are typically associated with LoRa?

(lxxvii) Differentiate between LoRa and NB-IoT.

(lxxix) How is the spread spectrum used for enhancing the efficiency of LoRa?

(lxxx) What is LoRaWAN? How is it different from LoRa?

(lxxxi) Differentiate between LoRaWAN and Sigfox.

(lxxxii) Describe the network topology of LoRa.

(lxxxiii) What are the modes of existence of NB-IoT?

(lxxxiv) How does NB-IoT make use of existing redundant GSM/CDMA bands?

(lxxxv) How does NB-IoT ensure high data handling capacities?

(lxxxvi) How does IEEE 802.11g achieve higher data rates?

(lxxxvii) What is the typical data transmission rate of IEEE 802.11n?

(lxxxviii) What is WLAN?

(lxxxix) Differentiate between WiFi and Bluetooth.

(xc) Differentiate between WiFI and Zigbee.

References

[1] Sikora, A. and V. F. Groza. 2005. "Coexistence of IEEE 802. 15.4 with Other Systems in the 2.4 GHz-ISM-Band." In 2005 *IEEE Instrumentation and Measurement Technology Conference Proceedings* 3: 1786–1791. IEEE.

[2] 802.15.4-2015 - IEEE Standard for Low-Rate Wireless Networks, IEEE. https://standards.ieee.org/standard/802_15_4-2015.html.

[3] Zigbee Alliance. https://zigbee.org/zigbee-for-developers/zigbee-3-0/.

[4] Farahani, S. 2011. *ZigBee Wireless Networks and Transceivers*. Newnes.

[5] Thread. https://www.threadgroup.org/.

[6]   OpenThread: An Open Foundation for the Connected Home. https://openthread.io/.

[7]   Petersen, S. and S. Carlsen. 2011. "WirelessHART vs. ISA100. 11a: The Format War Hits the Factory Floor." *IEEE Industrial Electronics Magazine* 5(4): 23–34.

[8]   Garfinkel, S. and H. Holtzman. 2006. "Understanding RFID Technology." *RFID*, pp. 15–36.

[9]   Coskun, V., B. Ozdenizci, and K. Ok. 2013. "A Survey on Near Field Communication (NFC) Technology." *Wireless Personal Communications* 71(3): 2259–2294.

[10]   Weyn, M., G. Ergeerts, L. Wante, C. Vercauteren, and P. Hellinckx. 2013. "Survey of the DASH7 Alliance Protocol for 433 MHz Wireless Sensor Communication." *International Journal of Distributed Sensor Networks* 9(12): 870430.

[11]   Badenhop, C. W., S. R. Graham, B. W. Ramsey, B. E. Mullins, and L. O. Mailloux. 2017. "The Z-Wave Routing Protocol and its Security Implications." *Computers & Security* 68: 112–129.

[12]   Weightless specification. http://www.weightless.org/about/weightless-specification.

[13]   Lauridsen, M., H. Nguyen, B. Vejlgaard, I. Z. Kovács, P. Mogensen, and M. Sorensen. 2017. "Coverage Comparison of GPRS, NB-IoT, LoRa, and SigFox in a 7800 km^2 Area." In *Proceedings of IEEE 85th Vehicular Technology Conference* (VTC Spring) (pp. 1–5). IEEE.

[14]   Augustin, A., J. Yi, T. Clausen, and W. Townsley. 2016. "A Study of LoRa: Long Range & Low Power Networks for the Internet of Things." *Sensors* 16(9): 1466.

[15]   Sinha, R. S., Y. Wei, and S. H. Hwang. 2017. "A Survey on LPWA Technology: LoRa and NB-IoT." *Ict Express* 3(1): 14–21.

[16]   IEEE 802.11TM WIRELESS LOCAL AREA NETWORKS – The Working Group for WLAN Standards. http://www.ieee802.org/11/.

[17]   Bluetooth specifications. https://www.bluetooth.com/specifications/.

Chapter 8

IoT Communication Technologies

Learning Outcomes

After reading this chapter, the reader will be able to:

- List common communication protocols in IoT
- Identify the salient features and application scope of each communication protocol
- Understand the terminologies and technologies in IoT communication
- Determine the requirements associated with each of these communication protocols in real-world solutions
- Determine the most appropriate communication protocol for their IoT implementation

8.1 Introduction

Having covered the various connectivity technologies for IoT in the previous chapter, this chapter specifically focuses on the various intangible technologies that enable communication between the IoT devices, networks, and remote infrastructures. We organize the various IoT communication protocols according to their usage into six groups: 1) Infrastructure protocols, 2) discovery protocols, 3) data protocols, 4) identification protocols, 5) device management protocols, and 6) semantic protocols. These protocols are designed to enable one or more of the functionalities and features associated with various IoT networks and implementations such as routing, data management, event handling, identification, remote management, and interoperability. Figure 8.1 outlines the distribution of these IoT communication protocol groups [3].

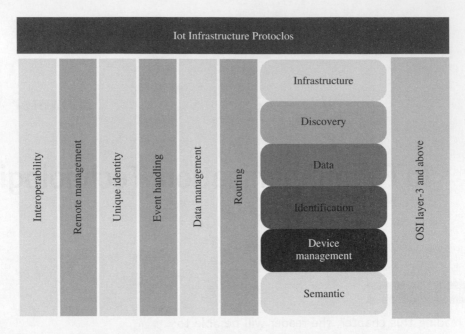

Figure 8.1 Various IoT communication protocol groups

Before delving into the various IoT communication protocols, we outline some of the essential terms associated with IoT networks that are indirectly responsible for the development of these communication protocols.

8.1.1 Constrained nodes

Constrained nodes is a term associated with those nodes where regular features of Internet-communicating devices are generally not available. These drawbacks are often attributed to the constraints of costs, size restrictions, weight restrictions, and available power for the functioning of these nodes. The resulting restrictions of memory and processing power often limit the usage of these nodes. For example, most of these nodes have a severely limited layer 2 capability and often lack full connectivity features and broadcasting capabilities. While architecting their use in networks and networked applications, these nodes require special design considerations. The issues of energy optimization and bandwidth utilization are dominant work areas associated with these nodes [1].

8.1.2 Constrained networks

Constrained networks [2], [1] are those networks in which some or all of the constituent nodes are limited in usage aspects due to the following constraints:

- limited processing power resulting in restrictions on achieving smaller duty cycles.

- low data rates and low throughput.

- asymmetric links and increased packet losses.

- restrictions on supported packet sizes due to increased packet losses.

- lack of advanced layer 3 functions such as multicasting and broadcasting.

- limited temporal device reachability from outside the network due to the inclusion of sleep states for power management in the devices.

8.1.3   Types of constrained devices

Constrained devices can be divided into three distinct classes according to the device's functionalities:

- **Class 0**: These devices are severely constrained regarding resources and capabilities. The barely feasible memory and processing available in these classes of devices do not allow for direct communication to the Internet. Even if the devices manage to communicate to the Internet directly, the mechanisms in place for ensuring the security of the device are not supported at all due to the device's reduced capabilities. Typically, this class of device communicates to the Internet through a gateway or a proxy.

- **Class 1**: These devices are constrained concerning available code space and processing power. They can primarily talk to the Internet, but cannot employ a regular full protocol stack such as HTTP (hyper text transfer protocol). Specially designed protocols stacks such as CoAP (common offer acceptance portal) can be used to enable Internet-based communication with other nodes. Compared to Class 0 devices, Class 1 devices have a comparatively increased power budget, which is attributed to the increased functionalities it supports over Class 0 devices. This class of devices does not need a gateway for accessing the Internet and can be armed with security features for ensuring safer communication over the Internet.

- **Class 2**: These devices are functionally similar to regular portable computers such as laptops and PDAs (personal digital assistants). They have the ability and capacity to support full protocol stacks of commonly used protocols such as HTTP, TLS, and others. However, as compared to the previous two classes of devices, these devices have a significantly higher power budget.

8.1.4   Low power and lossy networks

Low power and lossy networks (LLNs) typically comprise devices or nodes with limited power, small usable memory space, and limited available processing resources [4]. The network links between the devices in this network may be composed of low power Wi-Fi or may be based on the IEEE 802.15.4. The physical layers of the devices comprising LLNs are characterized by high variations in delivery rates, significant packet losses, and other similar behavior, which makes it quite unreliable, and often compromises network stability. However, LLNs have found extensive use in application areas such as industrial automation and monitoring, building automation, smart healthcare, smart homes, logistics, environment monitoring, and energy management.

8.2   Infrastructure Protocols

The protocols covered in this section are hugely dependent on the network and the network infrastructure for its operation. This section covers eight popular IoT-based communication technologies: Internet Protocol Version 6 (IPv6), Lightweight On-demand Ad hoc Distance vector Routing Protocol–Next Generation (LOADng), Routing Protocol for Low-Power and Lossy Networks (RPL), IPv6 over Low-Power Wireless Personal Area Networks (6LoWPAN), Quick UDP Internet Connection (QUIC), micro IP (uIP), nanoIP, and Content-Centric Networking (CCN).

8.2.1   Internet protocol version 6 (IPv6)

The Internet Protocol Version 6 or IPv6, as it is commonly known, is a resultant of the developments on and beyond IPv4 due to fast depleting address ranges in IPv4. The IPv4 was not designed to handle the needs of the future Internet systems, making it cumbersome and wasteful to use for IoT-based applications. The needs of massive scalability and limited resources gave rise to IPv6, which was developed by the IETF (Internet Engineering Task Force); it is also termed as the Internet version 2 [5].

Similar to IPv4, IPv6 also works on the OSI layer 3 (network layer). However, in contrast to IPv4 (which is 32 bits long and offers around 4,294,967,296 addresses), IPv6 has a massive logical address range (which is 128 bits long). Additional features in IPv6 include auto-configuration features, end-to-end connectivity, inbuilt security measures (IPSec), provision for faster routing, support for mobility, and many others. These features not only make IPv6 practical for use in IoT but also makes it attractive for a majority of the present-day and upcoming IoT-based deployments. Interestingly, as IPv6 was designed entirely from scratch, it is not backward compatible; it cannot be made to support IPv4 applications directly. Figure 8.2 shows the differences between IPv4 and IPv6 packet structures.

Some of the important features of IPv6 are as follows:

(i)   **Larger Addressing Range**: IPv6 has roughly four times more addressable bits than IPv4. This magnanimous range of addresses can accommodate the address requirements for any number of connected or massively networked devices in the world.

(ii)   **Simplified Header Structure**: Unlike IPv4, the IPv6 header format is quite simple. Although much bigger than the IPv4 header, the IPv6 header's increased

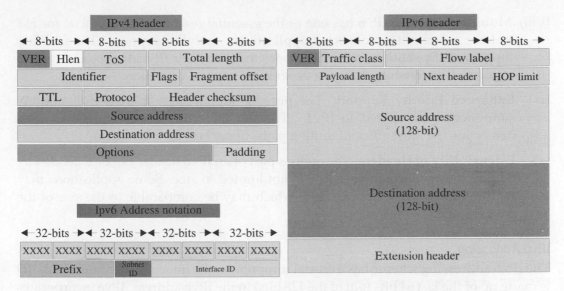

Figure 8.2 Differences between IPv4 and IPv6 packets and the IPv6 address notation

size is mainly attributed to the increased number of bits needed for addressing purposes.

(iii) **End-to-End Connectivity**: Unlike IPv4, the IPv6 paradigm allows for globally unique addresses on a significantly massive scale. This scheme of addressing enables packets from a source node using IPv6 to directly reach the destination node without the need for network address translations en route (as is the case with IPv4).

(iv) **Auto-configuration**: The configuration of addresses is automatically done in IPv6. It supports both stateless and stateful auto-configuration methods and can work even in the absence of DHCP (dynamic host configuration protocol) servers. This mechanism is not possible in IPv4 without DHCP servers.

(v) **Faster Packet Forwarding**: As IPv6 headers have all the seldom-used optional fields at the end of its packet, the routing decisions by a router are taken much faster, by checking only the first few fields of the header.

(vi) **Inbuilt Security**: IPv6 supports inbuilt security mechanisms (IPSec) that IPv4 does not directly support. IPv4 security measures were attained using separate mechanisms in conjunction with IPv4. The present-day version of IPv6 has security as an optional feature.

(vii) **Anycast Support**: Multiple networking interfacesare assigned the same IPv6 addresses globally; these addresses are known as anycast addresses. This mechanism enables routers to send packets to the nearest available destination during routing.

(viii) **Mobility Support**: IPv6 has one of the essential features that is crucial for IoT and the modern-day connected applications: mobility support. The mobility support of IPv6 allows for mobile nodes to retain their IP addresses and remain connected, even while changing geographic areas of operation.

(ix) **Enhanced Priority Support**: The priority support system in IPv6 is entirely simplified as compared to IPv4. The use of traffic classes and flow labels determine the most efficient routing paths of packets for the routers.

(x) **Extensibility of Headers**: The options part of an IPv6 header can be extended by adding more information to it; it is not limited in size. Some applications may require quite a large options field, which may be comparable to the size of the packet itself.

IPv6 Addressing

The IPv6 addressing scheme has a crucial component: the interface identifier (IID). IID is made up of the last 64 bits (out of the 128 bits) in the IPv6 address. IPv6 incorporates the MAC (media access control) address of the system for IID generation. As a device's MAC address is considered as its hardware footprint and is globally unique, the use of MAC makes IID unique too. The IID is auto-configured by a host using IEEE's extended unique identifier (EUI-64) format. Figure 8.2 illustrates the IPv6 addressing notation. IPv6 supports three types of unicasting: Global unicast address (GUA), link local address (LL), and unique local address (ULA).

The GUA is synonymous with IPv4's static addresses (public IP). It is globally identifiable and uniquely addressable. The global routing prefix is designated by the first (most significant) 48 bits. The first three bits of this routing prefix is always set to 001; these three bits are also the most significant bits of this prefix. In contrast, LLs are auto-configured IPv6 addresses, whose communication is limited to within a network segment only (under a gateway or a router). The first 16 bits of LL addresses are fixed and equals FE80 in hexadecimal. The subsequent 48 bits are set to 0. As these addresses are not routable, the LLs' scope is restricted to within the operational purview of a router or a gateway. Finally, ULAs are locally global and unique. They are meant for use within local networks only. Packets from ULAs are not routed to the Internet. The first half of an ULA is divided into four parts and the last half is considered as a whole. The four parts of the first part are the following: Prefix, local bit, global ID, and subnet ID, whereas the last half contains the IID. ULA's prefix is always assigned as FD in hexadecimal (1111 110 in binary). If the least significant bit in this prefix is assigned as 1, it signifies locally assigned addresses.

IPv6 Address Assignment

Any node in an IPv6 network is capable of auto-configuring its unique LL address. Upon assigning an IPv6 address to itself, the node becomes part of many multicast groups that are responsible for any communication within that segment of the network. The node then sends a neighbor solicitation message to all its IPv6 addresses.

If no reply is received in response to the neighbor solicitation message, the node assumes that there is no duplicate address in that segment, and its address is locally unique. This mechanism is known as duplicate address detection (DAD) in IPv6. Post DAD, the node configures the IPv6 address to all its interfaces and then sends out neighbor advertisements informing its neighbors about the address assignment of its interfaces. This step completes the IPv6 address assignment of a node.

IPv6 Communication

An IPv6 configured node starts by sending a router solicitation message to its network segment; this message is essentially a multicast packet. It helps the node in determining the presence of routers in its network segment or path. Upon receiving the solicitation message, a router responds to the node by advertising its presence on that link. Once discovered, the router is then set as that node's default gateway. In case the selected gateway is made unavailable due to any reason, a new default gateway is selected using the previous steps.

If a router upon receiving a solicitation message determines that it may not be the best option for serving as the node's gateway, the router sends a redirect message to the node informing it about the availability of a better router (which can act as a gateway) within its next hop.

IPv6 Mobility

A mobile IPv6 node located within its home link uses its home address for routing all communication to it. However, when the mobile IPv6 node goes beyond its home link, it has to first connect to a foreign link for enabling communication. A new IPv6 address is acquired from the foreign link, which is also known as the mobile node's care-of-address (CoA). The mobile node now binds its CoA to its home agent (a router/gateway to which the node was registered in its home segment). This binding between the CoA and the home agent is done by establishing a tunnel between them. Whenever the node's home agent receives a correspondence message, it is forwarded to the mobile node's CoA over the established tunnel. Upon receiving the message from a correspondent node, the mobile node may choose not to reply through its home agent; it can communicate directly to the correspondent node by setting its home address in the packet's source address field. This mechanism is known as route optimization.

> ### Check yourself
>
> IPv6 header structure, IPv6 extension header types, Neighbor discovery using IPv6, Mobility in IPv6

8.2.2 LOADng

LOADng stands for Lightweight On-demand Ad hoc Distance vector Routing Protocol – Next Generation. This protocol is inspired by the AODV (Ad hoc On-Demand Distance Vector) routing protocol, which is primarily a distance vector routing scheme [6]. Figure 8.3 illustrates the LOADng operation. Unlike AODV, LOADng was developed as a reactive protocol by taking into consideration the challenges of Mobile Ad hoc Networks (MANETs). The LOADng process starts with the initiation of the action of route discovery by a LOADng router through the generation of route requests (RREQs), as illustrated in Figure 8.3(a). The router forwards packets to its nearest connected neighbors, each of which again forwards the packets to their one-hop neighbors. This process is continued until the intended destination is reached. Upon receiving the RREQ packet, the destination sends back a route reply (RREP) packet toward the RREQ originating router (Figure 8.3(b)). In continuation, route error (RERR) messages are generated and sent to the origin router if a route is found to be down between the origin and the destination.

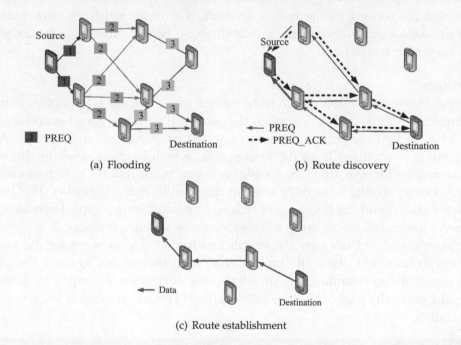

(a) Flooding

(b) Route discovery

(c) Route establishment

Figure 8.3 The LOADng routing mechanism

To summarize the operation of LOADng, a router performs the following tasks:

- Bi-directional network route discovery between a source and a destination.
- Route establishment and route maintenance between the source and the destination only when data is to be sent through the route.

- Generation of control and signaling traffic in the network only when data is to be transferred or a route to the destination is down.

Operational Principle

A LOADng router transmits an RREQ over all of its LOADng interfaces whenever a data packet from a local data source is received by it for transmission to a destination whose routing entry (a tuple) is not present with it. Figure 8.3(a) shows the flooding operation, where each LOADng's forward interfaces are numbered separately. Considering that it takes three hops to discover the destination from the source LOADng node, the individual forward interfaces are numbered from 1 to 3. The RREQ encodes the destination address received from the local source through the packet. The routing set managing the routing entries at each LOADng router updates or inserts an entry (with information of the originating address, and the immediate neighbor LOADng router) upon receiving an RREQ. This also works to enable a record of the reverse route between the source and destination (Figure 8.3(b)). The received RREQ initiates the checking of the destination address so that if the packets are intended for a local interface of a LOADng router, an RREP is sent back using the reverse route. In case the destination address is not local, it is forwarded to other LOADng interfaces in a hop-by-hop unicast manner through flooding.

When an RREP is received, it is recorded in the routing entry as the forward path toward the origin of the RREP along with the LOADng router that forwarded the message. The route metrics are additionally updated using RREQ and RREP messages. The LOADng determines the desired metric to be used (Figure 8.3(c)).

> Check yourself
>
> AODV routing, MANETs

8.2.3   RPL

RPL stands for routing protocol for low-power and lossy networks (LLN) and is designed for IPv6 routing. It follows a distance vector based routing mechanism [7]. The protocol aims to achieve a destination-oriented directed acyclic graph (DODAG). The network DODAG is formed based on an objective function and a set of network metrics. The DODAG built by RPL is a logical routing topology which is built over a physical network. The logical topology is built using specific criteria set by network administrators. The most optimum path (best path) is calculated from the objective function, a set of metrics, and constraints. The metrics in RPL may be expected transmission values (ETX), path latencies, and others. Similarly, the constraints in RPL include encryption of links, the presence of battery-operated nodes, and others. In general, the metrics are either minimized or maximized, whereas the constraints need to be minimized. The objective function dictates the rules for the formation of the

DODAG. Interestingly, in RPL, a single node in the mesh network may have multiple objective functions. The primary reason for this is attributed to the presence of different network traffic path quality requirements that need separate addressal within the same mesh network. Using RPL, a node within a network can simultaneously join more than one RPL instance (graphs). This enables RPL to support QoS-aware and constraint-based routing. An RPL node can also simultaneously take on multiple network roles: leaf node, router, and others. Figure 8.4 shows the RPL mechanism with different intra-mesh addressing arising due to different requirements of network and objective functions. The RPL border router, which is also the RPL root (in the illustrated figure), handles the intra-mesh addressing.

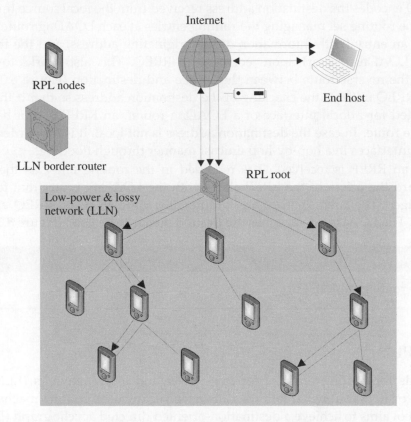

Figure 8.4 RPL information flow mechanism with different intra-mesh addressing and paths

RPL Instances

There are two instances associated with RPL: global and local. Global RPL instances are characterized by coordinated behavior and the possibility of the presence of more than one DODAG; they have a long lifetime. In contrast, local RPL instances are characterized by single DODAGs. The local RPL DODAG's root is associated directly with the DODAG-ID. The RPL instance ID is collaboratively and unilaterally allocated; it is divided between global and local RPL instances. Even the RPL control and data messages are tagged with their corresponding RPL instances using RPL instance IDs to avoid any ambiguity in operations.

> **Check yourself**
>
> Directed acyclic graphs (DAG), destination oriented directed acyclic graph (DODAG), vector based routing

8.2.4 6LoWPAN

6LoWPAN allows low power and constrained devices/nodes to connect to the Internet. 6LoWPAN stands for IPv6 over low power wireless personal area networks. As the name suggests, it enables IPv6 support for WPANs, which are limited concerning power, communication range, memory, and throughput [8]. 6LoWPAN is designed to be operational and straightforward over low-cost systems, and extend IPv6 networking capabilities to IEEE 802.15.4-based networks. Popular uses of this protocol are associated with smart grids, M2M applications, and IoT. 6LoWPAN allows constrained IEEE 802.15.4 devices to accommodate 128-bit long IPv6 addresses. This is achieved through header compression, which allows the protocol to compress and retro-fit IPv6 packets to the IEEE 802.15.4 packet format.

6LoWPAN networks can consist of both limited capability (concerning throughput, processing, memory, range) devices—called reduced function devices (RFD)—and devices with significantly better capabilities, called full function devices (FFD). The RFDs are so constrained that for accessing IP-based networks, they have to forward their data to FFDs in their personal area network (PAN). The FFDs yet again forward the forwarded data from the RFD to a 6LoWPAN gateway in a multi-hop manner. The gateway connects the packet to the IPv6 domain in the communication network. From here on, the packet is forwarded to the destination IP-enabled node/device using regular IPv6-based networking.

6LoWPAN Stack

The 6LoWPAN stack rests on top of the IEEE 802.15.4 PHY and MAC layers, which are generally associated with low rate wireless personal area networks (LR-WPAN). The choice of IEEE 802.15.4 for the base layer makes 6LoWPAN suitable for low power LR-WPANs. The network layer in 6LoWPAN enabled devices (layer 3) serves as an adaptation layer for extending IPv6 capabilities to IEEE 802.15.4 based devices. Figure 8.5 shows the 6LoWPAN packet structure.

- **PHY and MAC layers**: The PHY layer consists of 27 wireless channels, each having their separate frequency band and varying data rates. The MAC layer defines the means and methods of accessing the defined channels and use them for communication. The 6LoWPAN MAC layer is characterized by the following:

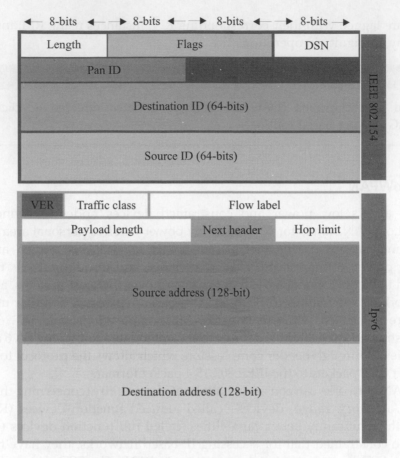

Figure 8.5 6LoWPAN packet structure

(i)   Beaconing tasks for device identification. These tasks include both beacon generation and beacon synchronization.

(ii) Channel access control is provided by CSMA/CA.

(iii) PAN membership control functions. Membership functions include association and dissociation tasks.

- **Adaptation layer**: As mentioned previously, 6LoWPAN accommodates and retro-fits the IPv6 packet to the IEEE 802.15.4 packet format. The challenge presented to 6LoWPAN is evident from the fact that IPv6 requires a minimum of 1280 octets for transmission. In contrast, IEEE 802.15.4 can support a maximum of only 1016 octets (127 bytes): 25 octets for frame overheads and 102 octets for payload. Additional inclusion of options in the IEEE 802.15.4 frame, such as security in the headers, leaves only 81 octets for IPv6 packets to use, which is insufficient. Even out of these available 81 octets, the IPv6 header reserves 40 octets for itself, 8 octets for UDP (user datagram protocol), and 20 octets for TCP (transmission control protocol), which are added in the upper layers. This leaves only 13 octets available at the disposal of the upper layers and the data itself. The 6LoWPAN adaptation layer between the MAC and the network layers takes care of these issues through the use of header compression, packet forwarding, and packet fragmentation.

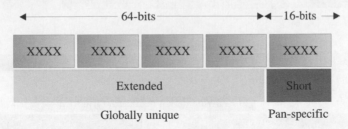

Figure 8.6 6LoWPAN address format

- **Address Format**: The 6LoWPAN address format is made up of two parts: 1) the short (16-bit) address and 2) the extended (64-bit) address. The short address is PAN specific and is used for identifying devices within a PAN only, which makes its operational scope highly restricted and valid within a local network only. In contrast, the globally unique extended address is valid globally and can be used to identify devices, even outside the local network uniquely. Figure 8.6 illustrates the 6LoWPAN address format.

Encapsulation Header Formats

The encapsulation headers, as the name suggests, defines methods and means by which 6LoWPAN encapsulates the IPv6 payloads within IEEE 802.15.4 frames. Figure 8.7 outlines the various header types associated with 6LoWPAN. 6LoWPAN has three encapsulation header types associated with it: dispatch, mesh addressing, and fragmentation. This system is similar to the IPv6 extension headers. The headers are identified by a *header type* field placed in front of the headers. The dispatch header type is used to initiate communication between a node and a destination node. The mesh addressing header is used for multi-hop forwarding by providing support for layer two forwarding of messages. Finally, the fragmentation header is used to fit large payloads to the IEEE 802.15.4 frame size.

Check yourself

LR-WPAN, WPAN, Beaconing

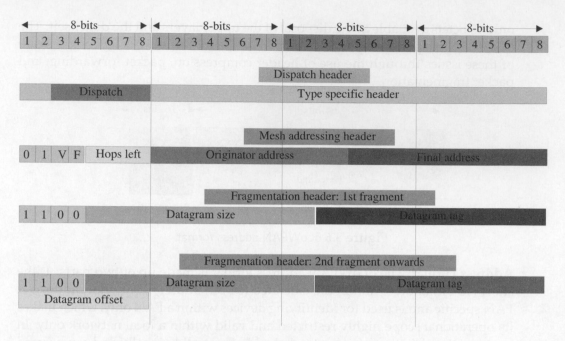

Figure 8.7 6LoWPAN header structures

8.2.5 QUIC

Quick UDP internet connections (QUIC) was developed (and still undergoing developments) to work as a low-latency and independent TCP connection [9]. The aim behind the development of this protocol is to achieve a highly reduced latency (almost zero round-trip-time) communication scheme with stream and multiplexing support like the SPDY protocol developed by Google. Figure 8.8 illustrates the differences between the positions of the various functionalities in QUIC and regular HTTP protocols.

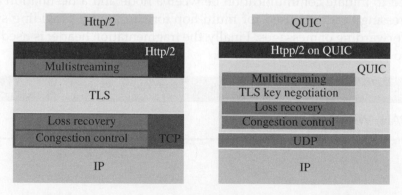

Figure 8.8 Differences between HTTP and QUIC protocols

The connection latency in QUIC is reduced by reducing the number of round trips incurred during connection establishment in TCP, such as those for handshaking, data requests, and encryption exchanges. This is achieved by including session negotiation information in the initial packet itself. The QUIC servers further enhance this compression by publishing a static configuration record corresponding to the connections. Clients synchronize connection information through cookies received from QUIC servers.

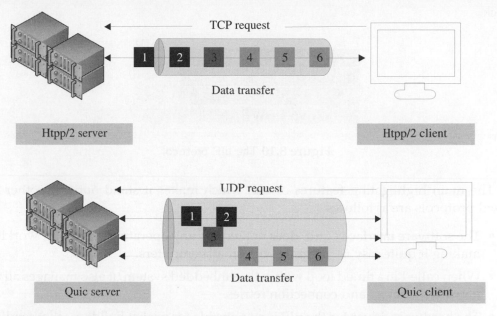

Figure 8.9 Differences between stream of packets over HTTP and QUIC protocols

QUIC uses advanced techniques such as packet pacing and proactive speculative retransmission to avoid congestion. Proactive speculative retransmission sends copies of most essential packets, which contain initial negotiation for encryption and error correction. The additional speedup is achieved using compression of data such as headers, which are generally redundant and repetitive. This feature enables QUIC connections to make multiple secured requests within a single congestion window, which would not have been possible using TCP. Figure 8.9 shows the difference in regular streaming of packets over HTTP and the improved performance of HTTP-over-QUIC during packet streaming. The use of UDP and multiple transmission paths significantly speeds up the performance of streaming over QUIC as compared to regular HTTP-based packet streaming.

Check yourself

QUIC use cases, SPDY protocol, TCP congestion control mechanism

8.2.6   Micro internet protocol (uIP)

The micro-IP (uIP) protocol is developed to extend the TCP/IP protocol stack capabilities to 8-bit and 16-bit microcontrollers [10]. uIP is an open-source protocol developed by the Swedish Institute of Computer Science (SICS). The low code space and memory requirements of uIP make it significantly useful for networking low-cost and low-power embedded systems. uIP now features a full IPv6 stack, which was developed jointly by Atmel, Cisco, and SICS. Figure 8.10 shows the micro-IP protocol stack.

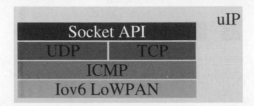

Figure 8.10 The uIP protocol

The main highlighting features of uIP, which makes it stand out from other IP-based protocols are as follows:

- The software interface of uIP does not require any operating system for working, making it quite easy to integrate with small computers.

- When called in a timed loop within the embedded system, it also manages all the network behavior and connection retries.

- The hardware driver for the uIP is responsible for packet builds, packet sending; it may also be used for response reception for the packets sent.

- uIP uses a minimal packet buffer (packet buffer = 1) in contrast to normal IP protocol stacks. This makes uIP suitable for low-power operations.

- The packet buffer is used in a half-duplex manner so that the same buffer can be repurposed for use in transmission and reception.

- Unlike regular TCP/IP protocols, uIP does not store data in buffers in case there is a need for retransmission. In the event of retransmission of packets, the previous data has to be reproduced and is recalled from the application code itself.

- Unlike conventional IP-based protocols, where a task is dedicated for each connection to a distant networked device/node, uIP stores connections in an array, and serves each connection sequentially through subroutine calls to the application for sending data.

8.2.7 Nano internet protocol (nanoIP)

The nano Internet protocol or NanoIP was designed to work with embedded devices, specifically sensor devices, by enabling Internetworking amongst these devices [11]. The concept of nanoIP enables wireless networking among low-power sensor devices, which is address-based, without incurring the overheads associated with the TCP/IP protocol stacks and mechanisms. Figure 8.11 shows the nano-IP TCP and UDP protocol stacks. The nanoIP is made up of two two transport mechanisms: nanoUDP and nanoTCP. These two transport mechanisms are analogous to the conventional UDP (unreliable transport protocol) and TCP (reliable transport protocol), respectively. NanoTCP even supports packet retransmissions and flow control, just like regular TCP. Instead of logical addressing, nanoIP uses hardware (MAC) addresses of devices for networking. The supported port range is 256 each for source and destination nodes, which is the allowable limit for an 8-bit port representation. With the advent of the nanoIP, several associated protocols have also come up, such as nHTTP and nPing.

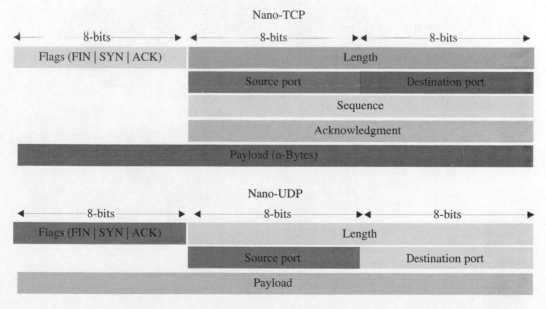

Figure 8.11 The nano-IP TCP and UDP protocols

> **Check yourself**
>
> nHTTP working, nPing working

8.2.8 Content-centric networking (CCN)

The content-centric networking (CCN) paradigm [12] is more commonly known as information-centric networking (ICN). Other names associated with this paradigm are named data networking (NDN) and publish–subscribe networking (PSIRP). CCN enables communication by defining and adhering to the concept of uniquely named data. This networking paradigm, unlike conventional networking approaches, is independent of location, application, and storage requirements. CCN is anchorless, which enables mobility and focuses on in-network caching for operations. These measures extend the features of data and communication efficiency, enhances scalability, and robustness, even in constrained and challenged networks. Figure 8.12 shows the operation of a typical CCN paradigm. Users can access cached content from multiple content generating sources by accessing data from trusted content servers, which also enable security of the data (not the communication channel).

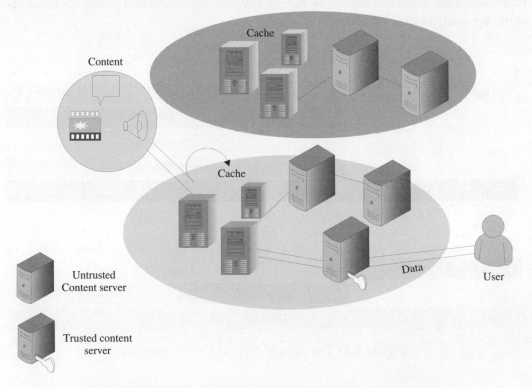

Figure 8.12 Content centric networking operation and its scope

In CCN, a forwarder checks a named request through hierarchical prefix matching (typically, longest prefix match) with a forwarding information base (FIB). A binary comparison is performed for prefix matching. The CCN request is a hierarchical sequence of network path segments. The FIB matching is then used to forward the named request to the appropriate network or network segment, which can respond to the issued request. The forwarder has to additionally determine the reverse path from the responder to the requester. All these operations are carried out without specifically binding a request to a network end point. The FIB at each CCN router stores information in a table, which is updated by a routing mechanism. Although the path segments and names are theoretically unbounded, they are restricted by the routing protocol being used for practical reasons.

> **Check yourself**
>
> Examples of publish–subscribe networking (PSIRP) and named data networking (NDN)

> **Points to ponder**
>
> A sensor node is made up of a combination of sensor/sensors, a processor unit, a radio unit, and a power unit.

8.3 Discovery Protocols

The protocols and paradigms covered in this section are largely focused on the discovery of services and logical addresses. We cover three interesting discovery protocols in this section: 1) Physical web, 2) mDNS, and 3) universal plug and play (UPnP).

8.3.1 Physical web

The physical web was designed to provide its users with the ability to interact with physical objects and locations seamlessly. The information to the users can be in the form of regular web pages or even dynamic web applications [13].

Some examples in the context of the physical web include user-friendly buses, which can alert its users about various route-related information, smart home appliances that can teach new users how to use them, self-diagnostic robots and machinery in industries, smart pet tags which can provide information about the pet's owner and its home location, and many more. Figure 8.13 shows the outline of a physical web model. The main takeaway of this concept is the seamless integration of several standalone smart systems through the web to provide information on demand to its users.

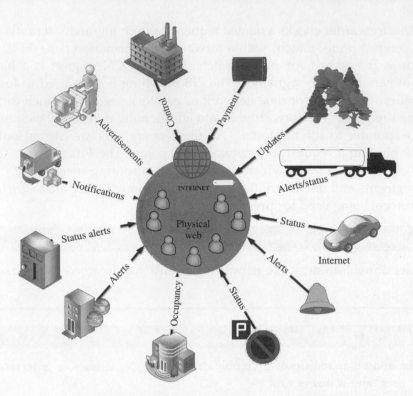

Figure 8.13 The physical web model

The physical web broadcasts a list of URLs within a short radius around it so that anyone in range can see the available URLs and interact with them. This paradigm is primarily built upon Bluetooth low energy (BLE), which is used to broadcast the content as beacons. The primary requirement of any supporting beacons to function in the physical web and broadcast URLs is their ability to support the Eddystone protocol specification. BLE was primarily chosen for the physical web due to its ubiquity, efficiency, and extended battery life of several years.

URLs are one of the core principles of the web and can be either flexible or decentralized. These URLs allow any application to have a presence on the web and enables the digital presence of an object or thing. As of now, physical web deployments have been undertaken in public spaces, and any device with a physical web scanner can detect these URLs. The use of URLs extends the benefits of the web security model to the physical web. Features such as secured login, secure communication over HTTPS (HTTP over secure socket layer), domain obfuscation, and others can be easily integrated with the physical web.

Check yourself

Eddystone protocol specification

8.3.2   Multicast DNS (mDNS)

The multicast domain name system or mDNS is explicitly designed for small networks and is analogous to regular DNS (domain name system), which is tasked with the resolution of IP addresses [14]. Interestingly, this system is free from any local name server from an operational point of view. However, it can work with regular DNS systems as it is a zero-configuration service. It uses multicast UDP for resolving host names. An mDNS client initiates a multicast query on the IP network, which asks a remote host to identify itself. The mDNS cache in the associated network subnet is updated with the multicast response received from the target. A node can give up its claim to a domain name by setting the time-to-live (TTL) field to zero in its response packet to an mDNS query. Some popular implementations of mDNS include the Apple Bonjour service and the networked printer discovery service in Windows 10 operating system from Microsoft. The main drawback of mDNS is its host name resolution reach to a top-level domain only.

> Check yourself
>
> DNS, DNS query, DNS response

8.3.3   Universal plug and play (UPnP)

Universal plug and play or UPnP encompasses a set of networking protocols aimed at service discovery and the establishment of functional network-based data sharing and communication services [15]. In brief, it is mainly used for enabling dynamic connections of devices to computing platforms. This paradigm is termed plug and play as the devices attaching to a computer network can configure themselves and update their hosts about their working configurations over a network. The UPnP is backed by a forum of many consumer electronics vendors and industries and is managed by the Open Connectivity Foundation. As UPnP is primarily designed for non-enterprise devices, its scope includes the discovery and intercommunication between networked devices such as mobiles, printers, access points, gateways, televisions, and other regular commercial systems enabled with IP capabilities. Figure 8.14 outlines the underlying UPnP stack and the relative location of the various functionalities in the stack.

The present-day UPnP has been designed to run on IP enabled networks, and makes use of the networking services of HTTP, XML, and SOAP for data transfer, device descriptions, and event generation and monitoring. UPnP enables UDP-based HTTP device search requests and advertisements using multicasting. The responses to device requests are necessarily unicast. UPnP advertisements use UDP port 1900 for multicasting. The unnecessary overheads and traffic generated by UPnP systems and their multicast behavior make them unsuitable for enterprise systems. The main

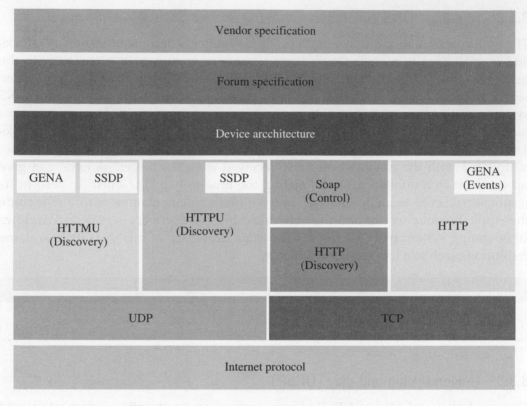

Figure 8.14 An outline of the basic UPnP stack

reason for this is because, on a large scale, the cost of this solution would be infeasible from an operational point of view.

> **Check yourself**
>
> Multicasting, Unicasting

8.4   Data Protocols

The protocols covered in this section are directly related to access, storage, and distribution of data through the IoT network. The data may be transferred between clients and servers as well as between brokers and subscribers in the IoT ecosystem. This section is further divided into seven parts: 1) MQTT, 2) MQTT-SN, 3) CoAP, 4) AMQP, 5) XMPP, 6) REST, and 7) websockets.

8.4.1 MQTT

Message queue telemetry transport or MQTT is a simple, lightweight publish–subscribe protocol, designed mainly for messaging in constrained devices and networks [16]. It provides a one-to-many distribution of messages and is payload-content agnostic. MQTT works reliably and flawlessly over high latency and limited bandwidth of unreliable networks without the need for significant device resources and device power. Figure 8.15 shows the working of MQTT. The MQTT paradigm consists of numerous clients connecting to a server; this server is referred to as a *broker*. The clients can have the roles of information publishers (sending messages to the broker) or information subscribers (retrieving messages from the broker). This allows MQTT to be largely decoupled from the applications being used with MQTT.

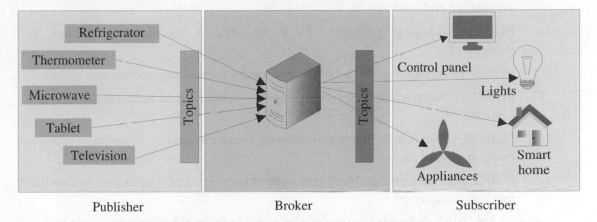

Figure 8.15 MQTT operation and its stakeholders

Operational Principle

MQTT is built upon the principles of hierarchical topics and works on TCP for communication over the network. Brokers receive new messages in the form of topics from publishers. A publisher first sends a control message along with the data message. Once updated in the broker, the broker distributes this topic's content to all the subscribers of that topic for which the new message has arrived. This paradigm enables publishers and subscribers to be free from any considerations of the address and ports of multiple destinations/subscribers or network considerations of the subscribers, and vice versa. In the absence of any subscribers of a topic, a broker normally discards messages received for that topic unless specified by the publisher otherwise. This feature removes data redundancies and ensures that maximally updated information is provided to the subscribers. It also reduces the requirements of storage at the broker. The publishers can set up default messages for subscribers in agreement with the broker if the publisher's connection is abruptly broken with the broker. This arrangement is referred to as the *last will and testament* feature of MQTT.

Multiple brokers can communicate in order to connect to a subscriber's topic if it is not present directly with the subscriber's primary broker.

MQTT's control message sizes can range between 2 bytes to 256 megabytes of data, with a fixed header size of 2 bytes. This enables the MQTT to reduce network traffic significantly. The connection credentials in MQTT are unencrypted and often sent as plain text. The responsibility of protecting the connection lies with the underlying TCP layer. The MQTT protocol provides support for 14 different message types, which range from connect/disconnect operations to acknowledgments of data. The following are the standard MQTT message types:

(i)   CONNECT: Publisher/subscriber request to connect to the broker.

(ii)   CONNACK: Acknowledgment after successful connection between publisher/ subscriber and broker.

(iii) PUBLISH: Message published by a publisher to a broker or a broker to a subscriber.

(iv)   PUBACK: Acknowledgment of the successful publishing operation.

(v)   PUBREC: Assured delivery component message upon successfully receiving publish.

(vi)   PUBREL: Assured delivery component message upon successfully receiving publish release signal.

(vii) PUBCOMP: Assured delivery component message upon successfully receiving publish completion.

(viii) SUBSCRIBE: Subscription request to a broker from a subscriber.

(ix)   SUBACK: Acknowledgment of successful subscribe operation.

(x)   UNSUBSCRIBE: Request for unsubscribing from a topic.

(xi)   UNSUBACK: Acknowledgment of successful unsubscribe operation.

(xii) PINGREQ: Ping request message.

(xiii) PINGRESP: Ping response message.

(xiv) DISCONNECT: Message for publisher/subscriber disconnecting from the broker.

MQTT Message Delivery QoS

MQTT's features and content delivery mechanisms are primarily designed for message transmission over constrained networks and through constrained devices. However, MQTT supports three QoS features:

- At most once: This is a best-effort delivery service and is largely dependent on the best delivery efforts of the TCP/IP network on which the MQTT is supported. It may result in message duplication or loss.

- At least once: This delivery service guarantees assured delivery of messages. However, message redundancy through duplication is a possibility.

- Exactly once: This delivery service guarantees assured message delivery. Additionally, this service also prevents message duplication.

> Check yourself
>
> MQTT clients and servers available online, implementing MQTT

8.4.2 MQTT-SN

The primary MQTT protocols heavily inspire MQTT for sensor networks or MQTT-SN; however, the MQTT-SN is robust enough to handle the requirements and challenges of wireless communications networks in sensor networks [17]. Typical features of MQTT-SN include low bandwidth usage, ability to operate under high link failure conditions; it is suitable for low-power, low-cost constrained nodes and networks. The major differences between the original MQTT and MQTT-SN include the following:

- The CONNECT message types are broken into three messages in which two are optional and are tasked with the communication of the testament message and testament topic to the broker.

- The topic name in the PUBLISH messages are replaced by topic identifiers, which are only 2 bytes long. This reduces the traffic generated from the protocol and enables the protocol to operate over bandwidth-constrained networks.

- A separate mechanism is present for topic name registration with the broker in MQTT-SN. After a topic identifier is generated for the topic name, the identifier is informed to the publisher/subscribers. This mechanism also supports the reverse pathway.

- In special cases in MQTT-SN, pre-defined topic identifiers are present that need no registration mechanism. The mapping of topic names and identifiers are known in advance to the broker as well as the publishers/subscribers.

- The presence of a special discovery process is used to link the publisher/subscriber to the operational broker's network address in the absence of a preconfigured broker address.

- The subscriptions to a topic, Will topic, and Will message are persistent in MQTT-SN. The publishers/subscribers can modify their Will messages during a session.

- Sleeping publishers/subscribers are supported by a keep-alive procedure, which is offline, and which helps buffer the messages intended for them in the broker until they wake up. This feature of MQTT-SN is not present in regular MQTT.

Figure 8.16 shows the two gateway types in MQTT-SN: 1) the transparent gateway and 2) the aggregating gateway. The MQTT-SN converts/translates MQTT and MQTT-SN traffic by acting as a bridge between these two network types. The transparent gateway (Figure 8.16(a)) creates as many connections to the MQTT broker as there are MQTT-SN nodes within its operational purview; whereas the aggregating gateway (Figure 8.16(b)) creates a single connection to the MQTT broker, irrespective of the number of MQTT-SN nodes under it.

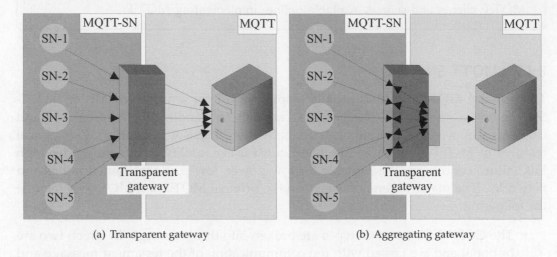

(a) Transparent gateway   (b) Aggregating gateway

Figure 8.16 The MQTT-SN types

8.4.3 CoAP

The constrained application protocol, or CoAP as it is more popularly known, is designed for use as a web transfer protocol in constrained devices and networks, which are typically low power and lossy [18]. The constrained devices typically have minimal RAM and an 8-bit processor at most. CoAP can efficiently work on such devices, even when these devices are connected to highly lossy networks with high packet loss, high error rates, and bandwidth in the range of kilobits.

CoAP follows a request–response paradigm for communication over these lossy networks. Additional highlights of this protocol include support for service discovery, resource discovery, URIs (uniform resource identifier), Internet media handling support, easy HTTP integration, and multicasting support, that too while maintaining low overheads. Typically, CoAP implementations can act as both clients and servers (not simultaneously). A CoAP client's request signifies a request for action from an identified resource on a server, which is similar to HTTP. The response sent by the server in the form of a response code can contain resource representations as well. However, CoAP interchanges are asynchronous and datagram-oriented over UDP. Figure 8.17 shows the placement of CoAP in a protocol stack. Packet traffic collisions are handled by a logical message layer incorporating the exponential back-off mechanism for providing reliability. The reliability feature of CoAP is optional. The two seemingly distinct layers of messaging (which handle the UDP and asynchronous messaging) and request-response (which handles the connection establishment) are part of the CoAP header.

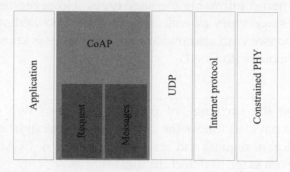

Figure 8.17 Position of the CoAP protocol in a stack

CoAP Features
The CoAP is characterized by the following main features:

(i) It has suitable web protocol for integrating IoT and M2M services in constrained environments with the Internet.

(ii) CoAP enables UDP binding and provides reliability concerning unicast as well as multicast requests.

(iii) Message exchanges between end points in the network or between nodes is asynchronous.

(iv) The limited packet header incurs significantly lower overheads. This also results in less complexity and processing requirements for parsing of packets.

(v) CoAP has provisions for URI and other content-type identifier support. CoAP additionally provides DTLS (datagram transport layer security) binding.

(vi) It has a straightforward proxy mechanism and caching capabilities, which is responsible for overcoming the effects of the lossy network without putting extra constraints on the low-power devices. The caching is based on the concept of the maximum age of packets.

(vii) The protocol provides a stateless mapping with HTTP. The server or receiving node does not retain information about the source of the message; rather, it is expected that the message packet carries that information with it. This enables CoAP's easy and uniform integration with HTTP.

CoAP Messaging

CoAP defines four messaging types: 1) Confirmable (CON), 2) non-confirmable (NON), 3) acknowledgment (ACK), and 4) reset. The method codes and the response codes are included in the messages being carried. These codes determine whether the message is a request message or a response message. Requests are typically carried in confirmable and non-confirmable message types. However, responses are carried in both of these message types as well as with the acknowledgment message. The transmission of responses with acknowledgment messages is known as piggybacking and is quite synonymous with CoAP.

Operational Principle

CoAP is built upon the exchange of messages between two or more UDP end points. Options and payload follow the compact 4-byte binary header in CoAP. This arrangement is typical of request and response messages of CoAP. A 2-byte message ID is used with each message to detect duplicates.

Whenever a message is marked as a CON message, it signifies that the message is reliable. In the event of delivery failure of a CON message, subsequent retries are attempted with exponential back-off until the receiving end point receives an ACK with the same message ID (Figure 8.18). In case the recipient does not have the resources to process the CON message, a RESET message is sent to the originator of the CON message instead of an ACK message.

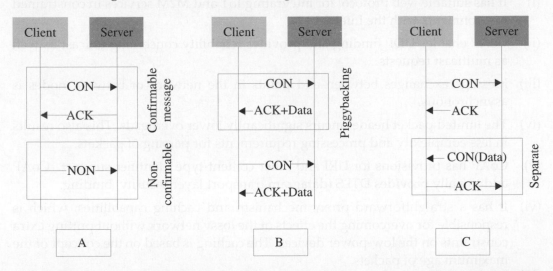

Figure 8.18 Various CoAP response–response models. (A): CON and NON messages, (B): Piggyback messages, and (C): Separate messages

Specific messages, which do not require reliable message transmission (such as rapid temporal readings of the environment from a sensor node), are sent as NON messages. NON messages do not receive an acknowledgment (Figure 8.18). However, the message ID associated with it prevents duplication. NON messages elucidate a

NON or CON response from a server, based on the settings and semantics of the application. If the receiver of the NON cannot process the message, a RESET message is sent to the originator of the NON message.

If a server fails to respond immediately to a request received by it in a CON message, an empty ACK response is sent to the requester to stop request retransmissions. Whenever the response is ready, a new CON message is used to respond to the previous request by the client. Here, the client then has to respond to the server using an ACK message. This scheme is known as a *separate response* (Figure 8.18).

The multicast support of CoAP over UDP results in multicast CoAP requests. The request and response semantics of CoAP is carried in the form of method and response codes in the CoAP messages itself. The options field of CoAP carries information about the requests and responses such as URI and MIME (multipurpose Internet mail extensions). The concept of tokens is used to match requests with their corresponding responses. The need for a token mechanism arose due to the asynchronous nature of the CoAP messaging. Similar to HTTP, CoAP uses GET, PUT, POST, and DELETE methods.

> **Check yourself**
>
> CoAP header fields, CoAP packet size

8.4.4 AMQP

AMQP or the advanced message queuing protocol is an open standard middleware at the application layer developed for message-oriented operations [19]. It tries to bring about the concept of interoperability between clients and the server by enabling cross-vendor implementations. Figure 8.19 shows the various components of AMQP and their relationships. An AMQP broker is tasked with maintaining message queues between various subscribers and publishers. The protocol is armed with features of message orientation, queuing, reliability, security, and routing. Both request–response and publish–subscribe methods are supported. AMQP is considered as a wire-level protocol. Here, the data format description is released on the network as a stream of bytes. This description allows AMQP to connect to anyone who can interpret and create messages in the same format. It also results in a level of interoperability where anyone with compliant or supporting means can make use of this protocol without any need for a specific programming language.

AMQP Features
AMQP is built for the underlying TCP and is designed to support a variety of messaging applications efficiently. It provides a wide variety of features such as flow-controlled communication, message-oriented communication, message delivery

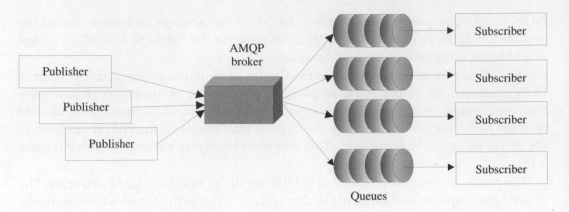

Figure 8.19 AMQP components and their relationships

guarantees (at most once, at least once, and exactly once), authentication support, and an optional SSL or TLS based encryption support. The AMQP is specified across four layers: 1) type system, 2) process to process asynchronous and symmetric message transfer protocol, 3) extensible message format, and 4) set of extensible messaging capabilities. In continuation, the primary unit of data in AMQP is referred to as a frame. These frames are responsible for the initiation of connections, termination of connections, and control of messages between two peers using AMQP. There are nine frame types in AMQP:

(i)   Open: responsible for opening the connection between peers.

(ii)   Begin: responsible for setup and control of messaging sessions between peers.

(iii)   Attach: responsible for link attachment.

(iv)   Transfer: responsible for message transfer over the link.

(v)   Flow: responsible for updating the flow control state.

(vi)   Disposition: responsible for updating of transfer state.

(vii)   Detach: responsible for detachment of link between two peers.

(viii) End: responsible for truncation of a session.

(ix)   Close: responsible for closing/ending a connection.

Operational Principle
The workings of AMQP revolve around the link protocol. A new link is initiated between peers that need to exchange messages by sending an ATTACH frame. A DETACH frame terminates the link between peers. Once a link is established, unidirectional messages are sent using the TRANSFER frame. Flow control is maintained by using a credit-based flow-control scheme, which protects a process from being overloaded by voluminous messages. Every message transfer state has to be mutually settled by both the sender and the receiver of the message. This

settlement scheme ensures reliability measures for messaging in AMQP. Any change in state and settlement of transfer is notified using a DISPOSITION frame. This allows for the implementation of various reliability guarantees. A session can accommodate multiple links in both directions. Unlike the link, a session is bidirectional and sequential. Upon initiation with a BEGIN frame, a session enables a conversation between peers. The session is terminated using an END frame. Multiple logically independent sessions can be multiplexed between peers over a connection. The OPEN frame initiates a connection and the connection is terminated by using a CLOSE frame.

8.4.5 XMPP

The extensible messaging and presence protocol, or XMPP, which was initially named as Jabber, is designed for message-oriented middlewares based on the extensible markup language (XML) [20]. XMPP was developed for instant messaging, maintenance of contacts, and information about network presence. Structured and extensible data between two networked nodes/devices can be exchanged in near real-time using this protocol. XMPP has found use in VOIP (voice-over Internet protocol) presence signaling, video and file transfers, smart grid, social networks, publish–subscribe systems, IoT applications, and others. The protocol, being open-source, has enabled a spurt of developments in various freeware as well as commercial messaging software. As XMPP follows a client–server architecture, peers in a network cannot talk directly to one another through XMPP. All communication between peers has to be routed through an XMPP server. The XMPP model is considered to be decentralized as anyone can host an XMPP server to which various clients can subscribe. Figure 8.20 shows the basic communication between the various XMPP stakeholders.

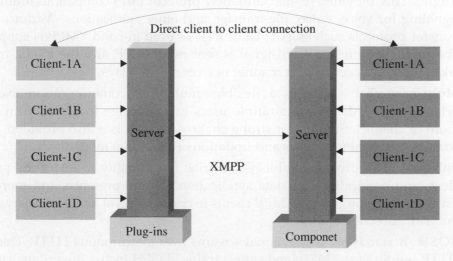

Figure 8.20 XMPP components

Operational Principle

A unique XMPP address, which is also referred to as a Jabber ID (JID), is assigned to every user on the network. The JID, similar to an email address, has a username and a domain name (user@domain.com). The domain name is mostly the IP address of the server hosting the XMPP service. XMPP allows its users to login from multiple devices by means of specifying resources. The resource is used to identify a user's clients/devices (home, mobile, work, laptop, and others), which is generally included in the JID by appending the JID with the resource name separated by a slash. A typical JID looks like this: user@domain.com/resource. Resources are prioritized using numerical IDs. Any message arriving at the default JID (without resource name) is forwarded to the resource with the highest priority (largest numerical ID value). Often JIDs without usernames are used for specific control messages and system messages, which are meant for the server. The use of JID in this mode—without an explicit IP address—allows XMPP to be used as an overlay network on top of multiple underlay networks.

XMPP Technologies

XMPP is an extensible, flexible, and diverse protocol; it has resulted in the development of a significant number of technologies based on it. Some key XMPP technologies include the following:

- **Core**: It deals with information about the core XMPP technologies for XML streaming over a network. The core includes the base XML layer for streaming, provides TLS-based encryption, imparts simple authentication and security layer (SASL) based authentication, informs about the availability of a network, provides UTF-8 support, and contact lists, which are presence enabled.

- **Jingle**: This provides session initiation protocol (SIP)-compatible multimedia signaling for voice, video, file transfer, and other applications. Various media transfer protocols such as TCP, UDP, RTP, or even in-band XMPP is supported. The Jingle session initiation signal is sent over XMPP, and the media transfer takes place in a peer-to-peer manner or over media relays.

- **Multi-user chat**: MUC is a flexible, multiparty communication exchange extension for XMPP. Here multiple users can exchange information in a chat room or channel. Support for strong chat room controls is also provided, which enables the banning of users and updation of chat room moderators.

- **Pub–sub**: This provides publish–subscribe functionality to XMPP by proving alerts and notifications for data syndication, vibrant presence, and more such features. Pub–sub enables XMPP clients to create topics at a pub–sub service and publish/subscribe to them.

- **BOSH**: It stands for bidirectional streams over synchronous HTTP. This is an HTTP binding for XMPP (and other) traffic. BOSH incurs lower latencies and lesser network bandwidth usage by doing away with HTTP polling. It is mainly used for the XMPP traffic exchange between clients and servers.

8.4.6 SOAP

SOAP or simple object access protocol is used for exchanging structured information in web services by making use of XML information set formatting over the application layer protocol (HTTP, SMTP) based transmission and negotiation of messages, as shown in Figure 8.21 [21]. This allows SOAP to communicate with two or more systems with different operating systems using XML, making it language and platform independent. The use of SOAP facilitates the messaging layer of the web services protocol stack.

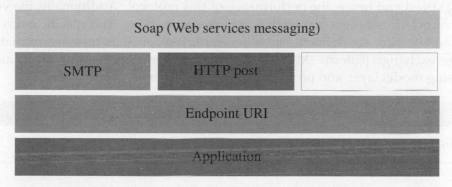

Figure 8.21 A representation of the position of the SOAP API in a stack

A SOAP application can send a request with the requisite search parameters to a server with web services enabled. The target server responds in a SOAP response format with the results of the search. The response from the server can be directly integrated with applications at the requester's end, as it is already in a structured and parsable format. Figure 8.22 illustrates the basic working of SOAP.

SOAP is made up of three broad components: 1) Envelope (which defines the structure of the message and its processing instructions), 2) encoding rules (which handles various datatypes arising out of the numerous applications), and 3) convention (which is responsible for web procedure calls and their responses). This messaging protocol extends the features of neutrality (can operate over any application layer protocol), independence (independent of programming models), and extensibility (features such as security and web service addressing can be extended) to its services. The use of SOAP with HTTP-based request–response exchanges does not require the modification of the communication and processing infrastructures. It can easily pass through network/system firewalls and proxies (similar to tunneling), as illustrated in Figure 8.22. However, the use of XML affects the

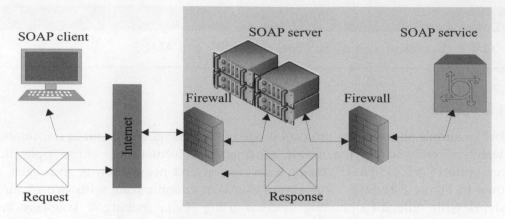

Figure 8.22 Working of SOAP

parsing speed and hence, the performance of this protocol. Additionally, the verbose nature of SOAP is not recommended for use everywhere. The specifications of the SOAP architecture are defined across several layers, such as message format layer, message exchange patterns (MEP) layer, transport protocol binding layer, message processing model layer, and protocol extensibility layer.

Check yourself

Limitations of SOAP, Protocols derived from SOAP

8.4.7 REST

Representational state transfer or REST encompasses a set of constraints for the creation of web services, mainly using a software architectural style [22]. The web services adhering to REST styles are referred to as RESTful services; these services enable interoperability between various Internet-connected devices. RESTful systems are stateless: the web services on the server do not retain client states. The use of stateless protocols and standards makes RESTful systems quite fast, reliable, and scalable. The reuse of components can be easily managed without hindering the regular operations of the system as a whole. Requesting systems can manipulate textual web resource representations by making use of this stateless behavior of REST. RESTful web services, in response to requests made to a resource's URI, mainly responds with either an HTML, XML, or JSON (JavaScript Object Notation) formatted payload. As RESTful services use HTTP for transfer over the network, the following four methods are commonly used: 1) GET (read-only access to a resource), 2) POST (for creating a new resource), 3) DELETE (used for removing a resource), and 4) PUT (used for updating an existing resource or creating a new one). Figure 8.23 represents the REST style and its components.

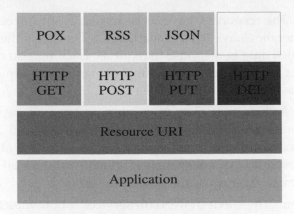

Figure 8.23 A representation of the REST style and its components

REST offers several advantages over regular web-based services. Enhanced network efficiency through the use of REST is ensured by an increase in the performance of interaction between components. Its use also enables a uniform and simple interface, easy live operational modification capabilities, reliability against component and data failures, portability of components, robust scalability, and support for a large number of components.

In REST, requests are used for identifying individual resources. As the resources can be represented in a variety of formats such as HTML, XML, JSON, and others, RESTful services can identify the individual resources from their representations, which allows them to modify, update, or delete these resources. The REST messages contain sufficient information in them to direct a parser on how to interpret the messages. REST client's can dynamically discover all web resources and actions associated with an initial URI. This enhances the dynamicity of applications using REST by avoiding the need to hard-code all clients with the information of the proper structure or dynamics of the web application.

RESTful systems are guided by six general constraints, which define and restrict the process of client–server interactions and requests–responses. These guidelines increase system performance, scalability, reliability, modifiability, portability, and visibility. All RESTful systems have to adhere to these six guidelines strictly:

(i) **Statelessness**: The statelessness of the client–server communication prevents the storage of any contextual information of the client on the server. Each client request has to be self-sufficient in informing its responders about its services and session state. This is done by including the possible links for new state transitions within the representation of each application state. Generally, upon detecting pending requests, the server infers that the client is in a state of transition.

(ii) **Uniform Interface**: Each part or component of a RESTful system must evolve independently as a result of the decoupling of architectures and its simplification.

(iii) **Cacheability**: The responses have to be implicitly, or in some cases, explicitly clear on whether they have to be cached or not. This helps the clients in retaining the most updated data in response to requests. Caching also reduces the number of client–server interactions, thereby improving the performance and scalability of the system as a whole.

(iv) **Client–server Architecture**: The user–interface interactions should be separate from data storage ones. This would result in enhanced portability of user interfaces across multiple platforms. The separation also allows for the independent evolution of components, which would result in scalability over the Internet across various organizational domains.

(v) **Layered System**: The client in RESTful services is oblivious to the nature of the server to which it is connected: an end point server or an intermediary server. The use of intermediaries also helps in improving the balancing of load and enhancing security measures and system scalability.

(vi) **Code on Demand**: This is an optional parameter. Here, the functionality of clients can be extended for a short period by the server. For example, the transfer of executable codes from compiled components.

Check yourself

Difference between REST and SOAP, evolution of REST

8.4.8 WebSocket

Websocket is an IETF (Internet Engineering Task Force)-standardized full-duplex communication protocol. Websockets (WS), an OSI layer seven protocol, enables reliable and full-duplex communication channels over a single TCP connection [23]. Figure 8.24 shows the position of a websocket layer in a stack. The WS relies on the OSI layer 4 TCP protocol for communication. Despite being different from the HTTP protocol, WS is compatible with HTTP and can work over HTTP ports 80 and 443, enabling support for network mechanisms such as the HTTP proxy, which is usually present during organizational Internet accesses through firewalls.

WS enables client–server interactions over the Web. Web servers and clients such as browsers can transfer real-time data between them without incurring many overheads. Upon establishment of a connection, servers can send content to clients without the clients requesting them first. Messages are exchanged over the established connection, which is kept open, in a standardized format. Support for WS is present in almost all modern-day browsers; however, the server must also include WS support for the communication to happen.

The full-duplex communication provided by WS is absent in protocols such as HTTP. Additionally, the use of TCP (which supports byte stream transfers) is also

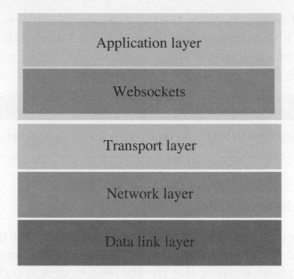

Figure 8.24 A representation of the position of websockets in a stack

enhanced by enabling it to provide message stream transfers using WS. Before the emergence of WS, comet channels were used for attaining full-duplex communication over port 80. However, comet systems were very complicated and incurred significant overheads, which made their utility limited for constrained application scenarios mainly associated with IoT.

Websocket (WS) and websocket secure (WSS) have been specified as uniform resource identifier (URI) schemes in the WS specification, which are meant for unencrypted and encrypted connections, respectively. The WS handshake process and the frames can be quickly inspected using browser development tools.

Operational Principle

A client initiates the WS connection process by sending a WS handshake request. In response, a WS server responds with a WS handshake response. As the servers have to incorporate both HTTP and WS connections on the same port, the handshaking is initiated by an HTTP request/response mechanism. Upon establishment of a connection between the client and server, the WS communication separates as a bi-directional protocol that is non-conformant with the HTTP protocol. The WS client sends an *update header* and a *sec-websocket-Key header*, which contains base64 encoded random bytes. The server responds to the client's request using a hash of the key included in the *Sec-WebSocket-Accept header*. This allows the WS to overcome a caching proxy's efforts to resend previous WS communication. A fixed string, 258EAFA5-E914-47DA-95CA-C5AB0DC85B11, is appended to the undecoded value from the *Sec-WebSocket-Key header* by a hashing function using the SHA (secure hash algorithm)-1, which is finally encoded using base64 encoding. Once the WS full-duplex connection is established, minimally framed data (small header and a payload), which may be data or text, can be exchanged. The WS transmissions or messages can be further split into multiple data frames whenever the full message length is not available during message transfer. This feature is occasionally exploited to include/multiplex several simultaneous streams, using simple extensions to the WS protocol. This multiplexing avoids the monopoly of a single large payload over the WS port.

> Check yourself
>
> Difference between regular client–server sockets and websockets

8.5   Identification Protocols

The surge of IoT devices and Things which are connected over the Internet, makes it significantly hard to identify each device securely. The number of connected things is rising exponentially; with this rise the need to design and develop protocols that can provide unique and distinguishable identifiers to so many Things becomes overwhelming. However, unified global efforts have come up with certain solutions to address the challenges regarding identification of Things, which keep on updating from time to time. Some of the commonly encountered ones are EPC, uCode, and URIs. This section outlines the various nuances associated with each of these methods.

8.5.1   EPC

EPC or the electronic product code identification system was designed to act as a universal identifier and provide unique identities accommodating all physical objects in the world [24]. The open standard and free EPCglobal Tag Data Standard defines the EPC structure. The official representation of EPC is an URI (uniform resource identifier) and referred to as the *pure identity URI*. Figure 8.25 illustrates the standard EPC representation. This representation is used for referring to physical objects in communicating information and business systems and application software. The standard also defines representations for EPC identifiers: tag encoding URI formats and formats for binary EPC identifier storage. In systems such as passive RFIDs that generally have low memory, the EPC binary identifier storage format plays a crucial role. The standard also provides EPC encoding and decoding rules to use URI and binary formats interchangeably seamlessly. Being a very flexible framework, external support for various coding schemes such as those used with barcodes is also possible with EPC. The EPC standard currently supports EPC identifiers, general identifiers, and seven types of identification keys from the GS1 Application Identifier system. As the EPC is not designed to be restricted for use only with RFID data carriers, the data carrier technology-agnostic behavior of EPC is further enhanced by the *pure identity URI*.

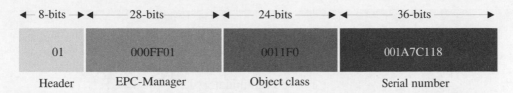

Figure 8.25 The EPC representation

8.5.2 uCode

Another identification number system, the uCode is designed to uniquely identify real-world things and objects whose information is digitally associated with the uCode system [25]. The uID center in Japan provides support for the uCode system. The uCode system can be used with any application, business processes, and technology (RFID, barcodes, matrix codes). uCode is application and technology independent and uses 128 bit codes for uniquely tagging/naming physical objects. The uCode provides 3.4×10^{38} unique codes for individually tagging objects. These features make uCode a crucial enabling technology for IoT. Figure 8.26 represents the working of uCode tags and its various stakeholders.

The uCode tags are generally grouped into five categories: 1) print tags, 2) acoustic tags, 3) active RF tags, 4) active infrared tags, and 5) passive RFID tags. In contrast to other identification systems, the uCode system has the following distinct features:

(i) It does not display product types, albeit it identifies individual objects. Existing codes identify products by individual vendors, making the possibility of identifier tag reuse a possibility, which is avoided in the uCode system.

(ii) In addition to physical objects, the uCode can be associated with places, concepts, and contents, enabling this system to identify such items universally.

(iii) Being application and business agnostic, the uCode system can be used across industries and organizations. The system provides a unique identification number, which does not carry any meaning or information about the tagged object/item. This enables the same system to be used seamlessly across organizations, industries, and product types.

(iv) uTRON, a ubiquitous security framework, which is incorporated with the ubiquitous ID architecture of the uCode system, makes it entirely secure and enables information privacy protection.

(v) The tag agnostic nature of the uCode system makes it possible for various systems such as RFIDs, and barcodes to store uCode information. This makes uCode highly ubiquitous and pervasive.

(vi) The uCode represents pure numbers and is devoid of any meaning or information related to the tagged item/object. This makes the reassignment of uCode tags quite robust and straightforward.

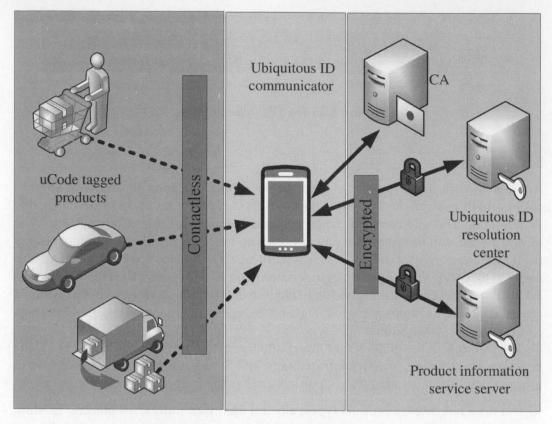

Figure 8.26 The operation of an uCode tag system

The ubiquitous ID architecture of the uCode system is made up of five distinct components: 1) uCode, 2) uCode tags, 3) ubiquitous communicators, 4) uCode resolution server, and 5) uCode information server. The operational process of reading a uCode is as follows:

(i) uCode tags are read using mobile phone cameras to identify the ucode.

(ii) An inquiry about the uCode is sent to the uCode resolution server from the mobile phone over the Internet.

(iii) The uCode resolution server returns information about the uCode to the mobile phone. The returned information contains the source of the read uCode information.

(iv) The ubiquitous communicator then acquires the contents and service information from the information providing source of the read uCode.

Just like the Internet DNS resolution mechanism, the uCode resolution system is hierarchically constructed. The three-tiered uCode resolution hierarchy has the root server at the top level. The uID center in Japan maintains the root server. The next level, the top level domain (TLD), is situated below the root server. As of now various

TLD servers are located around the globe in Japan, Finland, and a few other countries. Finally, the second level domain (SLD) is at the bottom of the hierarchy, below the TLD. The TLD and SLD servers are not restricted and can be added to the existing system.

> **Check yourself**
>
> Differences between EPC and uCode, Limitations of EPC, Limitations of uCode

8.5.3 URIs

One of the most common identifiers in use is the uniform resource identifier (URI). [26] The URI is used to identify individual resources only by using character strings distinctly. As with other protocols, the uniformity of this protocol is ensured by an agreed-upon set of syntax rules. These rules also allow for extensibility through the incorporation of separate hierarchical naming schemes such as "http://". URIs enable interaction with network-based resource representations through specific protocols, especially over the WWW. Some terms commonly derived from URIs are URLs and URNs. URLs or uniform resource locators are very commonly encountered during resource search over the Web or a network. URLs are generally referred to as web addresses and specify the location as well as the access mechanism for a remote resource. For example, "http://www.abc.xz/home/index" denotes the location of the resource at "/home/index", which is hosted at the domain "www.abc.xz", and can be accessed using HTTP. A less encountered form of URIs is the uniform resource name (URN), which identifies resources in particular namespaces only. URNs were initially designed to complement URLs. However, unlike URLs, URNs only identify resources and do not provide the location or method to access the identified resource. Figure 8.27 shows the typical URI format.

Figure 8.27 The representation of an URI link

8.6   Device Management

The need for device management protocols is vital given the rising number of applications of IoT in various application areas spread across the globe. In most of the cases, it is not possible to manage these devices or change their settings manually. Toward this goal, much work is being pursued in the domain of remote device management. We outline two of the most well-known device management protocols in this section.

8.6.1   TR-069

Owing to the rising need for remote management of customer premises equipment (CPE), the Broadband Forum defined the technical specifications for the application layer protocol for CPE over IP networks; these specifications are referred to as Technical Report 069 or TR-069 [27]. The TR-069 mainly focuses on the auto-configuration of Internet-connected devices using auto configuration servers (ACS). Within the premises of this report, the CPE WAN management protocol (CWMP) outlines the various support functions for CPE, which encompasses software and firmware management, status and performance report management, diagnostics, and auto-configuration. CWMP, a primarily SOAP/HTTP-based bi-directional protocol, which is also text-based, provides communication and management support between CPE and servers within a single framework. Devices connecting over the Internet such as routers, gateways, and end devices such as set-top boxes and VoIP devices fall under its purview. Figure 8.28 shows the main components of TR-069 and their relations between each other.

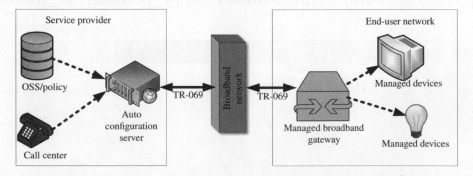

Figure 8.28 The various components of TR-069 and their inter-relations

The various functionalities of this protocol are as follows:

(i)   The commands between the CPE and the ACS during provisioning sessions are either HTTP or HTTPS based, where the ACS is the server and the CPE are clients.

(ii)   The provisioning session is responsible for the communications and operations between the CPE and ACS.

(iii) Session initiation is performed by the CPE through an "inform" message, to which the ACS indicates its readiness using an "inform response".

(iv)   In the subsequent stage, the CPE transmits orders to the ACS, which is invoked using a "transfer complete" message. An empty HTTP-request completes the transmission from the CPE to the ACS.

(v)   In response to the empty HTTP request, the HTTP response from the ACS to the device contains a CWMP (CPE WAN management protocol) request. An empty HTTP-response from the ACS indicates the completion of pending orders.

(vi)   Information security during transmission (login, password, and others) is handled using HTTPS and ACS certificate verification. Authentication of CPE is done based on a shared secret key between the CPE and ACS.

(vii) A time limit of 30 seconds is imposed on the start of the provisioning session after receiving device confirmation.

> **Points to ponder**
>
> The use of TR-069 for remote management of home networked devices and terminals is endorsed by various forums such as Home Gateway Initiative (HGI), Digital Video Broadcasting (DVB), and WiMAX Forum.

> **Check yourself**
>
> Security risks of CWMP, Data model of CWMP, multi-instance object handling

8.6.2 OMA-DM

The open mobile alliance (OMA) device management (DM) protocol is specified by the OMA working group and the data synchronization (DS) working group for remote device management of mobile devices, including mobile phones and tablets [28]. The management functions include provisioning, device configuration, software upgrades, fault management, and others. On the device end, any or all of these features may be implemented. The OMA-DM specification is designed for constrained devices with limited bandwidth, memory, storage, and processing. Data exchanges take place through SyncML, which is a subset of XML. OMA-DM supports both wired as well as wireless data transport (USB, RS-232, GSM, CDMA, Bluetooth, and others) over transport layer protocols such as WAP, HTTP, or OBEX.

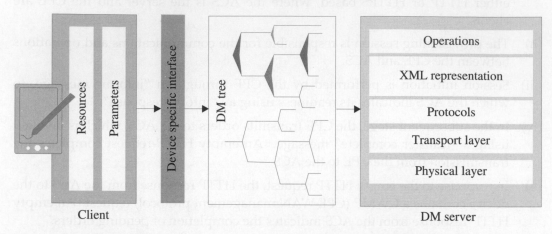

Figure 8.29 Communication between an OMA-DM client and a server

The OMA-DM follows a request–response communication model. The OMA-DM server asynchronously initiates the communication with the end device/client, which is generally in the form of a notification or alert message through WAP push or SMS. The client is meant to execute the command received from the server and reply with a message. More significant messages are generally broken down into chunks before transmission to the client. In terms of information security, authentication methods are built-in in this protocol, which prevents a client and a server from communicating until proper validation. Figure 8.29 shows the communication between a client and a server in OMA-DM.

Check yourself

Security mechanisms in OMA-DM commands

8.7   Semantic Protocols

The semantic protocols for IoT, which is a rapidly upcoming domain, focus on the meaning and logic behind data connectivity and formats. Examples include JSON-LD and the Web Thing model. Primarily designed to be cross-operable and modular, these protocols enhance the robustness and utility of IoT by incorporating the reach of the Web. As an example, the integration of semantic protocols such as JSON-LD with the Web Things model gives rise to the Semantic Web. The chapter on interoperability in this book discusses the challenges and developments in this domain.

8.7.1 JSON-LD

JavaScript object notation for linked data or JSON-LD is a lightweight protocol, which is designed for JSON-based encoding of linked data by seamlessly converting older JSON-based representations of data. The representations of the data are highly human-understandable and highly suitable for RESTful environments and unstructured data over the Web [29]. JSON-LD has an additional resource description framework (RDF) over and above the typical JSON model and is built to be contextual. This feature allows for the interoperability of JSON data over the Web. The contextual linking of the object properties of a JSON document follows a fixed ontology in JSON-LD through strategies such as tagging with a language by or forcibly assigning values to pre-defined groups/bins. Context embeddings in JSON-LD documents can be either direct or through the use of separate file references using HTTP link headers. Linked data allows for the existence of a network of machine-readable and standardized data over the Web, which can be parsed by starting at a singular piece of data and subsequently traversing the embedded links within it; this may lead to different locations across the Web.

A sample JSON-LD schema

```
1  <script type="application/ld+json">
2  {
3    "@context": "https :// schema.org",
4    "@type": "BlogPosting",
5    "mainEntityOfPage": {
6      "@type": "WebPage",
7      "@id": "www.xyz.com"
8    },
9    "headline": "Hello Readers",
10   "description": "This is a test",
11   "image": {
12     "@type": "ImageObject",
13     "url": "www.img1234.com",
14     "width": 696,
15     "height": 14
16   },
17   "author": {
18     "@type": "Person",
19     "name": "abc"
20   },
21   "publisher": {
22     "@type": "Organization",
23     "name": "CUP",
24     },
25   "datePublished": ""
26 }
27 </script>
```

8.7.2   Web thing model

The Web of Things (WoT) is another interoperability-driven initiative for achieving seamless Web-based uniformity for IoT devices. The main driving factor behind this initiative is to develop a unifying application layer-based framework for IoT which can provide URLs for the connected devices over the Web. This initiative aims to transform the traditionally predominant "Web of pages" to "Web of Things". As the current Web-based technologies and IoT-based integrations over the Web are vastly vendor-specific and use proprietary data formats, the cross-utilization of such technologies is seldom flawless. These drawbacks of the present technologies led to the need for a common syntactical vocabulary and API which will be able to induce ad hoc interoperability for IoT. The paradigms, such as "machine-to-machine communication", promote technological overhaul (most often complete technology replacement), without incorporating the existing technologies. In contrast, the WoT paradigm aims to integrate the existing Web with the various applications and systems already in place to fully utilize the infrastructural and technological leverage already present.

The following are the major sub-components of the WoT paradigm:

(i)   **Integration Patterns:** Dictates how the Things in IoT connect to the Web. It is mainly composed of three schemes: Direct connectivity, gateway based connectivity, and cloud-based connectivity.

(ii) **Web Things (WT) Requirements:** Provides guidelines and recommendations for handling various constraints and protocol implementations to enhance the seamless interaction between the WoT entities. A typical web-server is referred to as a Web Thing; it should also confirm to these recommendations.

(i)   **Web Thing Model:** Data exchange over the WoT ensues once a Web Thing is compliant. Additionally, in order to achieve context-awareness, this specification outlines RESTful web protocol, which has a defined set of payload syntax, data models, and resources. A fully compliant model is referred to as the Extended Web of Things model.

Summary

This chapter provided an outline of various communication technologies that are deemed as core technologies for developing IoT-based solutions. We initially explain the requirements and classification of IoT devices and communication types. We divide the various communication protocols under six heads based on their usability and functionalities: 1) Infrastructure, 2) discovery, 3) data, 4) identification, 5) device management, and 6) semantic. After this chapter, readers will be able to distinguish between various requirements and constraints associated with these protocols and select the best one amongst them according to their application's requirements.

Exercises

(i) What are the salient features of 6LoWPAN?

(ii) What is a WPAN?

(iii) Describe the addressing types in 6LoWPAN.

(iv) Describe the LOADng routing.

(v) Describe the RPL routing.

(vi) What are the different header types in 6LoWPAN?

(vii) What constitutes a low power lossy network (LLN)?

(viii) What is AMQP? Describe in detail.

(ix) What are the various message guarantees provided by AMQP? Explain each in detail.

(x) List some of the salient features of AMQP.

(xi) What are the frame types in AMQP?

(xii) Differentiate between OPEN, BEGIN and ATTACH frame types in AMQP.

(xiii) Differentiate between DETACH, END, and CLOSE frame types in AMQP.

(xiv) Differentiate between TRANSFER and FLOW frame types in AMQP.

(xv) What are BINDINGS in the context of AMQP?

(xvi) What are the various types of AMQP exchanges? Describe each.

(xvii) What are the popular applications of AMQP?

(xviii) Explain the working of MQTT

(xix) How is MQTT different from HTTP?

(xx) What are the various MQTT methods?

(xxi) What is SMQTT? How is it different from MQTT?

(xxii) List the salient features of MQTT.

(xxiii) List the salient features of XMPP.

(xxiv) Describe the XMPP protocol.

(xxv) Differentiate between structured and unstructured data.

(xxvi) What is XML?

(xxvii) What is BOSH? Explain in detail.

(xxviii) What is CORE? Explain in detail.

(xxix) What is Jingle? Explain in detail.

(xxx) What is Pub–Sub? Explain in detail.

(xxxi) List the significant limitations of XMPP.

(xxxii) List some of the popular uses of XMPP.

(xxxiii) What is CoAP?

(xxxiv) Describe the working of CoAP.

(xxxv) Explain the various messaging modes in CoAP.

(xxxvi) List the salient features of the CoAP protocol.

(xxxvii) What is REST?

(xxxviii) What are RESTful services?

(xxxix) Describe the LOADng protocol.

(xl) What is a DODAG?

(xli) Explain the mechanism of formation of a DODAG in RPL.

(xlii) Explain the working of RPL protocol.

(xliii) Illustrate the salient features of RPL.

(xliv) How is the global instance different from local instances in RPL?

(xlv) What is QUIC? How is the connection latency reduced in QUIC?

(xlvi) What is the purpose of publishing static configuration records in QUIC?

(xlvii) Highlight the various features of uIP.

(xlviii) What led to the development of nanoIP?

(xlix) How is the CCN paradigm different from traditional networking approaches?

(l) How is the Physical Web able to interact with physical objects and locations? What are its advantages?

(li) How is mDNS different from DNS?

(lii) What are some of the commonly used discovery protocols in IoT?

(liii) What features separate MQTT-SN from MQTT?

(liv) What are the main functional differences between transparent and aggregate gateways in MQTT-SN?

(lv) Differentiate between SOAP and REST.

(lvi) How does SOAP enable communication between two syntactically different devices/machines?

(lvii) What are the functional components of SOAP?

(lviii) What are the advantages of using REST over regular web-based services?

(lix) What are the various methods used in REST for transferring data over the network?

(lx) What is statelessness in the context of REST?

(lxi) How are websockets different from simple HTTP?

(lxii) Describe the working of websockets?

(lxiii) What is the functional mechanism for EPC in IoT?

(lxiv) What is uCode and how is it different from EPC?

(lxv) What are the various categories associated with uCode tags?

(lxvi) Describe the uCode resolution system.

(lxvii) What are URIs? How is it used for identifying individual resources?

(lxviii) How is auto-configuration over Internet-connected devices achieved using the auto configuration server?

(lxix) What are the various components of TR-069?

(lxx) What is OMA-DM?

(lxxi) How is OMA-DM functionally different from TR-069?

(lxxii) Differentiate between JSON-LD and XML.

(lxxiii) What is the Web Thing model? Illustrate its strengths and weaknesses.

References

[1] Bormann, C., M. Ersue and A. Keranen. 2014. "Terminology for Constrained Node Networks." https://tools.ietf.org/html/rfc7228.

[2] Annamalaisamy, Vijay. 2019. "Introduction to IoT Constrained Node Networks." https://www.hcltech.com/blogs/introduction-iot-constrained-node-networks.

[3] Postscapes. 2019. "IoT Standards and Protocols." https://www.postscapes.com/internet-of-things-protocols/.

[4] Vasseur, J. P., Cisco Systems, Internet Engineering Task Force (IETF). 2014. *RFC-7102* ISSN: 2070-1721. https://tools.ietf.org/html/rfc7102.

[5] Deering, S., R. Hinden. 1998. "Internet Protocol, Version 6 (IPv6) Specification, IETF." https://tools.ietf.org/html/rfc2460.

[6] Sobral, J., J. Rodrigues, R. Rabelo, K. Saleem, and V. Furtado. 2019. "LOADng-IoT: An Enhanced Routing Protocol for Internet of Things Applications over Low Power Networks." *Sensors* 19(1): 150.

[7] Winter, T. 2012. "RPL: IPv6 Routing Protocol for Low-Power and Lossy Networks, IETF." https://tools.ietf.org/html/rfc6550.

[8] Kushalnagar, N., G. Montenegro, and C. Schumacher. 2005. "IPv6 over Low-Power Wireless Personal Area Networks (6LoWPANs): Overview, Assumptions, Problem Statement, and Goals, IETF." https://datatracker.ietf.org/doc/rfc4919/.

[9] The Chromium Projects. 2015. "QUIC, a Multiplexed Stream Transport over UDP." Available online: https://www.chromium.org/quic.

[10] Dunkels, A. 2002. "uIP-A Free Small TCP/IP Stack." *The uIP* 1.

[11] Shelby, Z., J. Riihijärvi, O. Raivio, and O. Mähönen. 2003. "NanoIP: The Zen of Embedded Networking." *IEEE International Conference on Communications.*

[12] Franck, F., S. A. S. Alcatel Lucent. 2016. "Content-centric Networking." U. S. Patent 9,338,150.

[13] The Physical Web, Available online: https://google.github.io/physical-web/.

[14] Cheshire, S. and M. Krochmal. 2013. "Multicast dns." *RFC* 6762, February.

[15] Jeronimo, M. and J. Weast. 2003. *UPnP Design by Example* (Vol. 158). Intel Press.

[16] Banks, A. and R. Gupta. 2014. "MQTT Version 3.1. 1." *OASIS Standard* 29: 89.

[17] Stanford-Clark, A. and H. L. Truong. 2013. "MQTT for Sensor Networks (MQTT-SN) Protocol Specification." International Business Machines (IBM) Corporation version 1: 2.

[18] Shelby, Z., K. Hartke, and C. Bormann. 2014. "The Constrained Application Protocol (CoAP)." *IETF, RFC* 7252.

[19] Vinoski, S. 2006. "Advanced Message Queuing Protocol." *IEEE Internet Computing* (6): 87–89.

[20] Saint-Andre, P., K. Smith, R. Tronçon, and R. Troncon. 2009. *XMPP: The Definitive Guide.* "O'Reilly Media, Inc."

[21] Box, D., D. Ehnebuske, G. Kakivaya, A. Layman, N. Mendelsohn, H. F. Nielsen, S. Thatte, and D. Winer. 2000. "Simple Object Access Protocol (SOAP) 1.1."

[22] Battle, R. and E. Benson. 2008. "Bridging the Semantic Web and Web 2.0 with Representational State Transfer (REST)." *Web Semantics: Science, Services and Agents on the World Wide Web* 6(1): 61–69.

[23] Fette, I. 2011. "The Websocket Protocol."

[24] Song, B. and C. J. Mitchell. 2008. "RFID Authentication Protocol for Low-cost Tags." In *Proceedings of the First ACM Conference on Wireless Network Security.* ACM. 140–147.

[25] Ishikawa, C. 2012. "A URN Namespace for uCode."

[26] Berners-Lee, T., R. Fielding, and L. Masinter. 1998. "Uniform Resource Identifiers (URI): Generic Syntax."

[27] Broadband Forum, TR-069: CPE WAN Management Protocol. 2018. https://www.broadband-forum.org/download/TR-069_Amendment-6.pdf.

[28] Open Mobile Alliance, OMA Device Management Protocol. 2016. http://www.openmobilealliance.org/release/DM/V1_3-20160524-A/OMA-TS-DM_Protocol-V1_3-20160524-A.pdf.

[29] JSON-LD Working Group. 2018. "JSON for Linking Data." https://www.w3.org/2018/json-ld-wg/.

[30] Guinard Dominique. 2017. "The Web Thing Model", WEB OF THINGS INTEREST GROUP. https://www.w3.org/blog/wotig/2017/01/13/web-thing-model-member-submission/.

Chapter **9**

IoT Interoperability

Learning Outcomes

After reading this chapter, the reader will be able to:

- Understand the importance of interoperability in IoT
- List various interoperability types
- Identify the salient features and application scope of each interoperability type
- Understand the challenges associated with interoperability in IoT
- Comprehend the importance of real-world use of interoperability frameworks in IoT

9.1 Introduction

The introduction of billions of connected devices under the IoT environment, which may extend to trillions soon, has contributed massively to the evolution of interoperability. As more and more manufacturers and developers are venturing into IoT, the need for uniform and standard solutions is felt now more than ever before [1]. Figure 9.1 shows the various facets of interoperability in IoT. Interoperability is considered as the interface between systems or products—hardware, software, or middleware—designed in such a manner that the connecting devices can communicate, exchange data, or services with one another seamlessly irrespective of the make, model, manufacturer, and platform.

The urgency in the requirement for interoperability and interoperable solutions in IoT arose mainly due to the following reasons:

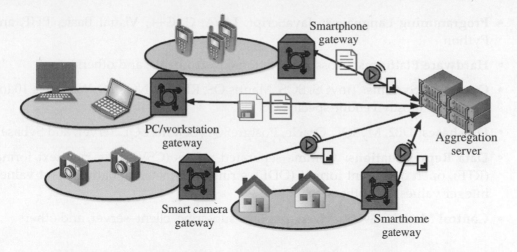

Figure 9.1 An illustration of the various facets of interoperability in IoT

(i) **Large-scale Cooperation:** There is a need for cooperation and coordination among the huge number of IoT devices, systems, standards, and platforms; this is a long-standing problem. Proprietary solutions are seldom reusable and economical in the long run, which is yet another reason for the demand for interoperability.

(ii) **Global Heterogeneity:** The network of devices within and outside the purview of gateways and their subnets are quite large considering the spread of IoT and the applications it is being adapted to daily. Device heterogeneity spans the globe when connected through the Internet. A common syntax, platform, or standard is required for unifying these heterogeneous devices.

(iii) **Unknown IoT Device Configuration:** Device heterogeneity is often accompanied by further heterogeneity in device configurations. Especially considering the global-scale network of devices, the vast combinations of device configurations such as data rate, frequencies, protocols, language, syntax, and others, which are often unknown beforehand, further raise the requirement of interoperable solutions.

(iv) **Semantic Conflicts:** The variations in processing logic and the way data is handled by the numerous sensors and devices making up a typical IoT implementation, makes it impossible for rapid and robust deployment. Additionally, the variations in the end applications and their supported platform configurations further add to the challenges.

The heterogeneity in IoT devices may arise due to several reasons. Some of the common ones are as follows:

- **Communication Potocols:** ZigBee(IEEE 802.15.4), Bluetooth (IEEE 802.15.1), GPRS, 6LowPAN, Wi-Fi (IEEE 802.11), Ethernet (IEEE 802.3), and Higher Layer LAN Protocols (IEEE 802.1)

- **Programming Languages:** JavaScript, JAVA, C, C++, Visual Basic, PHP, and Python

- **Hardware Platforms:** Crossbow, National Instruments, and others

- **Operating Systems:** TinyOS, SOS, Mantis OS, RETOS, NOOBS, Windows 10 IoT Core, and mostly vendor-specific OS

- **Databases:** DB2, MySQL, Oracle, PostgreSQL, SQLite, SQL Server, and Sybase

- **Data Representations:** Comma separated values (CSV), text, rich text format (RTF), open document format (ODF), strings, characters, floating-point values, integer values, and others

- **Control Models:** Event-driven, publish–subscribe, client–server, and others

9.1.1 Taxonomy of interoperability

The significant range of interoperable solutions that has been developed for IoT can be broadly categorized into the following groups:

(i) **Device:** The existence of a vast plethora of devices and device types in an IoT ecosystem necessitates device interoperability. Devices can be loosely categorized as low-end, mid-end, and high-end devices based on their processing power, energy, and communication requirements. Low-end devices are supposed to be deployed in bulk, with little or no chance of getting their energy supplies replenished, depending on the application scenario. These devices rely on low-power communication schemes and radios, typically accompanied by low-data rates. The interface of such devices with high-end devices (e.g., smartphones, tablets) requires device-level interoperability [2].

(ii) **Platform:** The variations in the platform may be due to variations in operating systems (Contiki, RIOT, TinyOS, OpenWSN), data structures, programming languages (Python, Java, Android, C++), or/and application development environment. For example, the Android platform is quite different from the iOS one, and devices running these are not compatible with one another [3].

(iii) **Semantic:** Semantic conflicts arise during IoT operations, mainly due to the presence of various data models (XML, CSV, JSON), information models (°C, °F, K, or different representations of the same physical quantity), and ontologies [4]. There is a need for semantic interoperability, especially in a WoT environment, which can enable various agents, applications, and services to share data or knowledge in a meaningful manner.

(iv) **Syntactic:** Syntactic interoperability is a necessity due to the presence of conflicts between data formats, interfaces, and schemas. The variation in the syntactical grammar between a sender and a receiver of information results in massive stability issues, redundancies, and unnecessary data handling efforts [5]. For example, a packet from a device has a format as *Header-Identifier-SensorA-SensorB-Footer*, whereas another device from a different manufacturer, but deployed for the same application has the data format as *Header-Identifier-SensorB-SensorA-Footer*. This change in position of sensor A and sensor B in the two packets creates syntactic errors, although they contain the same information.

(v) **Network:** The large range of connectivity solutions, both wired and wireless, at the disposal of developers and manufacturers of IoT devices and components, further necessitates network interoperability. Starting from the networks and sub-networks on the ground, to the uplink connectivity solutions, there is a need for uniformity or means of integrating to devices enable seamless and interoperable operations.

9.2 Standards

Toward enabling IoT interoperability, various technologies have been standardized and are recognized globally for incorporating consistent interoperability efforts worldwide across various industries, domains, and technologies. We list seven of the popular ones in this chapter.

9.2.1 EnOcean

EnOcean is a wireless technology designed for building automation systems, primarily based on the principle of energy harvesting [6]. Due to the robustness and popularity of EnOcean, it is being used in domains such as industries, transportation, logistics, and homes. As of 2012, EnOcean was adopted as a wireless standard under ISO/IEC 14543-3-10, providing detailed coverage of the physical, data link, and networking layers. EnOceanbased devices are batteryless. They use ultra-low power consuming electronics along with micro energy converters to enable wireless communication among themselves; the devices include networking components such as wireless sensors, switches, controllers, and gateways. The energy harvesting modules in EnOcean use micro-level variations and differences in electric, electromagnetic, solar, or other forms of energy to transform the energy into usable energy through highly efficient energy converters. The wireless signals from the batteryless EnOcean sensors and switches, which are designed to be maintenance-free, can operate up to 30 meters in buildings and homes and up to 300 meters in the open. EnOcean wireless sensor modules wirelessly transmit their data to EnOcean system modules, as shown in Figure 9.2.

EnOcean is typically characterized by low data rates (of about 125 kbit/s) for wireless packets that are 14 bytes long. This reduces the energy consumption of the EnOcean devices. Additional features such as the transmission of RF (radio frequency) energy only during transmission of 1s in the binary encoded message further reduce

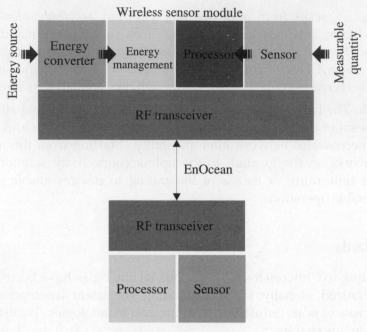

Figure 9.2 A representation of the major constituents of EnOcean devices

the energy consumption of these devices. Frequencies of 902 MHz, 928.35 MHz, 868.3 MHz, and 315 MHz are employed for transmission of messages in this technology.

> Check yourself
>
> EnOcean ultra-low power management, self-powered IoT

9.2.2 DLNA

The Digital Living Network Alliance (DLNA), previously known as the Digital Home Working Group (DHWG), was proposed by a consortium of consumer electronics companies in 2003 to incorporate interoperability guidelines for digital media sharing among multimedia devices such as smartphones, smart TVs, tablets, multimedia servers, and storage servers. Primarily designed for home networking, this standard relies majorly on WLAN for communicating with other devices in its domain and can easily incorporate cable, satellite, and telecom service providers to ensure data transfer link protection at either end. The inclusion of a digital rights management layer allows for multimedia data sharing among users while avoiding piracy of data. The consumers in DLNA, which may consist of a variety of devices such as TVs, phones, tablets, media players, PCs, and others, can view subscribable content without any additional add-ons or devices through VidiPath. Figure 9.3 shows the steps involved in a typical DLNA-based multimedia streaming application. As of 2019, DLNA has over a billion devices following its guidelines globally [7].

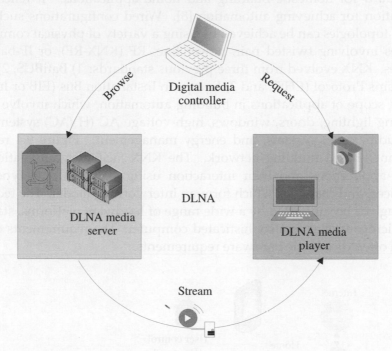

Figure 9.3 A representation of the various roles in a DLNA-based media streaming application

DLNA outlines the following key technological components, which enable interoperability guidelines for manufacturers [7].

(i) Network and Connectivity

(ii) Device and Service Discovery and Control

(iii) Media Format and Transport Model

(iv) Media Management, Distribution, and Control

(v) Digital Rights Management and Content Protection

(vi) Manageability

Check yourself

DLNA Home Network and Infrastructure devices and components, DLNA mobile infrastructure

9.2.3 Konnex

Konnex or KNX is a royalty-free open Home Automation Network (HAN) based wired standard for domestic building and home applications. It relies on wired communication for achieving automation [8]. Wired configurations such as a star, tree, or line topologies can be achieved by using a variety of physical communication technologies involving twisted pair, power line, RF (KNX-RF), or IP-based (KNX-net/IP) ones. KNX evolved from three previous standards: 1) BatiBUS, 2) European Home Systems Protocol (EHS), and 3) European Installation Bus (EIB or Instabus). It has a broad scope of applications in building automation, which involve tasks such as controlling lighting, doors, windows, high-voltage AC (HVAC) systems, security systems, audio/video systems, and energy management. Figure 9.4 represents a typical Konnex-based building network. The KNX facilitates automation through distributed applications and their interaction using standard data types, objects, logical devices, and channels, which form an interworking model. The technology is robust enough to be supported by a wide range of hardware platforms, starting from a simple microcontroller to a sophisticated computer. The requirements of building automation often dictate the hardware requirements.

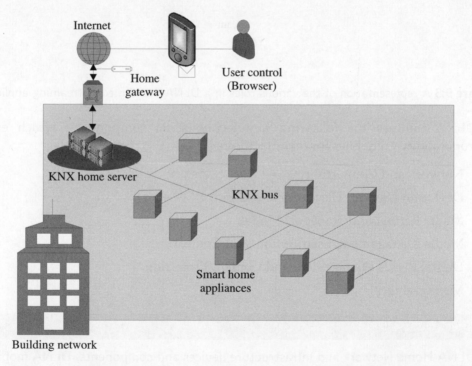

Figure 9.4 A representation of the Konnex network

The KNX architecture consists of sensors (temperature, current, light), actuators (motors, switches, solenoids, valves), controllers (implementable logic), and other system devices and components (couplers). Typically, the KNX uses a twisted pair bus for communication, which is channeled through the building/home alongside the electrical wiring. Using a 16-bit address bus, KNX can accommodate 57375 devices. A KNXnet/IP installation allows the integration of KNX sub-networks via IP. A system interface component is used for loading application software, system topology, and operating software onto the devices, after which the devices can be accessed over LAN or phone networks. This feature also allows for the centralized as well as distributed control of systems remotely. KNX has three different configuration modes according to device categories.

(i) Automatic mode (A mode): Typically used for auto-configurable devices, and is generally installed by the end users.

(ii) Easy mode (E mode): Devices require initial training for installation, where the configuration is done as per the user's requirements; the device behavior is pre-programmed using E mode.

(iii) System mode (S mode): Some devices generally require specialists to install; the system mode is used for this. The devices do not have a default behavior but can be used for deploying complex building automation systems.

> **Points to ponder**
>
> KNX is an approved standard under International standards (ISO/IEC 14543-3), European standards (CENELEC EN 50090 and CEN EN 13321-1), US standards (ANSI/ASHRAE 135), and China Guobiao (GB/T 20965)

> **Check yourself**
>
> KNX architecture, KNX addressing, KNX use cases

9.2.4 UPnP

The Universal Plug and Play (UPnP) was designed primarily for home networks as a set of protocols for networking devices such as PCs, printers, mobile devices, gateways, and wireless access points. UPnP can discover the presence of other UPnP devices on the network, as well as establish networks amongst them for communication and data sharing [9]. Whenever they are connected to a network, UPnP devices can establish working configurations with other devices. As of 2016, UPnP is managed by the Open Connectivity Forum (OCF). The underlying assumption of UPnP is the presence of an IP network over which it uses HTTP to share events, data, actions, and service/device descriptions through a device-to-device networking arrangement. Device search and advertisements are multicast through HTTP over UDP (HTTPMU) over port 1900. The responses are returned in a unicast manner through HTTP over UDP (HTTPU). UPnP is based on established protocols and architectures such as TCP/IP protocol suite, HTTP, XML, and SOAP. UPnP is a distributed and open standard. Devices controlled by UPnP are handled by UPnP control points (CPs). The networked UPnP devices are designed to dynamically join networks, obtain IP addresses, advertise its presence and capabilities, and detect the presence and capabilities of other neighboring and networked devices through a process known as zero configuration networking.

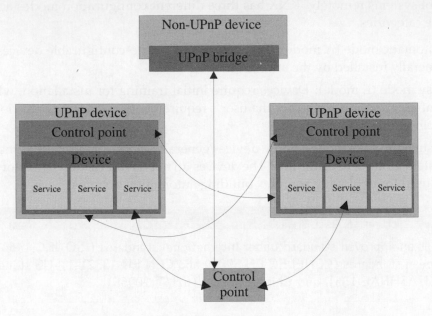

Figure 9.5 A representation of the UPnP operation

UPnP devices are typically characterized by a control point and service(s). The service(s) need to communicate with the control point for further instructions/execution. Figure 9.5 shows a typical UPnP operation. A central control point in a room can be used to control various UPnP services across a home. Non-UPnP devices can be easily integrated with the UPnP services through a bridge.

UPnP supports a range of IP supporting media such as Ethernet, IR, Bluetooth, Wi-Fi, FireWire, and others, without the need for individual device drivers. UPnP, being an OS and language independent protocol, typically uses web browsers for the user interface. Each UPnP device implements a DHCP (dynamic host configuration protocol) client and searches for a DHCP server during its first initiation in the network. These devices can also use a feature known as AutoIP to assign itself an IP address, in case a DHCP server is not available. The UPnP device then discovers the network through the simple service discovery protocol (SSDP), which advertises the device through the CPs (coordination protocols) on the network. The CP then retrieves the device's information through a location URL sent by the device. The device information is in the form of an XML schema using SOAP; it additionally contains a list of services: commands, actions, and actionable variables and parameters. To the control URL in the description, CPs use control messages to send actions to a device's service. Finally, if a device has a URL for presentation, the CP retrieves the contents, allowing a user to control or view the device and device status.

Check yourself

UPnP device discovery, UPnP protocol, Event notification

9.2.5 LonWorks

LonWorks or local operating network, as it was initially named, is a protocol developed by the Echelon Corp [10]. It was primarily developed for addressing the needs of networked control applications within buildings over physical communication media such as twisted pair, fiber optic cables, powerlines, and RF. The twisted pair uses differential Manchester encoding and has a data rate of 78 kbit/s, whereas the powerline is much slower and can have either 5.4 kbit/s or 3.6 kbit/s depending on the frequency of the power line. This protocol was standardized by ANSI (American National Standards Institute) as early as 1999 when it was known as LonTalk and was used for control networking. This protocol has been used in a variety of deployment areas such as the pneumatic braking system of trains, semiconductor equipment manufacturing, petrol station controls, and as a building automation standard. LonWorks extends backward compatibility support to its legacy installations through an IP-based tunneling standard (ISO/IEC 14908-4). Regular IP-based services can be readily used with LonWorks platforms or installations for UI or control level applications. Figure 9.6 illustrates a typical LonWorks network.

Initially, a LonTalk protocol node could only be installed using a custom-designed IC with an 8-bit processor; this IC was referred to as the "neuron chip". The neuron chip is a system on a chip and is essentially the soul of the LonWorks-based devices. There are two types of neuron chips based on the memory capabilities and packaging: 1) the 3120 and 2) the 3150. Presently, a significant number of LonWorks-based devices use the neuron chip, which is also accessible by general processors by porting to an IP-based or 32-bit chip. A neuron chip has three CPUs, one each for MAC processing, network processing, and application processing. The MAC processor is tasked with CRCs (cyclic redundancy checks), transmitting and receiving messages over the physical media, and confirming message destinations. The network processor deals with addressing, routing, acknowledgments, and other network layer tasks. Finally, the application processor is used for deploying custom applications which typically support 8-bit operations; it can also be used as a communication co-processor for high-end processors. The decoupling of processors based on tasks enables the robust and speedy performance of the neuron chips. Each neuron chip has three memory types available with it: 1) ROM, 2) RAM, and 3) EEPROM. The LonTalk,

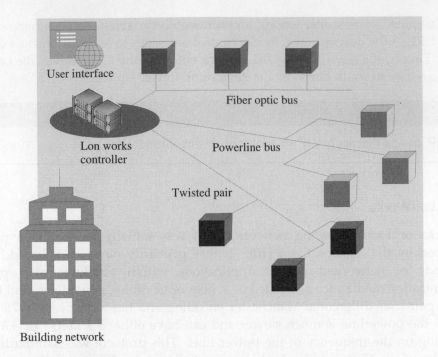

Figure 9.6 A representation of the LonWorks network

along with the OS and I/O libraries are typically programmed in the ROM during manufacturing.

9.2.6 Insteon

Insteon was developed as a home automation technology by Smartlabs in 2005 and marketed under its subsidiary Insteon. Insteon enables interoperability and automation among household devices such as lights, switches, thermostats, motion and gas sensors, and others through RF or powerline communication [11]. Insteon-connected devices act as peers and can independently perform network-based functions such as message transmission and reception by using a dual mesh network topology. These devices operating over the powerline have a frequency of 131.65 kHz; the devices use binary phase shift keying (BPSK), with a minimum receive signal level of 10 mV. In contrast, Insteon devices using RF operates over a frequency of 915 MHz; these devices use frequency shift keying for communication and Manchester codes for encoding data with a data rate of 4.56 kbit/s over ranges of approximately 120 m without obstructions. Figure 9.7 shows a typical home-based Insteon network.

Insteon networks can have 16 million+ unique IDs and can support 65,000+ devices. Each of these devices has a built-in engine, which has an 80 byte RAM and a 3 kbyte ROM. Application-specific requirements of Insteon devices such as lights and switches require 256 bytes of RAM and EEPROM, and 7 kbytes of flash memory. Insteon devices have an average data rate of 180 bit/s, using which a standard message of 10 bytes or extended message of 24 bytes is transmitted. Each Insteon message can accommodate up to 14 bytes of user data and contain a two-bit field meant for counting hops. Message originating nodes initialize this field value to 3, which is decremented by the number of times a node repeats the message during its transmission. Upon receiving a message, each device performs error detection and correction. Retransmission of erroneous messages in this manner enhances the reliability of Insteon technology. All devices transmit the same message simultaneously using PSK to ensure synchronicity with the powerline frequency. This ensures message integrity and strengthens the signal over the powerline.

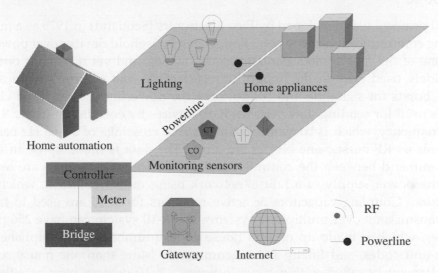

Figure 9.7 A representation of an Insteon network

The dual mesh/ dual band network topology of Insteon is named so mainly because, during operations over the RF band, interferences are mitigated by transmitting data over the powerline, and vice versa. As this is a peer-to-peer network, it can operate without the need for central controllers. Central controllers can be integrated with this technology to extend control operations over smartphones and tablets. As a security measure to avoid hijacking a neighbor's Insteon devices, Insteon requires users to have physical ownership of the devices they want to connect to their network and the respective device IDs (which is unique and similar to a MAC ID). The inbuilt firmware on the devices prevents Insteon devices from forming connections and identifying themselves to other devices until a button is physically pressed on them during their installation.

Points to ponder

Legacy Insteon chipsets are interoperable with X10 powerline messaging, but with reduced functionalities. Present-day initiatives have incorporated compatibility for certain functionalities of Insteon with Amazon Echo, Microsoft Cortana, Apple Watch, and the Google-owned NEST.

Check yourself

Insteon installation, Insteon functionalities with other platforms, Insteon use cases

---

9.2.7 X-10

The X-10 protocol was developed by Pico Electronics (Scotland) in 1975 as a means of achieving communication and automation among household devices over powerlines. It was one of the first home automation technologies, and yet it remains one of the most widely used even in the present day [12]. Data and controls are encoded as brief RF bursts for signaling and control over the powerlines. Household electrical wiring is used for sending data between X-10 devices by encoding it over a 120 kHz carrier frequency, which is transmitted during zero crossings of 50–60 Hz consumer AC signals as RF bursts, one bit per crossing. The data is made up of an address and a command between the controller and the device. X-10 signals are restricted within the power supply of a house/network using inductive filters, which act as attenuators. Coupling capacitors or active repeaters for X-10 are used to facilitate signal transmission over multiphase systems. An X-10 system can have 256 possible addresses, which is made up of 4-bit house codes (numbered from alphabets A to P), 4-bit unit codes, and finally, a 4-bit command. More than one house code can be simultaneously called within a single house. X-10 devices may be either one-way or two-way. One-way devices are typically very cheap and can only receive commands, whereas two-way devices are more expensive and can send as well as receive commands. These two-way devices are generally used as controllers. Figure 9.8 represents a typical X-10 setup and controller, which allows a user to connect to and control a variety of appliances and devices at home.

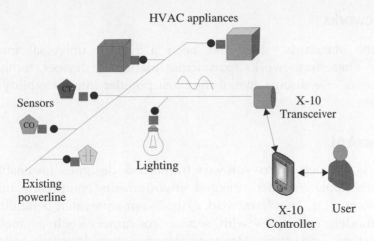

Figure 9.8 A representation of the X-10 network

A bit value of 1 is represented by a 1 ms burst of 120 kHz for a typical 60 Hz AC powerline at the zero crossing. The absence of a pulse follows this bit value. A 0 bit is represented by the absence of a 120 kHz burst at the zero crossings, followed by a pulse. The data rate for the X-10 protocol is typically around 20 bit/s, which includes retransmission time and control signals. Due to the meager data rates, X-10 commands are kept simple and have limited functionalities such as on/off. Each data frame in X-10 is transmitted twice, which although incurs redundancy also allows for reliable data transmission over noisy channels. A new command over the powerline is separated by at least six clear zero crossings from the previous command. An RF protocol is also defined under X-10 to accommodate wireless remotes and switches. This protocol operates over a 310 MHz channel in the US and a 433.92 MHz channel in Europe. The wireless data packets from X-10 devices communicate to a radio receiver, which acts as a bridge between the wireless devices and the powerline-based X-10 devices.

9.3   Frameworks

Similar to the standards, there has been a rise in universal interoperability frameworks. These frameworks span across platforms, devices, technologies, and application areas. We discuss five of the most popular interoperability frameworks in this chapter.

9.3.1   universAAL

UniversAAL is an open-source software framework designed for enabling runtime support in distributed service oriented environments comprising mainly of the system of systems [13]. This framework extends semantic interoperability by sharing compatible models/ontologies with service consumers such as mobile devices, embedded systems, and others. Managers, along with middleware, collectively form the universAAL platform. These managers are considered low-level applications and provide functional APIs (application programming interfaces) to final applications utilizing universAAL. Hardware such as sensors and actuators connect to the universAAL platform through exporters, which are specific for different technologies such as Zigbee, Konnex, and others.

The universAAL middleware is tasked with core coordination among the nodes within a peer-to-peer connectivity layout, referred to as the uSpace. The sharing of various universAAL communication semantics such as the shared ontological model, context, service interactions, and user interactions is performed in this uSpace, which creates a logical environment for enabling communications irrespective of the underlying device, technology, or network. The services or set of services run by a universAAL application is human/user-centric. A coordinator node is responsible for creating each uSpace, and subsequently keeping track of its status, and adding/deleting new nodes to it.

A container is responsible for supporting the middleware and the code and building rules under different environments such as Java environments, Android environments, and other embedded systems. As of now, universAAL supports only Bundles in OSGi (for embedded systems) and APKs in Android. The peering part handles various instances of middleware communication and interconnections. A UPnP-like connector is tasked with the discovery of universAAL nodes and multi-technology bridging.

The most crucial aspect of the middleware is the communication, which provides the logic for semantic information flow between the peers. This flow is enabled through purpose-specific buses to which various applications connect irrespective of the device, container, or peering technology. Buses have been defined for purposes such as context, service and user interactions, internal strategy handling, semantics, peer matchmaking, and others. The ontology model, encryption, and message parsing through message serialization are defined in a representation model. A uSpace gateway handles communication across different uSpaces by handling message exchanges and authentication between them.

> Check yourself
>
> Composition of universAAL ontologies, context sharing in universAAL, service handling in universAAL, user interaction in universAAL

9.3.2 AllJoyn

The AllJoyn is an open-source software initiative proposed by Qualcomm in 2011 that allows devices within this framework to communicate with other devices near its vicinity [14]. The flexible AllJoyn framework encourages proximal connectivity and even has the option of including cloud connectivity to it. It was subsequently signed over to the Linux Foundation under the aegis of the AllSeen Alliance, which was formed primarily to promote IoT interoperability. Major global consumer electronics corporations such as LG, Sony, Panasonic, Haier, Cisco, HTC, Microsoft, and many others are part of the AllSeen Alliance. In 2016, AllJoyn merged with IoTvity and joined the Open Connectivity Forum (OCF), which allowed various open-source projects to include it within their framework. The AllJoyn and IoTvity technologies are currently interoperable and backward compatible with one another.

The open-source AllJoyn software framework enables interoperability amongst connected devices and applications, resulting in the creation of dynamic proximal networks using a D-Bus message bus. The software framework and the core components of the system seamlessly discover, communicate, and collaborate irrespective of platform, product, brand, or connection types, although within the limitations of the collaborating brands only (which is quite large). As of now, communication is only through Wi-Fi, but it includes devices concerning smart homes, smart TVs, smart audio, gateways, and even automotive devices.

The AllJoyn framework follows a client–server model. The clients are often referred to as "consumer" and the server as the "producer". For example, in a smart home environment, a proximity sensor senses the presence of humans in the house and switches on appliances based on the occupancy of the house. If the house is empty, the appliances are turned off. Here, the proximity sensor is the consumer, and the appliance (maybe, a light) is a producer. In this framework, each producer is characterized by an introspection file, which is an XML schema of the producer's capabilities and functionalities. The requests for each producer are based on its introspection file. The framework's capabilities can be extended by incorporating other protocols with it through bridging. Complex functionalities such as simultaneous audio streams to multiple devices can also be executed using this framework.

Some of the core services provided by the AllJoyn framework include onboarding services (attaching a new device to the framework's Wi-Fi network), configuration service (configuring device attributes such as languages, passwords, and names), notification service (text/view-URL based audio and image notifications), control panel (remote app-based control of all connected devices), and common device model service (unified monitoring of IoT devices irrespective of vendors or manufacturers).

Check yourself

Device XML schema, AllJoyn source code, AllJoyn products and services

---

9.3.3   IoTivity

Similar to the AllJoyn, IoTivity is an open-source project which is sponsored by the OSF (Open Science Framework) and hosted by the Linux Foundation [15]. This framework was developed to unify billions of IoT devices, be it wired or wireless, across the Internet, to achieve a robust and interoperable architecture for smart and thin devices. IoTivity is interoperable and backward compatible with AllJoyn. This framework can connect across profiles ranging from consumer, health, enterprise, industrial, and even automotive.

The IoTivity framework uses CoAP at the application layer and is not bothered with the physical layer requirements of devices. However, the network layer of the connecting devices must communicate using IP. The connectivity technologies of IoTivity connecting devices can consist of Wi-Fi, Ethernet, Bluetooth Low Energy, Thread, Z-Wave, Zigbee, or other legacy standards.

The IoTivity architecture supports the following core functionalities: Discovery (finding devices in one's vicinity and offering services to them), data transmission (standardized message transmission between devices), device management, and data management.

Under the purview of the resource-bounded context in IoTivity's OCF (Open Connectivity Foundation) Native Cloud 2.0 framework, which aims to utilize and enhance the benefits of IoT for companies fully, a resource hosting server has to be accessible through the OCF's native cloud. A resource is an object, which consists of a type, associated data, resource relationships, and operational methods. A server can only publish discoverable resources (which can be found by other connected clients), once it is successfully connected, authenticated, and authorized. Clients can discover resources, either based on the resource type or server identifiers.

Check yourself

IoTivity services and functionalities, IoTivity source code, IoTivity use cases

---

9.3.4 Brillo and Weave

Google introduced its IoT framework in 2015 as Project Brillo. It is primarily designed as an operating system for IoT devices; it can be considered as a skinny version of Android, having a minimal footprint [16]. Brillo is currently Wi-Fi and BLE (Bluetooth low energy) enabled, with ongoing efforts for the addition of further low-power solutions such as Thread. As the framework is Android-based, it extends scalability in terms of rapid acceptance and portability. Brillo extends interoperability amongst devices and platforms from various vendors and manufacturers.

The underlying communication layer of Brillo is known as Weave. Weave provides a common language for devices such as phones to talk to the cloud. The Weave is the communications layer by which Things can talk to one another. It provides a common language so that devices can talk to one another, with the cloud and the phone. The Brillo framework extends interoperability and uniformity over a diverse range of applications such as smart farming devices, smart homes, smart parking systems, and others. Weave devices communicate over TCP or UDP, using either IPv4 or IPv6. Interestingly, Weave is an information schema for devices that defines device types, functionalities, and modes of communication.

The Weave stack comprises four core modules: Security manager, exchange manager, message layer, and fabric state. Weave provides some core functionalities: Bulk data exchange (file transfers), common (system status and error reports), data management, echo (network connectivity testing), security, service directory, and others. Secondary protocols built on top of the core protocols of Weave include alarm, device control, service provisioning, network provisioning, heartbeat, and others.

> Check yourself
>
> Brillo and Weave use cases

9.3.5 HomeKit

The HomeKit software framework is designed by Apple to work with its iOS mobile operating system for achieving a centralized device integrating and control framework [17]. It enables device configuration, communication, and control of smart home appliances. Home automation is achieved by incorporating room designs, items, and their actions within the HomeKit service. Users can interact with the framework using speech-based voice commands through Apple's voice assistant, Siri, or through external apps. Smart home devices such as thermostats, lights, locks, cameras, plugs, and others, spread over a house can be controlled by a single HomeKit interface through smartphones. HomeKit-enabled device manufacturers need to have an MFi program, and all devices were initially required to have an encryption coprocessor. Later, the processor-based encryption was changed to a software-based one.

Non-HomeKit devices can have the benefits of HomeKit through the use of HomeKit gateways and hubs.

HomeKit devices within a smart home securely connect to a hub either through Wi-Fi or Bluetooth. However, as the range of Bluetooth is severely limited, the full potential of the HomeKit may not be adequately exploited. This framework allows for individual as well as grouped control of connected devices based on scenarios. Features such as preconfigured devices settings can be collectively commanded using voice commands to Siri.

> **Points to ponder**
>
> The MFi program is Apple's licensing program for hardware/software/firmware developers. It stands for "Made For iPhone/iPad/iMAC".

> **Check yourself**
>
> HomeKit interfacing, HomeKit controls, HomeKit use case

Summary

This chapter introduces the concept of interoperability in the context of IoT architectures, frameworks, and application domains. We initially outline the taxonomy of interoperability to give the readers a perspective of the challenges and the present-day solutions or attempts to solve these challenges. We outline the various standardization efforts to address interoperability issues in different domains. Further, at the end of this chapter, we also provide a brief description of the different interoperability-enabling frameworks that are under development by various corporations across the globe.

Exercises

(i)   Differentiate between semantic and syntactic interoperability.

(ii)   What are the various types of interoperability encountered in IoT environments?

(iii) What is meant by the heterogeneity of IoT devices in the context of interoperability?

(iv)   How is device interoperability different from platform interoperability?

(v)   Describe the following standards:

    (a) EnOcean

    (b) DLNA

      (c) Konnex

      (d) LonWorks

      (e) UPnP

      (f) X-10

      (g) Insteon

(vi) How does EnOcean use energy harvesting for its operations?

(vii) What is LonTalk?

(viii) What is a neuron chip in the context of LonWorks?

(ix) How is X-10 different from DLNA?

(x) How is the UniversAAL framework different from the Alljoyn framework?

(xi) How is Brillo different from Weave?

References

[1] Al-Fuqaha, A., M. Guizani, M., Mohammadi, M., Aledhari, and Ayyash. 2015. "Internet of Things: A Survey on Enabling Technologies, Protocols, and Applications." *IEEE Communications Surveys and Tutorials* 17(4): 2347–2376.

[2] Aloi, G., G. Caliciuri, G. Fortino, R. Gravina, P. Pace, W. Russo, and C. Savaglio. 2017. "Enabling IoT Interoperability through Opportunistic Smartphone-based Mobile Gateways." *Journal of Network and Computer Applications* 81: 74–84.

[3] Bröring, A., S. Schmid, C. K. Schindhelm, A. Khelil, S. Käbisch, D. Kramer, D. Le Phuoc, J. Mitic, D. Anicic, and E. Teniente. 2017. "Enabling IoT Ecosystems through Platform Interoperability." *IEEE Software* 34(1): 54–61.

[4] Kiljander, J., A. D'elia, F. Morandi, P. Hyttinen, J. Takalo-Mattila, A. Ylisaukko-Oja, J. P. Soininen, and T. S. Cinotti. 2014. "Semantic Interoperability Architecture for Pervasive Computing and Internet of Things." *IEEE Access* 2: 856–873.

[5] Bandyopadhyay, S., M. Sengupta, S. Maiti, and S. Dutta. 2011. "Role of Middleware for Internet of Things: A Study." *International Journal of Computer Science and Engineering Survey* 2(3): 94–105.

[6] The EnOcean Alliance. https://www.enocean-alliance.org/.

[7] The Digital Living Networking Alliance. https://www.dlna.org/.

[8] Konnex. https://www.konnex.group/en/.

[9] Cheng, D. Y., Philips North America LLC. 2002. "UPnP Enabling Device for Heterogeneous Networks of Slave Devices." U. S. Patent Application 09/742,278.

[10] Echelon, "Introduction to the LonWorks Platform." https://www.echelon.com/assets/blt893a8b319e8ec8c7/078-0183-01B_Intro_to_LonWorks_Rev_2.pdf.

[11] Insteon: The Technology. https://www.insteon.com/technology.

[12] X10 Basics. https://www.x10.com/x10-basics.html.

[13] UniversAAL IoT. https://www.universaal.info/.

[14] AllJoyn Open Source Project. https://openconnectivity.org/developer/reference-implementation/alljoyn/.

[15] IoTivity. https://iotivity.org/.

[16] Brillo/Weave Part 1: High Level Introduction. https://events.static.linuxfound.org/sites/events/files/slides/Brillo%20and%20Weave%20-%20Introduction$_v$3$_1$.pdf.

[17] HomeKit. https://developer.apple.com/homekit/.

PART THREE
ASSOCIATED IOT TECHNOLOGIES

Chapter **10**

Cloud Computing

After reading this chapter, the reader will be able to:

- Understand the concept of cloud computing and its features
- Understand virtualization, different cloud models, and service-level agreements (SLAs)
- Identify the salient features of various cloud computing models
- Understand the concept of sensor-clouds

10.1   Introduction

Sensor nodes are the key components of Internet of Things (IoT). These nodes are resource-constrained in terms of storage, processing, and energy. Moreover, in IoT, the devices are connected and communicate with one another by sharing the sensed and processed data. Handling the enormous data generated by this large number of heterogeneous devices is a non-trivial task. Consequently, cloud computing becomes an essential building block of the IoT architecture. This chapter aims at providing an extensive overview of cloud computing. Additionally, *Check yourself* will help the learner to learn different concepts are related to cloud computing.

Cloud computing is more than traditional network computing. Unlike network computing, cloud computing comprises a pool of multiple resources such as servers, storage, and network from *single/multiple* organizations. These resources are allocated to the end users as per requirement, on a payment basis. In cloud computing architecture, an end user can request for customized resources such as storage space, RAM, operating systems, and other software to a *cloud service provider* (*CSP*) as shown in Figure 10.1. For example, a user can request for a *Linux* operating system for running an application from a CSP; another end user can request for *Windows 10* operating system from the same CSP for executing some application. The cloud services are accessible from anywhere and at any time by an authorized user through Internet connectivity.

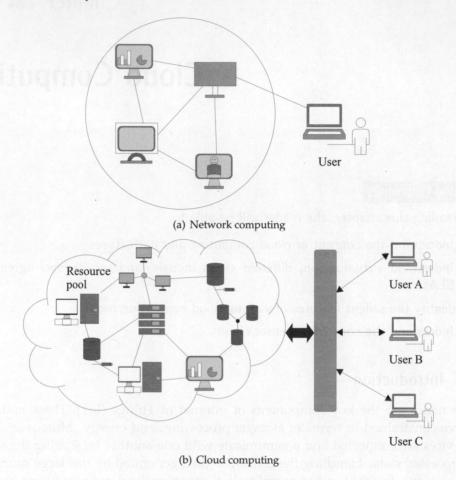

(a) Network computing

(b) Cloud computing

Figure 10.1 Network computing versus cloud computing

Points to ponder

- *Gmail*, *Facebook*, and *Twitter* are examples of cloud computing applications.

- Currently, many companies such as *Amazon Web Service* and *Microsoft Azure* provide cloud services.

Cloud computing comprises a shared pool of computing resources, which are accessible dynamically, ubiquitously, and on-demand basis by the users. This shared pool of resources includes networks, storage, processor, and servers. These resources are accessible by multiple users through a regular command-line terminal at the same or different time instants. The services of cloud computing are based on the *pay-per-use* model. The concept is the same as paying utility bills based on consumption. In cloud computing, a user pays for the cloud services as per the duration of their resource usage. On the other hand, there is a CSP, that provides cloud services to end user organizations.

10.2 Virtualization

The key concept of cloud computing is *virtualization*. The technique of sharing a single resource among multiple end user organizations or end users is known as virtualization. In the virtualization process, a physical resource is logically distributed among multiple users. However, a user perceives that the resource is unlimited and is dedicatedly provided to him/her. Figure 10.2(a) represents a traditional desktop, where an application (App) is running on top of an OS, and resources are utilized only for that particular application. On the other hand, multiple resources can be used by different end users through virtualization software, as shown in Figure 10.2(b). Virtualization software separates the resources logically so that there is no conflict among the users during resource utilization.

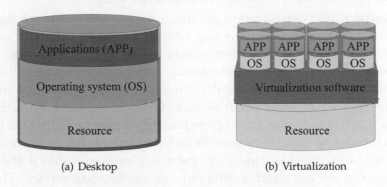

(a) Desktop   (b) Virtualization

Figure 10.2 Traditional desktop versus virtualization

10.2.1 Advantages of virtualization

With the increasing number of interconnected heterogeneous devices in IoT, the importance of virtualization also increases. In IoT, a user is least bothered about where the data from different heterogeneous devices are stored or processed for a particular application. Users are mainly concerned for their services. Typically, there are different software such as *VMware*, which enable the concept of virtualization.

With the increasing importance of cloud computing, different organizations and individuals are using it extensively. Moreover, there is always a risk of system crash at any instant of time. In such a scenario, cloud computing plays a vital role by keeping backups through virtualization. Primarily, there are two entities in a cloud computing architecture: *end users* and *CSP*. Both end users and CSP are benefited in several aspects through the process of virtualization. The major advantages, from the perspective of the end user and CSP, are as follows:

(i)   **Advantages for End Users**

  (a)   **Variety:** The process of virtualization in cloud computing enables an end user organization to use various types of applications based on the requirements. As an example, suppose John takes up still photography as a hobby. His resource-limited PC can barely handle the requirements for a photo editing software, say *X-photoeditor*. In order to augment his PC's regular performance, he uninstalls the *X-photoeditor* software and purchases a cloud service, which lets him access a virtual machine (VM). In his VM, he installs the *X-photoeditor* software, by which he can edit photos efficiently and, most importantly, without worrying about burdening his PC or running out of processing resources. After six months, John's interest in his hobby grows and he moves on to video-editing too. For editing his captured videos, he installs a video editing software, *Y-videoeditor*, in his VM and can edit videos efficiently. Additionally, he has the option of installing and using a variety of software for different purposes.

  (b)   **Availability:** Virtualization creates a logical separation of the resources of multiple entities without any intervention of end users. Consequently, the concept of virtualization makes available a considerable amount of resources as per user requirements. The end users feel that there are unlimited resources present dedicatedly for him/her. Let us suppose that Jane uses a particular email service. Her account has been active for over ten years now; however, it offers limited storage of 2 GB. Due to the ever-accumulating file attachments in different emails, her 2 GB complimentary space is exhausted. However, there is a provision that if she pays $100 annually, she can attach additional space to her mail service. This upgrade allows her to have more storage at her disposal for a considerable time in the future.

  (c)   **Portability:** Portability signifies the availability of cloud computing services from anywhere in the world, at any instant of time. For example, a person flying from the US to the UK still has access to their documents, although they cannot physically access the devices on which the data is stored. This has been made possible by platforms such as Google Drive.

(d) **Elasticity:** Through the concept of virtualization, an end user can scale-up or scale-down resource utilization as per requirements. We have already explained that cloud computing is based on a pay-per-use model. The end user needs to pay the amount based on their usage. For example, Jack rents two VMs in a cloud computing infrastructure from a CSP. VM1 has the *Ubuntu* operating system (OS), on which Jack is simulating a network scenario using *Network Simulator-2* (*NS2*). VM2 has *Windows 10* OS, on which he is running a *MATLAB* simulation. However, after a few days, Jack feels that his VM2 has served its purpose and is no longer required. Consequently, he releases VM2 and, after that, he is only billed for VM1. Thus, Jack can scale-up or scale-down his resources in cloud computing, which employs the concept of virtualization.

(ii) **Advantages for CSP**

(a) **Resource Utilization:** Typically, a CSP in a cloud computing architecture procures resources on their own or get them from third parties. These resources are distributed among different users dynamically as per their requirements. A segment of a particular resource provided to a user at a time instant, can be provided to another user at a different time instant. Thus, in the cloud computing architecture, resources can be re-utilized for multiple users.

(b) **Effective Revenue Generation:** A CSP generates revenue from the end users based on resource utilization. As an example, today, a user **A** is utilizing storage facility from a particular CSP. The user will release the storage after a few days when his/her requirement is complete. The CSP earns some revenue from user A for the utilization of the allocated storage facility. In the future, the CSP can provide the same storage facility to a different user, **B**. Again, the CSP can generate revenue from user B for his/her storage utilization.

Check yourself

VMware, hypervisor, virtual machine

10.2.2 Types of virtualization

Based on the requirements of the users, we categorized virtualization as shown in Figure 10.3.

(i) **Hardware Virtualization:** This type of virtualization indicates the sharing of hardware resources among multiple users. For example, a single processor appears as many different processors in a cloud computing architecture.

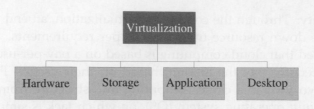

Figure 10.3 Types of virtualization

Different operating systems can be installed in these processors and each of them can work as stand-alone machines.

(ii) **Storage Virtualization:** In storage virtualization, the storage space from different entities are accumulated virtually, and seem like a single storage location. Through storage virtualization, a user's documents or files exist in different locations in a distributed fashion. However, the users are under the impression that they have a single dedicated storage space provided to them.

(iii) **Application Virtualization:** A single application is stored at the cloud end. However, as per requirement, a user can use the application in his/her local computer without ever actually installing the application. Similar to storage virtualization, in application virtualization, the users get the impression that applications are stored and executed in their local computer.

(iv) **Desktop Virtualization:** This type of virtualization allows a user to access and utilize the services of a desktop that resides at the cloud. The users can use the desktop from their local desktop.

> Check yourself
>
> Server virtualization, Para virtualization, User virtualization

10.3   Cloud Models

As per the National Institute of Standards and Technology (NIST) [1] and Cloud Computing Standards Roadmap Working Group, the cloud model can be divided into two parts: (1) *Service model* and (2) *Deployment model* as shown in Figure 10.4. Further the service model is categorized as: *Software-as-a-Service* (*SaaS*), *Platform-as-a-Service* (*PaaS*), and *Infrastructure-as-a-Service* (IaaS). On the other hand, the deployment model is further categorized as: *Private cloud*, *Community cloud*, *Public cloud*, and *Hybrid cloud*.

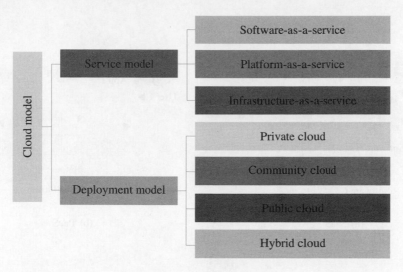

Figure 10.4 Cloud model

(i) **Service Model**

The service model is depicted in Figure 10.5.

(a) **Software-as-a-Service (SaaS):** This service provides access to different software applications to an end user through Internet connectivity. For accessing the service, a user does not need to purchase and install the software applications on his/her local desktop. The software is located in a cloud server, from where the services are provided to multiple end users. SaaS offers scalability, by which users have the provision to use multiple software applications as per their requirements. Additionally, a user does not need to worry about the update of the software applications. These software are accessible from any location. One example of SaaS is *Microsoft Office 365*.

(b) **Platform-as-a-Service (PaaS):** PaaS provides a computing platform, by which a user can develop and run different applications. The cloud user need not go through the burden of installing and managing the infrastructure such as operating system, storage, and networks. However, the users can develop and manage the applications that are running on top of it. An example of PaaS is *Google App Engine*.

(c) **Infrastructure-as-a-Service (IaaS):** IaaS provides infrastructure such as storage, networks, and computing resources. A user uses the infrastructure without purchasing the software and other network components. In the infrastructure provided by a CSP, a user can use any composition of the operating system and software. An example of IaaS is *Google Compute Engine*.

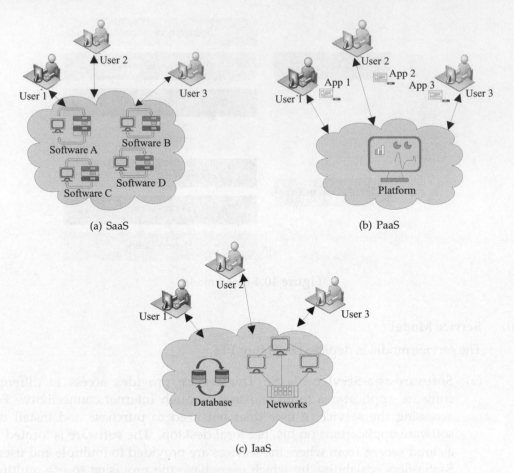

Figure 10.5 Service models

(ii)   Deployment Model

(a) **Private Cloud:** This type of cloud is owned explicitly by an end user organization.   The internal resources of the organization maintain the private cloud.

(b) **Community Cloud:** This cloud forms with the collaboration of a set of organizations for a specific community.   For a community cloud, each organization has some shared interests.

(c) **Public Cloud:** The public cloud is owned by a third party organization, which provides services to the common public. The service of this cloud is available for any user, on a payment basis.

(d) **Hybrid Cloud:** This type of cloud comprises two or more clouds (private, public, or community).

10.4 Service-Level Agreement in Cloud Computing

The most important actors in cloud computing are the end user/customer and CSP. Cloud computing architecture aims to provide optimal and efficient services to the end users and generate revenue from them as per their usage. Therefore, for a clear understanding between CSP and the customer about the services, an agreement is required to be made, which is known as *service-level agreement* (*SLA*). An SLA provides a detailed description of the services that will be received by the customer. Based on the SLA, a customer can be aware of each and every term and condition of the services before availing them. An SLA may include multiple organizations for making the legal contract with the customers.

10.4.1 Importance of SLA

An SLA is essential in cloud computing architecture for both CSP and customers. It is important because of the following reasons:

- **Customer Point of View**: Each CSP has its SLA, which contains a detailed description of the services. If a customer wants to use a cloud service, he/she can compare the SLAs of different organizations. Therefore, a customer can choose a preferred CSP based on the SLAs.

- **CSP Point of View**: In many cases, certain performance issues may occur for a particular service, because of which a CSP may not be able to provide the services efficiently. Thus, in such a situation, a CSP can explicitly mention in the SLA that they are not responsible for inefficient service.

10.4.2 Metrics for SLA

Depending on the type of services, an SLA is constructed with different metrics. However, a few common metrics that are required to be included for constructing an SLA are as follows:

(i) **Availability:** This metric signifies the amount of time the service will be accessible for the customer.

(ii) **Response Time:** The maximum time that will be taken for responding to a customer request is measured by response time.

(iii) **Portability:** This metric indicates the flexibility of transferring the data to another service.

(iv) **Problem Reporting:** How to report a problem, whom and how to be contacted, is explained in this metric.

(v) **Penalty:** The penalty for not meeting the promises mentioned in the SLA.

10.5   Cloud Implementation

10.5.1   Cloud simulation

With the rapid deployment of IoT infrastructure for different applications, the requirement for cloud computing is also increasing. It is challenging to estimate the performance of an IoT system with the cloud before real implementation. On the other hand, real deployment of the cloud is a complex and costly procedure. Thus, there is a requirement for simulating the system through a cloud simulator before real implementation. There are many cloud simulators that provide pre-deployment test services for repeatable performance evaluation of a system. Typically, a cloud simulator provides the following advantages to a customer:

- Pre-deployment test before real implementation
- System testing at no cost
- Repeatable evaluation of the system
- Pre-detection of issues that may affect the system performance
- Flexibility to control the environment

Currently, different types of cloud simulators are available. A few cloud simulators are listed here:

(i)   CloudSim

  (a)   Description: CloudSim [3] is a popular cloud simulator that was developed at the University of Melbourne. This simulator is written in a Java-based environment. In CloudSim, a user is allowed to add or remove resources dynamically during the simulation and evaluate the performance of the scenario.

  (b)   Features: CloudSim has different features, which are listed as follows:

    (1)   The CloudSim simulator provides various cloud computing data centers along with different data center network topologies in a simulation environment.

    (2)   Using CloudSim, virtualization of server hosts can be done in a simulation.

    (3)   A user is able to allocate virtual machines (VMs) dynamically.

    (4)   It allows users to define their own policies for the allocation of host resources to VMs.

    (5)   It provides flexibility to add or remove simulation components dynamically.

    (6)   A user can stop and resume the simulation at any instant of time.

(ii) CloudAnalyst

    (a) Description: CloudAnalyst [4] is based on CloudSim. This simulator provides a graphical user interface (GUI) for simulating a cloud environment, easily. The CloudAnalyst is used for simulating large-scale cloud applications.

    (b) Features:

        (1) The CloudAnalyst simulator is easy to use due to the presence of the GUI.

        (2) It allows a user to add components and provides a flexible and high level of configuration.

        (3) A user can perform repeated experiments, considering different parameter values.

        (4) It can provide a graphical output, including a chart and table.

(iii) GreenCloud

    (a) Description: GreenCloud [2] is developed as an extension of a packet-level network simulator, NS2. This simulator can monitor the energy consumption of different network components such as servers and switches.

    (b) Features:

        (1) GreenCloud is an open-source simulator with user-friendly GUI.
        (2) It provides the facility for monitoring the energy consumption of the network and its various components.
        (3) It supports the simulations of cloud network components.
        (4) It enables improved power management schemes.
        (5) It allows a user to manage and configure devices, dynamically, in simulation.

10.5.2 An open-source cloud: OpenStack

For the real implementation of cloud, there are various open-source cloud platforms available such as *OpenStack*, *CloudStack*, and *Eucalyptus*. Here, we will discuss the OpenStack platform briefly. The OpenStack [12] is free software, which provides a cloud IaaS to users. A user can easily use this cloud with the help of a GUI-based web interface or through the command line. OpenStack supports a vastly scalable cloud system, in which different pre-configured software suites are available. The service components of OpenStack along with their functions are depicted in Table 10.1.

Features of OpenStack

(i) OpenStack allows a user to create and deploy virtual machines.

(ii) It provides the flexibility of setting up a cloud management environment.

Table 10.1 Components in OpenStack

Component	Function
Nova	Compute
Neutron	Networking
Cinder	Block storage
Keystone	Identity
Glance	Image
Swift	Object storage
Horizon	Dashboard
Trove	Database
Sahara	Elasticmap reduce
Manila	Shared file system
Designate	DNS
Searchlight	Search
Barbican	Key manager

(iii) OpenStack supports an easy horizontal scaling: dynamic addition or removal of instances for providing services to multiple numbers of users.

(iv) This cloud platform allows the users to access the source code and share their code to the community.

10.5.3 A commercial cloud: Amazon web services (AWS)

Besides the open-source cloud, there are various commercial cloud infrastructures available in the market. Few of the popular commercial cloud infrastructures are *Amazon Web Services (AWS)*, *Microsoft Azure*, and *Google App Engine*. In this section, we will discuss the different features of AWS [13]. A user can launch and manage server instances in AWS. Typically, a web interface is used to handle the instances. Additionally, AWS provides different APIs (application programming interfaces), tools, and utilities for users. Like other commercial clouds, Amazon AWS follows the *pay-per-use* model. This cloud infrastructure provides a virtual computing environment, where different configurations, such as CPU, memory, storage, and networking capacity are available.

Features of AWS

(i) It provides flexibility to scale and manage the server capacity.

(ii) AWS provides control to OS and deployment software.

(iii) It follows the pay-per-use model.

(iv) The cloud allows a user to establish connectivity between the physical network and private virtual network

(v) The developer tools in this cloud infrastructure help a user for fast development and deployment of the software.

(vi) AWS provides excellent management tools, which help a user to monitor and automate different components of the cloud.

(vii) The cloud provides *machine learning* facilities, which are very useful for data scientists and developers.

(viii) For extracting meaning from data, analytics play an important role. AWS also provides a data analytics platform.

Check yourself

CloudStack cloud, Eucalyptus cloud, Amazon cloud

10.6 Sensor-Cloud: Sensors-as-a-Service

In this chapter, we have already discussed different services of cloud computing, which include SaaS, PaaS, and IaaS. Now, we will explore a new concept known as *Sensors-as-a-Service* (*Se-aaS*) in a sensor-cloud architecture [5]. Virtualization of resources is the backbone of cloud computing. Similarly, in a sensor-cloud, *virtualization of sensors* plays an essential role in providing services to multiple users. Typically, in a sensor-cloud architecture, multiple users receive services from different asensor nodes, simultaneously. However, the users remain oblivious to the fact that a set of sensor nodes is not dedicated solely to them for their application requirements. In reality, a particular sensor may be used for serving multiple user applications, simultaneously. The main aim of sensor-cloud infrastructure is to provide an opportunity for the common mass to use Wireless Sensor Networks (WSNs) on a payment basis. Similar to cloud computing, sensor-cloud architecture also follows the *pay-per-use* model.

10.6.1 Importance of sensor-cloud

The sensor-cloud infrastructure is based on the concept of cloud computing, in which a user application is served by a set of homogeneous or heterogeneous sensor nodes. These sensor nodes are selected from a common pool of sensor nodes, as per the requirement of user applications. Using the sensor-cloud infrastructure, a user receives data for an application from multiple sensor nodes without owning them. Unlike sensor-cloud, if a user wants to use traditional WSN for a certain application, he/she has to go through different pre-deployment and post-deployment hurdles. Figures 10.6 depicts the usage of sensor nodes using traditional WSN and sensor-cloud infrastructure. With the help of a case study, we will discuss the advantages of sensor-cloud over traditional WSN.

Case Study: John is a farmer, and he has a significantly vast farmable area with him. As manual supervision of the entire field is very difficult, he has planned to deploy a WSN in his farming field. Before purchasing the WSN, he has to decide which sensors should be used in his fields for sensing the different agricultural parameters. Additionally, he has to decide the type and number of other components such as an electronics circuit board and communication module required along with the sensors. As there are numerous vendors, it is challenging for him to choose the correct (in terms of quality and cost) vendor, as well as the sensor owner from whom the WSN will be procured. He finally decides the type of sensors along with the other components that are required for monitoring his agricultural field. Now, John faces the difficulty of optimally planning the sensor node deployment in his fields. After going through these hurdles, he decides on the number of sensor nodes that are required for monitoring his field. Finally, John procures the WSNs from a vendor. After procurement, he deploys the sensor nodes and connects different components. As WSN consists of different electronic components, he has to maintain the WSN after its deployment. After three months, as his requirement of agricultural field monitoring is completed, he removes the WSN from the agricultural field. Six months later, John plans to use the WSN that was deployed in the agricultural field for home surveillance. As the agriculture application is different from the home surveillance application, the sensor required for the system also changes. Thus, John has to go through all the steps again, including maintenance, deployment, and hardware management, for the surveillance system. Thus, we observe that the users face different responsibilities for using a WSN for an application. In such a situation, if sensor-cloud architecture is present, John can easily use WSNs for his application on a rental basis. Moreover, through the use of sensor-cloud, John can easily switch the application without any manual intervention. On the other end, service providers of the sensor-cloud infrastructure may serve multiple users with the same sensors and earn profit.

Check yourself

Difference between sensor-cloud and virtual sensor network (VSN)

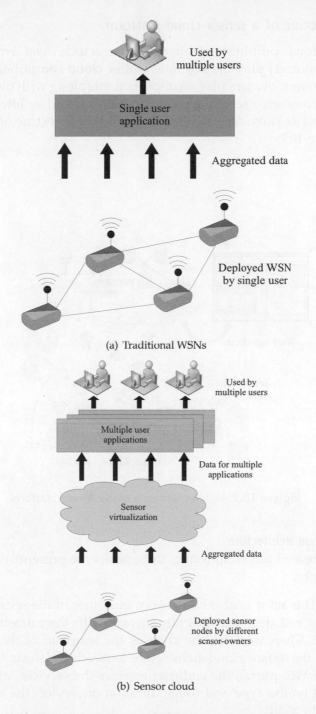

(a) Traditional WSNs

(b) Sensor cloud

Figure 10.6 Traditional WSN versus sensor-cloud

10.6.2 Architecture of a sensor-cloud platform

In a traditional cloud computing architecture, two actors, *cloud service provider* (*CSP*) and *end users* (customer) play the key role. Unlike cloud computing, in sensor-cloud architecture, the sensor owners play an important role along with the service provider and end users. However, a service provider in sensor-cloud architecture is known as a sensor-cloud service provider (SCSP). The detailed architecture of a sensor-cloud is depicted in Figure 10.7.

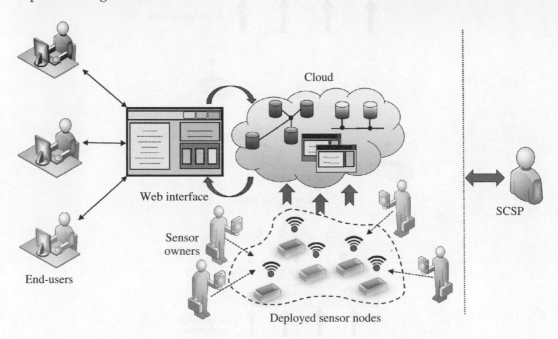

Figure 10.7 Architecture of a sensor-cloud platform

Actors in sensor-cloud architecture

Typically, in a sensor-cloud architecture, three actors are present. We briefly describe the role of each actor.

(i) **End User:** This actor is also known as a customer of the sensor-cloud services. Typically, an end user registers him/herself with the infrastructure through a Web portal. Thereafter, he/she chooses the template of the services that are available in the sensor-cloud architecture to which he/she is registered. Finally, through the Web portal, the end user receives the services, as shown in Figure 10.7. Based on the type and usage duration of service, the end user pays the charges to the SCSP.

(ii) **Sensor Owner:** We have already discussed that the sensor-cloud architecture is based on the concept of Se-aaS. Therefore, the deployment of the sensors is essential in order to provide services to the end users. These sensors in a sensor-cloud architecture are owned and deployed by the sensor owners, as depicted in Figure 10.7. A particular sensor owner can own multiple homogeneous or heterogeneous sensor nodes. Based on the requirements of the users, these sensor nodes are virtualized and assigned to serving multiple applications at the same time. On the other hand, a sensor owner receives rent depending upon the duration and usage of his/her sensor node(s).

(iii) **Sensor-Cloud Service Provider (SCSP):** An SCSP is responsible for managing the entire sensor-cloud infrastructure (including management of sensor owners and end users handling, resource handling, database management, cloud handling etc.), centrally. The SCSP receives rent from end users with the help of a pre-defined pricing model. The pricing scheme may include the infrastructure cost, sensor owners' rent, and the revenue of the SCSP. Typically, different algorithms are used for managing the entire infrastructure. The SCSP receives the rent from the end users and shares a partial amount with the sensor owners. The remaining amount is used for maintaining the infrastructure. In the process, the SCSP earns a certain amount of revenue from the payment of the end users.

Check yourself

Pricing scheme for sensor-cloud [6, 9], Caching in sensor-cloud [11], data center scheduling for sensor-cloud, Big sensor-cloud [8], Mobile sensor-cloud [7]

Sensor-Cloud Architecture from Different Viewpoints

We explore the sensor-cloud architecture from two view points: (i) User organizational view and (ii) real architectural view [5]. Different views of sensor-cloud architecture are shown in Figure 10.8.

(i) **User Organizational View:** This view of sensor-cloud architecture is simple. In a sensor-cloud, end users interact with a Web interface for selecting templates of the services. Thereafter, the services are received by the end users through the Web interface. In this architecture, an end user is unaware of the complex processes that are running at the back end.

(ii) **Real Architectural View:** The complex processing of sensor-cloud architecture is visualized through this view. The processes include sensor allocation, data extraction from the sensors, virtualization of sensor nodes, maintenance of the infrastructure, data center management, data caching, and others. For each process, there is a specific algorithm or scheme.

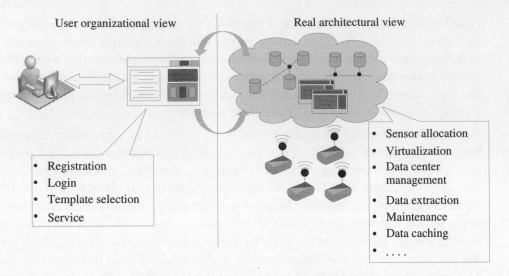

User organizational view

- Registration
- Login
- Template selection
- Service

Real architectural view

- Sensor allocation
- Virtualization
- Data center management
- Data extraction
- Maintenance
- Data caching
- . . . .

Figure 10.8 Sensor-cloud architecture

Summary

This chapter covered different aspects of cloud computing which would help the reader to visualize its requirement in IoT. Additionally, we discussed commercial cloud platforms along with simulation tools. We explored the different aspects of the OpenStack cloud platform, which would help a learner to implement and use it. Finally, we concluded the chapter with a newly explored concept known as sensor-cloud, which deals with sensors-as-a-service.

Exercises

(i)   What are the advantages of cloud computing?

(ii)   With an example, explain how software-as-a-service is different from platform-as-a-service?

(iii) What is an SLA? Why it is important in cloud computing?

(iv)   Differentiate between scalability and elasticity.

(v)   What is an Amazon Machine Image?

(vi)   What are the differences between modular and containerized data centers?

(vii) What is the relationship between IoT and cloud computing?

(viii) What is a sensor-cloud? Why do we use sensor-cloud?

(ix)   Differentiate among different cloud deployment models.

References

[1]   Hogan, Michael, Fang Liu, Annie Sokol, Jin Tong. 2011. *NIST Cloud Computing Standards Roadmap*. National Institute of Standards and Technology (NIST), U. S. Department of Commerce.

[2]   Kliazovich, D., P. Bouvry, Y. Audzevich, and S. U. Khan. 2010. "GreenCloud: A Packet-Level Simulator of Energy-Aware Cloud Computing Data Centers." In *Proceedings of the IEEE Global Telecommunications Conference (GLOBECOM), December, 2010.*

[3]   Calheiros, Rodrigo N., Rajiv Ranjan, Anton Beloglazov, César A. F. De Rose, and Rajkumar Buyya. 2011. "CloudSim: A Toolkit for Modeling and Simulation of Cloud Computing Environments and Evaluation of Resource Provisioning Algorithms." *Software: Practice and Experience* 41(1): 23–50.

[4]   Wickremasinghe, Bhathiya, Rodrigo N. Calheiros, and Rajkumar Buyya. 2010. "CloudAnalyst: A CloudSim-Based Visual Modeller for Analysing Cloud Computing Environments and Applications." In *24th IEEE International Conference on Advanced Information Networking and Applications, AINA* 446–452.

[5]   Misra, S., S. Chatterjee, and M. S. Obaidat. 2017. "On Theoretical Modeling of Sensor Cloud: A Paradigm Shift From Wireless Sensor Network." *IEEE Systems Journal* 11: 1084–1093.

[6]   Roy, A., S. Misra, and P. Dutta. 2019. "Dynamic Pricing for Sensor-Cloud Platform in the Presence of Dumb Nodes." *IEEE Transactions on Cloud Computing*. doi: 10.1109/TCC.2019.2950396.

[7]   Roy, A., S. Misra, and Lakshya. 2019. "OPTIVE: Optimal Configuration of Virtual Sensor in Mobile Sensor-Cloud." *IEEE Wireless Communications and Networking Conference (WCNC), Marrakesh, Morocco* 1–6. doi: 10.1109/WCNC.2019.8885626.

[8]   Chatterjee, S., A. Roy, S. K. Roy, S. Misra, M. Bhogal, and R. Daga. 2019. "Big-Sensor-Cloud Infrastructure: A Holistic Prototype for Provisioning Sensors-as-a-Service." *IEEE Transactions on Cloud Computing*. doi: 10.1109/TCC.2019.2908820.

[9]   Chakraborty, A., A. Mondal, A. Roy, and S. Misra. 2018. "Dynamic Trust Enforcing Pricing Scheme for Sensors-as-a-Service in Sensor-Cloud Infrastructure." *IEEE Transactions on Services Computing*. doi: 10.1109/TSC.2018.2873763.

[10]   Roy, C., A. Roy, and S. Misra. 2018. "DIVISOR: Dynamic Virtual Sensor Formation for Overlapping Region in IoT-based Sensor-cloud." *IEEE Wireless Communications and Networking Conference (WCNC), Barcelona.* 1–6. doi: 10.1109/WCNC.2018.8377221.

[11]   Roy, A., S. Misra, and S. Ghosh. 2018. "QoS-Aware Dynamic Caching for Destroyed Virtual Machines in Sensor-Cloud Architecture." In *Proceedings of the 19th International Conference on Distributed Computing and Networking (ACM ICDCN), Varanasi, India.*

[12]   OpenStack. https://www.openstack.org/.

[13]   Amazon Web Services. https://aws.amazon.com/.

Fog Computing and Its Applications

After reading this chapter, the reader will be able to:

- Understand the concept of Fog computing and its features

- List the salient features of a Fog computing architecture

- Identify the requirements of a Fog computing architecture

- Understand the importance of Fog computing through real-life use cases and identify its application scope in IoT

11.1 Introduction

In the Internet of Things (IoT), billions of Things connect through wired or wireless connectivity. These Things or devices produce a huge volume of data, which are typically transmitted to the cloud for analysis. Sometimes, the process of transmitting the data to the cloud and analyzing the data may consume a significant amount of time, which is undesirable for time-critical applications, such as healthcare. Therefore, providing real-time services is a major challenge in cloud computing. The concept of *Fog* computing was introduced considering the high latencies involved in cloud computing. This chapter aims to provide an insight into the fog computing architecture and how it reduces the latency of data processing and transmission.

In the IoT architecture, physical devices transmit data to the cloud. After specific processing of these data, service is provided to the end user applications. Fog computing follows a distributed architecture, which enables processes to execute near the edge of the devices to avoid service latency. A fog layer is an intermediate layer between the physical IoT devices and the cloud. The term, fog computing, was coined by Cisco.

In Figure 11.1, we depict the difference between the basic architecture of cloud and fog computing. In traditional cloud architecture, all data from the devices are transmitted to the cloud directly and then, processed in order to receive the final end user services. These end user services may be in the form of results of analytics, visualizations, and other such processed information. In fog computing architecture, time-sensitive data from different devices are transmitted to the fog devices at the fog layer. Further, data are processed to serve an end user application. Moreover, based on the requirements, the data are transmitted to the cloud layer for long-term storage.

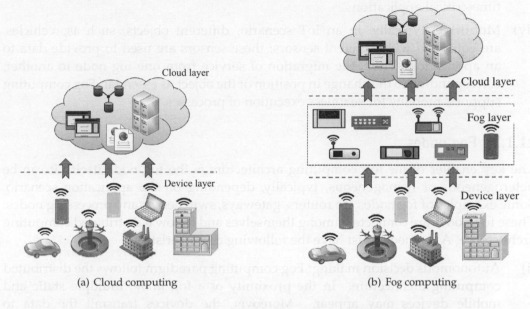

Figure 11.1 Difference between cloud and fog computing

11.1.1 Essential characteristics in fog computing

A fog computing platform resides between cloud and physical IoT devices; it provides processing, storage, and networking services. The challenges of a fog computing platform are as follows:

(i) **Location Awareness:** The fog nodes operate near the edge of the physical IoT devices. Therefore, these fog nodes must be aware of their actual locations and how far they are located from the physical devices. A fog node collaborates with its neighboring fog nodes to provide the end user services.

(ii) **Heterogeneity and Interoperability:** The fog platform consists of a sufficient number of heterogeneous fog nodes. These nodes have different capabilities of processing, storing, and networking. Therefore, in fog architecture, it is essential to process the data of a physical device with the most suitable fog nodes. In order to serve applications, a set of fog nodes need to work collaboratively.

Therefore, in such a situation, communication among these fog nodes is essential. As the fog nodes are heterogeneous, the communication protocols may also differ from one to another. Moreover, these fog nodes are manufactured by different vendors. Consequently, for proper communication among these fog nodes, interoperability is an essential characteristic of fog computing.

(iii) **Low Latency:** For a time-critical application in a traditional cloud computing architecture, latency affects the quality of service. Therefore, fog computing architecture is introduced to process services at the edge itself to avoid latency in time-critical applications.

(iv) **Mobility:** Typically, in an IoT scenario, different objects, such as, vehicles, are equipped with different sensors; these sensors are used to provide data to an application. Thus, the migration of service from one fog node to another, corresponding to the change in position of the object, is essential. Fog computing supports mobility for seamless execution of processes.

11.1.2 Fog nodes

The key enabler of the fog computing architecture is the fog node itself. It can be heterogeneous or homogeneous, typically, depending on the application scenario. Some examples of fog nodes are routers, gateways, switches, or any processing nodes. These fog nodes are connected among themselves and follow a distributed computing architecture. A fog node must have the following characteristics:

(i) Autonomous decision making: Fog computing paradigm follows the distributed computing architecture. In the proximity of a fog node, multiple static and mobile devices may appear. Moreover, the devices transmit the data to the nearest fog node or the most capable fog node, based on certain pre-defined optimization algorithms. Within the fog architecture, fog nodes work collaboratively to provide a service. Consequently, for service migration from one fog node to another, a fog node must select/switch to another appropriate fog node intelligently and autonomously.

(ii) Programmability: An essential characteristic of a fog node is programmability. In fog computing architecture, physical devices transmit the data directly to a fog node for providing a service. Moreover, a fog node must process the data immediately, after receiving it from any physical device. Thus, the fog nodes must be programmable so that a pre-defined program can be installed and executed inside the fog device as per requirement.

(iii) Heterogeneity: Typically, heterogeneity is introduced as different vendors manufacture the various constituents of fog devices. The devices at the fog layer may be of different types and include components such as routers, switches, gateways, and other processing devices. Based on the accessibility and requirements, these devices are used for provisioning different services.

(iv) Network-enabled: For a particular application, a set of fog nodes work together. These fog nodes may be located at different physical locations. For establishing communication among these fog nodes, the network becomes an essential component. Therefore, fog devices need to be network-enabled.

11.1.3 Fog node deployment model

As per NIST (National Institute of Standards and Technology), the deployment model of fog nodes are categorized into four parts [2], which are similar to traditional cloud computing.

(i) Private fog node: These fog nodes are allocated dedicatedly for a single user organization as shown in Figure 11.2(a). For example, a hospital uses a private fog node for analyzing the physiological data of the patients. These fog nodes are managed either by the organization itself or by a third party.

(ii) Community fog node: These fog nodes are used for a set of functionally similar organizations or a community. As an example, a set of hospitals use fog nodes for analyzing the physiological data of the patients. In this case, only those hospitals that are associated with the deployed fog nodes can access the services. The community fog nodes are owned and maintained by single or multiple organizations. The use of community fog nodes is shown in Figure 11.2(b).

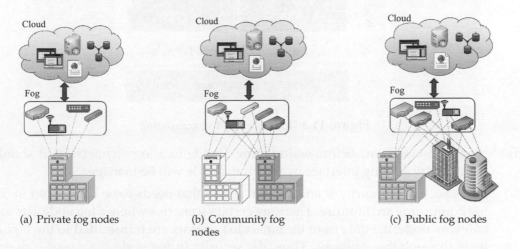

(a) Private fog nodes   (b) Community fog nodes   (c) Public fog nodes

Figure 11.2 Difference between cloud and fog computing

(iii) Public fog node: The fog nodes that are accessible and used by multiple functionally different organizations are known as public fog nodes. For example, a fog node can be used by a hospital at a time instant, and at a later instant of time, the same fog node can be used by a road transportation organization. A public fog node is owned and managed by multiple organizations. The use of public fog nodes is depicted in Figure 11.2(c).

(iv) Hybrid fog node: A hybrid fog node consists of single or multiple public, community, or public fog nodes, which are logically combined to provide services to multiple organizations.

11.2 View of a Fog Computing Architecture

In this section, we will discuss the perspective of *OpenFog Consortium Architecture Working Group* [1] on fog computing architecture. The view of fog architecture is of three folds: (i) Node view, (ii) system view, and (iii) software view.

11.2.1 Node view

This view is the bottom-most in the architecture of fog computing. In a fog computing architecture, certain aspects need to be considered as shown in Figure 11.3.

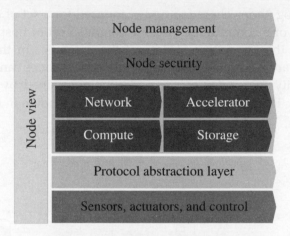

Figure 11.3 Node view in fog computing

(i) Node management: Before assimilating a node in a fog architecture, it should support the existing interface by which the node will be managed.

(ii) Node security: Security is an essential aspect that needs to be considered in the fog computing architecture. There are certain scenarios where a fog node acts as a gateway node; the data from the smart IoT sensors are transmitted to the higher level through this gateway. Thus, the security in fog node is a major concern, which needs to be addressed.

(iii) Network: Fog computing is used for reducing the service delay of an application. Thus, a fog node must communicate with other nodes and smart IoT devices efficiently with the least possible delay. This will also enable the fog computing network to support time-sensitive networking (TSN).

(iv) Accelerators: In order to increase the speed of service accessibility, fog applications use accelerators.

(v) Computation: The fog nodes must have the computational capabilities for taking instant and quick decisions for certain applications.

(vi) Storage: The fog node should have adequate storage facilities associated with it. Additionally, this storage must be reliable and meet the performance requirements of the applications.

(vii) Protocol abstraction layer: Traditionally, available smart IoT devices may not be able to communicate with the fog devices directly. However, a protocol abstraction layer logically connects smart IoT devices with fog components at the fog layer.

(viii) Sensors, Actuators, and Control: These smart devices lie at the bottom of the fog architecture. One or multiple fog nodes can connect to multiple smart IoT devices. Typically, these devices have less processing capabilities and limited connectivity capabilities, such as, Wi-Fi, Zigbee, and Ethernet.

11.2.2 System view

The system view consists of two hardware-level views that are indispensable for a fog computing architecture. A diagram of the system view is depicted in Figure 11.4.

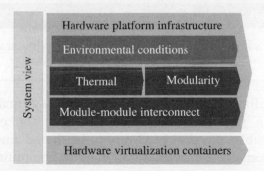

Figure 11.4 System view of fog computing

(i) Hardware Platform Infrastructure: Sometimes, a fog computing architecture deployment is required in an application area with harsh environments, such as, those with high temperatures and heavy rainfall. Moreover, the fog nodes may also be deployed over an area where the protection of these fog devices from physical attack and theft is the utmost requirement. The infrastructure view discusses the essential requirements of the hardware platform infrastructure. Additionally, this view sheds some light on the mechanical support needed by the internal components in fog architecture. In the hardware platform infrastructure view, the following conditions are required to be considered:

- Environmental conditions: Internationally, there are different environmental safety and responsibility standards to which a fog architecture must comply.

Many applications, such as industrial, military, and commercial, have a higher range of operating temperatures. With the increasing range of temperature, the fog nodes may fail to work properly. Thus, it must be ensured that fog computing architectures operate well under such harsh operating conditions. Similarly, the fog nodes must work under any environmental condition.

- Deployment: Sometimes, a fog node may be required for deployment over certain areas with high thermal effects. In such high-temperature conditions, the fog nodes should not require any additional cooling systems, such as a fan. Thus, during the deployment of a fog-based solution, the issue of thermal effects and other such possibilities must also be taken into account.

- Modularity: The fog architecture should be modular in such a way that the other required components can be, robustly and easily, configured with it.

- Module–Module Interconnection: The interconnectivity among different modules is necessary for a fog computing architecture. These interconnecting medium between any two modules can be wired. This connecting medium is also referred to as its fabric.

Check yourself
International safety and environmental responsibility standards: UL, CSA, ROHS

(ii) Hardware Virtualization and Containers: Hardware virtualization is an essential component of a fog computing architecture. The concept of hardware virtualization enables physical hardware to be shared among multiple entities. The concept of containers enables isolation in fog computing architecture.

11.2.3 Software view

This view consists of four layers: node management, software backplane, application service, and application support. A diagram of the software view is depicted in Figure 11.5.

(i) Node management: This layer is associated with the management of nodes and the connections among them. The software and hardware are required to be managed properly so that they can provide the expected services. The operating system and support configuration are managed with certain software. For monitoring different components of a fog architecture, operational management, such as generating alarm, is crucial. Node management also includes security related issues, such as key and identity management. Availability management is an essential component of node management. Upon failure of a working

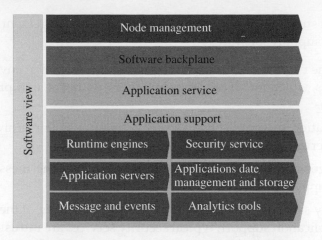

Figure 11.5 Software view of fog computing

piece of hardware, alternate hardware needs to be arranged. Similarly, upon encountering major disruptions in software, the VMs must be regenerated.

(ii)   Software backplane: This layer coordinates the communications among all the levels, such as between smart IoT devices to fog architecture, one fog architecture to another, and between fog and cloud architecture. All the security-related aspects (such as confidentiality, integrity, and non-repudiation) should be taken care of by the software backplane. Sometimes, multiple fog architectures are required for serving certain applications. In such a situation, service discovery is essential to work cooperatively within the fog architecture. A new fog node should broadcast its presence to the existing fog architecture so that the new fog node can participate in serving an application. Therefore, node discovery is a vital component of the software backplane.

(iii) Application services: The application services include fog connector service, core service, supporting service, analytics service, integration service, and user interface service. In the application services layer, the fog connector service works above the protocol abstraction layer. The key role of a fog connector is to convert the data received from smart devices into a common data structure and to transmit those to the core services. Further, core services provide suitable data to the services. The responsibilities of the supporting layer are data logging, scheduling, and service registration. In fog computing, reactive and predictive capabilities are essential features. The analytics service is responsible for performing both reactive and predictive operations close to the edge. The intergeneration services enable a fog node to retrieve data from another fog node with the correct format, place, and time. The user interface (UI) is mostly useful for displaying the collected and merged data at the fog node. Additionally, it displays the results, analytics, status, and operation in a fog node.

(iv) Application support: The main role of application management is to manage the application support software.

- Runtime engines: These engines provide the execution environment for the services, which includes VMs, containers, platform, and program language libraries.

- Application servers: These servers are responsible mainly for hosting the microservices.

- Messages and events: These primarily support the messages and event-based applications.

- Security service: This service looks after the security of the applications. In the application support, typical security services are intrusion detection and prevention, and packet inspection.

- Application data management and storage: Management and storage of data are important aspects of fog computing. This layer handles the data management and storage issues, including database and cache.

- Analytics tools and frameworks: These layers take care of different analytical tools such as *Hadoop*.

11.3   Fog Computing in IoT

Typically, an IoT infrastructure consists of a large number of smart devices, which produce a considerable volume and variety of data. The analysis of data is essential for serving an application successfully. Thus, in such a situation, the role of fog computing is crucial.

11.3.1   Importance of fog computing

IoT devices can be distributed across a large area for monitoring different environmental parameters. Traditionally, data generated from the smart IoT devices are analyzed in the cloud. During the transmission of data from the smart IoT devices to the cloud, an additional delay is incurred in serving an application. In such a scenario, the use of fog computing is pertinent. In a fog computing architecture, a large geographical application area can be divided logically into multiple sub-areas. A set of fog devices are responsible for serving these sub-areas. Therefore, in fog computing architecture, the data from the smart IoT devices are analyzed and processed in one of the fog devices in a particular sub-area, which may also be restricted geographically. As in this architecture, the data is analyzed near the edge, the latency of data transmission reduces as compared to the traditional cloud computing architecture.

11.3.2 Time sensitiveness in fog computing

In a fog infrastructure, based on the time sensitiveness, data are categorized into three types (Figure 11.6): (i) Extremely time-sensitive data, (ii) Moderately time-sensitive data, and (iii) Non-time-sensitive data.

(i) Extremely time-sensitive data: These data are required to be analyzed within a fraction of a second, and the decision has to be made immediately in an application. The nearest best possible fog nodes handle these data. However, after analyzing the data, the summary is transferred to the cloud.

(ii) Moderately time-sensitive data: These data are comparatively less sensitive than extremely time-sensitive data; they may need to be analyzed within a minute. The aggregate fog nodes take the responsibilities of analyzing the moderately time-sensitive data through the nearest fog node of the smart IoT device. Moreover, a fog aggregate node takes the decision and acts on the smart IoT devices. The summary of the data analysis, decisions, and actions are sent to the cloud for future reference.

(iii) Non-time-sensitive data: These data are not time-sensitive, and can wait for analysis for hours, weeks, and months. The non-time-sensitive data are analyzed in the cloud itself.

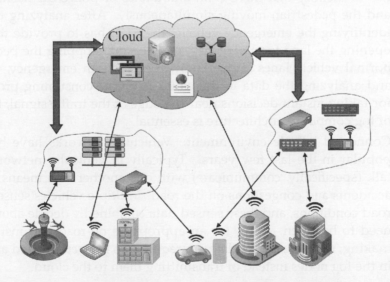

Figure 11.6 Time sensitiveness in data

11.4 Selected Applications of Fog Computing

We have already learned that fog computing is essential and useful for time-sensitive applications. There are numerous application areas of fog computing. However, in this section, we will discuss a few selected applications of fog computing in real-life scenarios.

(i)   Smart road transportation system (SRTS): Typically, in an SRTS, different sensors are attached to different actors of a transport system, such as vehicles, signals, and roads. These sensors sense the value of different parameters that are related to the transportation system. However, to build an SRTS, stand-alone sensors are not enough. The requirement of quickly analyzing and making an instant decision are essential components. In such a scenario, fog computing plays a crucial role. SRTS can be divided into two sub-application areas: (a) smart traffic signal system and (b) connected vehicle environment.

   (a)   Smart traffic signal system (STSS): These systems are typically used to handle on-road running traffic, the running traffic and avoid accidents. The use of fog computing in STSS is critical in order to fulfill its intention. For example, let there be a camera-enabled STSS on the junction of four roads. In the road, there is an emergency vehicle lane, pedestrian lane, and normal vehicle lane. Suddenly the camera on the STSS captures the flashing light of an emergency vehicle. However, on the same road, the lane is already open for crossing the road by a pedestrian. In such an emergency, the STSS has to identify that there is an ambulance or police car flashing the light and the pedestrian moving simultaneously. After analyzing the data and identifying the emergency vehicle, the STSS has to provide the signal for opening the lane for emergency vehicles and stopping the pedestrian and normal vehicle lanes immediately. For such an emergency, transmitting and analyzing the data in the cloud is a time-consuming process. Thus, for taking instant decisions near the edge of the traffic signal, the presence of fog computing architecture is essential.

   (b)   Connected vehicle environment: Vehicular networks have become very popular in the last few years. Typically in vehicular networks, vehicles talk (specifically, communicate) with one another as a means of avoiding accidents and congestions on the road. Multiple vehicles sense the current road conditions, and these sensed data combinedly decide about the action need to be taken. Thus, for an appropriate on-road and instant decision making, multiple sensor data are required to be accumulated and analyzed in the fog nodes instead of transmitting them to the cloud.

Check yourself

Fog Computing in Industrial Internet of Things and Industry 4.0 [4]

(i)   Healthcare: The traditional cloud computing architecture consumes a significant amount of time for analyzing a huge volume and variety of healthcare data. The emergence of fog computing in the healthcare industry results in real-time analysis and decision making of healthcare data. Recently, Tata Consultancy Services (TCS) developed a fog computing platform for remote patient monitoring. In the TCS fog platform, the healthcare data is analyzed in a decentralized fashion [3]. This fog platform is accessible at any time from anywhere. The platform provides an alert to the user when a particular physiological parameter crosses a pre-defined threshold/value. The TCS fog platform provides PaaS to its users.

(iii) Mining industries: Currently, several expensive instruments, such as autonomous vehicles and autonomous drilling systems, are used in mining industries. Further, various risks, such as emission of poisonous gases, sliding, and water logging, are some of the dangers that are typically associated with the mining industry [4]. Therefore, different sensors and actuators are deployed in the mines to sense the surroundings and collect data. These data need to be rapidly analyzed before taking any actual action, such as drilling. For this data analysis and instantaneous decision-making action, fog computing plays a crucial role.

(iv) Product advertisement: Currently, fog computing is also used for advertising certain products. A very recent literature [4] highlights the applicability of fog computing for product advertisement. As an example, consider a person who is a regular customer of a supermarket. This customer, being a regular, purchases groceries periodically. When this customer enters the supermarket, a camera associated with the fog computing architecture captures his/her image to identify the customer. With the help of previous data, which are already stored, and from previous records, the fog computing architecture can predict the customer's purchases. Based on that, the system can send notifications about newly launched products and freshly stocked grocery to the customer's smartphone.

Summary

This chapter explored the basic concepts of fog computing, which would help a new learner to understand fog architecture. The most important enabler, fog nodes, of the fog computing is discussed along with its deployment model. Additionally, different views of fog computing and their different components are elaborately explored. A few real implementations of fog computing architecture are also discussed in this chapter.

Exercises

(i) How is fog computing different from cloud computing?

(ii) List the characteristics of a fog node.

(iii) How can fog computing be used in a smart city?

(iv) What is the role of the "protocol abstraction layer" of a fog node?

(v) What are roles of software backplane in the software view of a fog computing architecture?

(vi) What do you mean by time sensitiveness in fog computing?

(vii) How is the fog computing architecture useful in a hospital scenario/environment?

(viii) Why is autonomous decision making important in fog computing?

(ix) What do you mean by community fog node?

References

[1] OpenFog Consortium Architecture Working Group, OpenFog Reference Architecture for Fog Computing, February 2017. https://www.iiconsortium.org.

[2] Iorga, Michaela, Larry Feldman, Robert Barton, Michael J. Martin, Nedim Goren, Charif Mahmoudi. 2017. "The NIST Definition of Fog Computing." National Institute of Standards and Technology (NIST), U. S. Department of Commerce.

[3] Fog Computing Platform for Remote Patient Monitoring. https://www.tcs.com/content/dam/tcs/pdf/Industries/life-sciences-and-healthcare/solution-brochure/Fog%20Computing%20Platform%20for%20Remote%20Patient%20Monitoring.pdf.

[4] Aazam, M. and S. Zeadally and K. A. Harras. 2018. "Deploying Fog Computing in Industrial Internet of Things and Industry 4.0." *IEEE Transactions on Industrial Informatics* 14: 4674–4682.

PART FOUR
IOT CASE STUDIES AND FUTURE TRENDS

Chapter 12

Agricultural IoT

After reading this chapter, the reader will be able to:

- Relate to the applicability of IoT in real scenarios
- List the salient features of agricultural IoT
- Understand the requirements, challenges, and advantages in implementing IoT in agriculture
- Relate to the appropriate use of various IoT technologies through real-life use cases on IoT-based leaf area index assessment and an IoT-based irrigation system

12.1 Introduction

Currently, IoT-enabled technologies are widely used for increasing crop productivity, generating significant revenue, and efficient farming. The development of the IoT paradigm helps in precision farming. Agricultural IoT systems perform crop health monitoring, water management, crop security, farming vehicle tracking, automatic seeding, and automatic pesticide spraying over the agricultural fields. In an IoT-based agricultural system, different sensors necessarily have to be deployed over agricultural fields, and the sensed data from these sensors need to be transmitted to a centralized entity such as a server, cloud, or fog devices. Further, these data have to be processed and analyzed to provide various agricultural services. Finally, a user should be able to access these services from handheld devices or computers. Figure 12.1 depicts a basic architecture of an agricultural IoT.

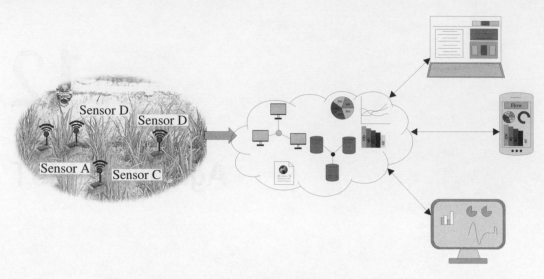

Figure 12.1 Architecture of agricultural IoT

12.1.1   Components of an agricultural IoT

The development of an agricultural IoT has helped farmers enhance crop productivity and reduce the overhead of manual operations of the agricultural equipment in the fields. Different components such as analytics, drone, cloud computing, sensors, hand-held devices, and wireless connectivity enable agricultural IoT as depicted in Figure 12.2.

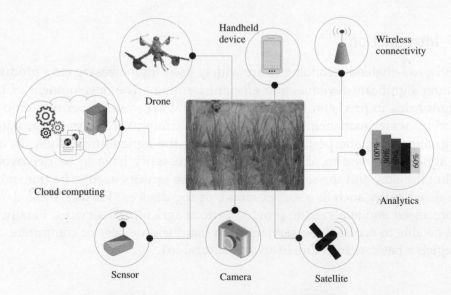

Figure 12.2 Components of agricultural IoT

The different components of an agricultural IoT are discussed as follows:

- Cloud computing: Sensors such as the camera, devices to measure soil moisture, soil humidity, and soil pH-level are used for serving different agricultural applications. These sensors produce a huge amount of agricultural data that need to be analyzed. Sometimes, based on the data analysis, action needs to be taken, such as switching on the water pump for irrigation. Further, the data from the deployed sensors are required to be stored on a long-term basis since it may be useful for serving future applications. Thus, for agricultural data analysis and storage, the cloud plays a crucial role.

- Sensors: In previous chapters, we already explored different types of sensors and their respective requirements in IoT applications. We have seen that the sensors are the major backbone of any IoT application. Similarly, for agricultural IoT applications, the sensors are an indispensable component. A few of the common sensors used in agriculture are sensors for soil moisture, humidity, water level, and temperature.

- Cameras: Imaging is one of the main components of agriculture. Therefore, multispectral, thermal, and RGB cameras are commonly used for scientific agricultural IoT. These cameras are used for estimating the nitrogen status, thermal stress, water stress, and crop damage due to inundation, as well as infestation. Video cameras are used for crop security.

- Satellites: In modern precision agriculture, satellites are extensively used to extract information from field imagery. The satellite images are used in agricultural applications to monitor different aspects of the crops such as crop health monitoring and dry zone assessing over a large area.

- Analytics: Analytics contribute to modern agriculture massively. Currently, with the help of analytics, farmers can take different agricultural decisions, such as estimating the required amount of fertilizer and water in an agricultural field and estimating the type of crops that need to be cultivated during the upcoming season. Moreover, analytics is not only responsible for making decisions locally; it is used to analyze data for the entire agricultural supply chain. Data analytics can also be used for estimating the crop demand in the market.

- Wireless connectivity: One of the main components of agricultural IoT is wireless connectivity. Wireless connectivity enables the transmission of the agricultural sensor data from the field to the cloud/server. It also enables farmers to access various application services over handheld devices, which rely on wireless connectivity for communicating with the cloud/server.

- Handheld devices: Over the last few years, e-agriculture has become very popular. One of the fundamental components of e-agriculture is a handheld device such as a smartphone. Farmers can access different agricultural information, such as soil and crop conditions of their fields and market tendency, over their smartphones. Additionally, farmers can also control different field equipment, such as pumps, from their phones.

- **Drones:** Currently, the use of drones has become very attractive in different applications such as surveillance, healthcare, product delivery, photography, and agriculture. Drone imaging is an alternative to satellite imaging in agriculture. In continuation to providing better resolution land mapping visuals, drones are used in agriculture for crop monitoring, pesticide spraying, and irrigation.

An agricultural food chain (agri-chain) represents the different stages that are involved in agricultural activity right from the agricultural fields to the consumers. Figure 12.3 depicts a typical agricultural food chain with the different operations that are involved in it. Additionally, the figure depicts the applications of different IoT components required for performing these agricultural operations. In the agri-chain, we consider *farming* as the first stage. In farming, various operations, such as seeding, irrigation, fertilizer spreading, and pesticide spraying, are involved. For performing these operations, different IoT components are used. As an example, for monitoring the soil health, soil moisture and temperature sensors are used; drones are used for spraying pesticides; and through wireless connectivity, a report on on-field soil conditions is sent directly to a users' handheld device or cloud. After farming, the next stage in the agri-chain is *transport*. Transport indicates the transfer of crops from the field to the local storage, and after that, to long-term storage locations. In transport, smart vehicles can automatically load and unload crops. The global

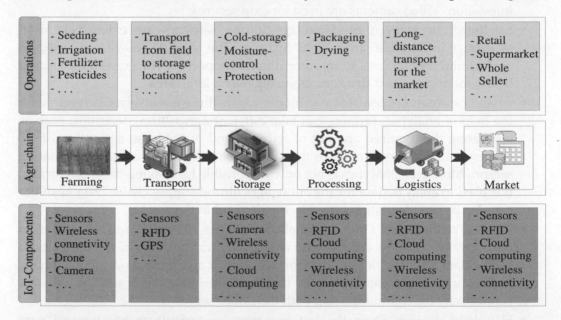

Figure 12.3 Use of IoT components in the agricultural chain

positioning system (GPS) plays an important role by tracking these smart devices, and radio frequency identification (RFID) is used to collect information regarding the presence of a particular container of a crop at a warehouse. Storage is one of the important operations in the agri-chain. It is responsible for storing crops on a long-term basis. Typically, cold storage is used for preserving the crops for a long time and providing them with the necessary climatic and storage conditions and protection. In the storage, cameras are used to keep a check and protect the harvested crops. The camera feeds are transferred through wireless connectivity to a remote server or a cloud infrastructure. Moreover, the amount and type of crops stored in a storage location are tracked and recorded with the help of sensors and cloud computing. For pushing the crops into the market, processing plays a crucial role in an agri-chain. Processing includes proper drying and packaging of crops. For drying and packaging, different sensors are used. Packaging is the immediate operation prior to pushing the crop into the market. Thus, it is essential to track every package and store all the details related to the crops in the cloud. Logistics enables the transfer of the packed crops to the market with the help of smart vehicles. These smart vehicles are equipped with different sensors that help in loading and unloading the packed crop autonomously. Additionally, GPS is used in these smart vehicles for locating the position of the packed crops at any instant and tracking their whereabouts. All the logistical information gets logged in the cloud with the help of wireless connectivity. Finally, the packed items reach the market using logistical channels. From the market, these items are accessible to consumers. The details of the sale and purchase of the items are stored in the form of records in the cloud.

12.1.2 Advantages of IoT in agriculture

Modern technological advancements and the rapid developments in IoT components have gradually increased agricultural productivity. Agricultural IoT enables the autonomous execution of different agricultural operations. The specific *advantages* of the agricultural IoT are as follows:

(i) Automatic seeding: IoT-based agricultural systems are capable of autonomous seeding and planting over the agricultural fields. These systems significantly reduce manual effort, error probability, and delays in seeding and planting.

(ii) Efficient fertilizer and pesticide distribution: Agricultural IoT has been used to develop solutions that are capable of applying and controlling the amount of fertilizers and pesticides efficiently. These solutions are based on the analysis of crop health.

(iii) Water management: The excess distribution of water in the agricultural fields may affect the growth of crops. On the other hand, the availability of global water resources is finite. The constraint of limited and often scarce usable water resources is an influential driving factor for the judicious and efficient distribution of agricultural water resources. Using the various solutions available for agricultural IoT, water can be distributed efficiently, all the while, increasing field productivity and yields. The IoT-enabled agricultural systems are capable of monitoring the water level and moisture in the soil, and accordingly, distribute the water to the agricultural fields.

(iv) Real-time and remote monitoring: Unlike traditional agriculture, in IoT-based farming, a stakeholder can remotely monitor different agricultural parameters, such as crop and soil conditions, plant health, and weather conditions. Moreover, using a smart handheld device (e.g., cellphone), a farmer can actuate on-field farming machinery such as a water pump, valves, and other pieces of machinery.

(v) Easy yield estimation: Agricultural IoT solutions can be used to record and aggregate data, which may be spatially or temporally diverse, over long periods. These records can be used to come up with various estimates related to farming and farm management. The most prominent among these estimates is crop yield, which is done based on established crop models and historical trends.

(vi) Production overview: The detailed analysis of crop production, market rates, and market demand are essential factors for a farmer to estimate optimized crop yields and decide upon the essential steps for future cropping practices. Unlike traditional practices, IoT-based agriculture acts as a force multiplier for farmers by enabling them to have a stronger hold on their farming as well as crop management practices, and that too mostly autonomously. Agricultural IoT provides a detailed product overview on the farmers' handheld devices.

12.2 Case Studies

In this section, we discuss a few case studies that will provide an overview of real implementation of IoT infrastructure for agriculture.

12.2.1 In-situ assessment of leaf area index using IoT-based agricultural system

In this case study, we focus on an IoT-based agricultural system developed by Bauer et al. [1]. The authors focus on the in-situ assessment of the leaf area index (LAI), which is considered as an essential parameter for the growth of most crops. LAI is a dimensionless quantity which indicates the total leaf area per unit ground area. For determining the canopy (the portion of the plant, which is above the ground) light, LAI plays an essential role.

Architecture
The authors integrated the hardware and software components of their implementation in order to develop the IoT-based agricultural system for LAI assessment. One of the important components in this system is the wireless sensor network (WSN), which is used as the LAI assessment unit. The authors used two types of sensors: (i) ground-level sensor (G) and (ii) reference sensor (R). These sensors are used to measure photosynthetically active radiation (PAR). The distance between the two types of sensors must be optimal so that these are not located very far from one another. In this system, the above-ground sensor (R) acts as a cluster head while the other sensor nodes (Gs) are located below the canopy. These Gs and R connect and

form a star topology. A solar panel is used to charge the cluster head. The system is based on IoT architecture. Therefore, a cluster head is attached to a central base station, which acts as a gateway. Further, this gateway connects to an IoT infrastructure. The architecture of the system is depicted in Figure 12.4.

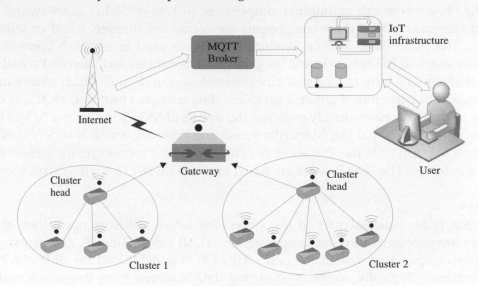

Figure 12.4 System architecture

Hardware

For sensing and transmitting the data from the deployment fields to a centralized unit, such as a server and a cloud, different hardware components are used in the system. The commercial off-the-shelf (COTS) TelosB platform is used in the system. The TelosB motes are equipped with three types of sensors: temperature, humidity, and light sensors. With the help of an optical filter and diffuser accessory on the light sensors, the PAR is calculated to estimate the LAI. The system is based on the cluster concept. A Raspberry-Pi is used as a cluster head, which connects with four ground sensor motes. The Raspberry-Pi is a tiny single board, which works as a computer and is used to perform different operations in IoT. Humidity and wet plants intermittently cause attenuation to the system, which is minimized with the help of forward error coding (FEC) technique.

The real deployment of the LAI assessment system involves various environmental and wild-life challenges. Therefore, for reliable data delivery, the authors take the redundant approach of using both wired and wireless connectivity. In the first deployment generation, USB power supply is used to power-up the sensors motes. Additionally, the USB is used for configuring the sensor board and accessing the failure as per requirement. In this setup, a mechanical timer is used to switch off the sensor nodes during the night. In the second deployment generation, the cluster is formed with wireless connectivity. The ground sensor motes consist of external antennas, which help to communicate with the cluster head. A Raspberry-Pi with long-term evolution (LTE) is used as a gateway in this system.

Communication

The LAI system consists of multiple components, such as WSN, IoT gateway, and IoT-based network. All of these components are connected through wired or wireless links. The public land mobile network (PLMN) is used to establish connectivity between external IoT networks and the gateway. The data are analyzed and visualized with the help of a farm management information system (FMIS), which resides in the IoT-based infrastructure. Further, a prevalent data transport protocol: MQTT, is used in the system. We have already explored the details of MQTT in Chapter 8. MQTT is a very light-weight, publish/subscribe messaging protocol, which is widely used for different IoT applications. The wireless LAN is used for connecting the cluster head with a gateway. The TelosB motes are based on the IEEE 802.15.4 wireless protocol.

Software

Software is an essential part of the system by which different operations of the system are executed. In order to operate the TelosB motes, TinyOS, an open-source, low-power operating system, is used. This OS is widely used for different WSN applications. Typically, in this system, the data acquired from the sensor node is stored with a timestamp and sequence number (SN). For wired deployments (the first generation deployment), the sampling rate used is 30 samples/hour. However, in the wireless deployment (the second generation), the sampling rate is significantly reduced to 6 samples/hour. The TinyOS is capable of activating low-power listening modes of a mote, which is used for switching a mote into low-power mode during its idle state. In the ground sensor, TelosB motes broadcast the data frame, and the cluster head (Raspberry-Pi) receives it. This received data is transmitted to the gateway. Besides acquiring ground sensor data, the Raspberry-Pi works as a cluster head. In this system, the cluster head can re-boot any affected ground sensor node automatically.

IoT Architecture

The MQTT broker runs in the Internet server of the system. This broker is responsible for receiving the data from the WSN. In the system, the graphical user interface (GUI) is built using an Apache server. The visualization of the data is performed at the server itself. Further, when a sensor fails, the server informs the users. The server can provide different system-related information to the smartphone of the registered user.

12.2.2 Smart irrigation management system

In precision agriculture, the regular monitoring of different agricultural parameters, such as water level, soil moisture, fertilizers, and soil temperature are essential. Moreover, for monitoring these agricultural parameters, a farmer needs to go to his/her field and collect the data. Excess water supply in the agricultural field can damage the crops. On the other hand, insufficient water supply in the agricultural field also affects the healthy growth of crops. Thus, efficient and optimized water supply in the agricultural field is essential.

This case study highlights a prototype of an irrigation management system [2], developed at the Indian Institute of Technology Kharagpur, funded by the Government of India. The primary objective of this system is to provide a Web-based platform to the farmer for managing the water supply of an irrigated agricultural field. The system is capable of providing a farmer-friendly interface by which the field condition can be monitored. With the help of this system, a farmer can take the necessary decision for the agricultural field based on the analysis of the data. However, the farmer need not worry about the complex background architecture of the system. It is an affordable solution for the farmers to access the agricultural field data easily and remotely.

Architecture

The architecture of this system consists of three layers: Sensing and actuating layer, remote processing and service layer, and application layer. These layers perform dedicated tasks depending on the requirements of the system. Figure 12.5 depicts the architecture of the system. The detailed functionalities of different layers of this system are as follows:

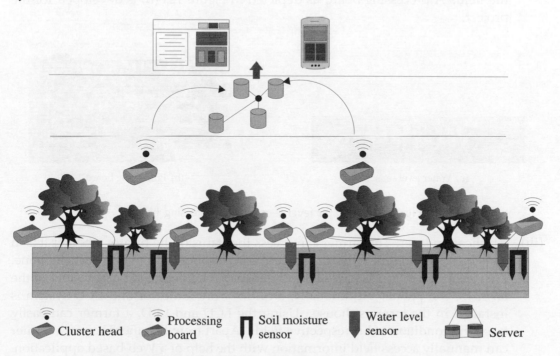

Figure 12.5 Architecture: Smart irrigation management system

(i)   Sensing and Actuating layer: This layer deals with different physical devices, such as sensor nodes, actuators, and communication modules. In the system, a specially designated sensor node works as a cluster head to collect data from other sensor nodes, which are deployed on the field for sensing the value of soil moisture and water level. A cluster head is equipped with two communication module: *ZigBee* (IEEE 802.15.4) and *General Packet Radio Service* (*GPRS*). The communication between the deployed sensor nodes and the cluster head takes place with the help of ZigBee. Further, the cluster heads use GPRS to transmit data to the remote server. An *electrically erasable programmable read-only memory* (*EEPROM*), integrated with the cluster head, stores a predefined threshold value of water levels and soil moisture. When the sensed value of the deployed sensor node drops below this predefined threshold value, a solenoid (pump) activates to start the irrigation process. In the system, the standard EC-05 soil moisture sensor is used along with the water level sensor, which is specifically designed and developed for this project. A water level sensor is shown in Figure 12.6(a).

(ii)   Processing and Service layer: This layer acts as an intermediate layer between the sensing and actuating layer and the application layer. The sensed and process data is stored in the server for future use. Moreover, these data are accessible at any time from any remote location by authorized users. Depending on the sensed values from the deployed sensor nodes, the pump actuates to irrigate the field. A processing board as depicted in Figure 12.6(b) is developed for the project.

(a) Water level sensor   (b) Processing board

Figure 12.6 Water level sensor and processing board

(iii) Application layer: The farmer can access the status of the pump, whether it is in switch on/off, and the value of different soil parameters from his/her cell phone. This information is accessible with the help of the integrated GSM facility of the farmers' cell phone. Additionally, an LED array indicator and LCD system is installed in the farmers' house. Using the LCD and LED, a farmer can easily track the condition of his respective fields. Apart from this mechanism, a farmer can manually access field information with the help of a Web-based application. Moreover, the farmer can control the pump using his/her cell phone from a remote location.

Deployment

The system has been deployed and experimented in two agricultural fields: (i) an agricultural field at the Indian Institute of Technology Kharagpur (IIT Kharagpur), India, and (ii) Benapur, a village near IIT Kharagpur, India. Both the agricultural fields were divided into 10 equal sub-fields of $3 \times 3m^2$. In order to examine the performance, the system was deployed at over 4 sub-fields. Each of these sub-fields consists of a solenoid valve, a water level sensor, and a soil moisture sensor, along with a processing board. On the other hand, the remaining six sub-fields were irrigated through a manual conventional irrigation process. The comparison analysis between these six and four fields summarily reports that the designed system's performance is superior to the conventional manual process of irrigation.

Summary

This chapter explored the applications of IoT in the domain of agriculture. Further, the chapter helps the reader to visualize the importance of IoT in the various links of the agricultural food chain. A case study on a very important aspect of agriculture, leaf area index assessment, was explored in this chapter. This case study gives a detailed idea about the system along with basic knowledge of the hardware used in it. Another real deployed system of irrigation management is discussed in this chapter. This case study is beneficial for the learner to understand the importance of IoT architecture in the irrigation process of agriculture.

Exercises

(i) List the type of sensors which can be used for agricultural IoT.

(ii) Explain two use cases where drones can be used for agricultural IoT.

(iii) Design a scenario where we can use fog computing in agriculture.

(iv) How can agricultural IoT help in the efficient distribution of water in agricultural fields?

(v) What are the roles of the various IoT components in an agri-chain?

(vi) What are the advantages of agricultural IoT?

(vii) List a few communication modules used for agricultural IoT?

(viii) Design a case study to develop an IoT-based agricultural planter. In the case study, you should include the requirement analysis of different components and justify their usability in the planter.

(ix) What is the importance of satellites in agricultural IoT?

References

[1]   Bauer, J. and N. Aschenbruck. 2018. "Design and Implementation of an Agricultural Monitoring System for Smart Farming." In *Proceedings of IoT Vertical and Topical Summit on Agriculture - Tuscany (IOT Tuscany), May 2018.*

[2]   Roy, Sanku Kumar, Sudip Misra, Narendra Singh Raghuwanshi, and Amitava Roy. 2017. "A Smart Irrigation Management System using WSNs." Indian Patent File No.: 201731031610.

[3]   Roy, S. K., A. Roy, S. Misra, N. S. Raghuwanshi, and M. S. Obaidat. 2015. "AID: A Prototype for Agricultural Intrusion Detection using Wireless Sensor Network." In *Proceedings of the IEEE International Conference on Communications (ICC), London, 2015.* pp. 7059–7064.

Chapter **13**

Vehicular IoT

Learning Outcomes

After reading this chapter, the reader will be able to:

- Relate to the applicability of IoT in real scenarios
- List the salient features of vehicular IoT
- Understand the requirements, challenges, and advantages of implementing IoT in vehicles
- Relate to the appropriate use of various IoT technologies through a real-life case on vehicular IoT

13.1   Introduction

In this chapter, we discuss the application of IoT in connected vehicular systems. The use of connected vehicles is increasing rapidly across the globe. Consequently, the number of on-road accidents and mismanagement of traffic is also increasing. The increasing number of vehicles gives rise to the problem of parking. However, the evolution of IoT helps to form a connected vehicular environment to manage the transportation systems efficiently. Vehicular IoT systems have penetrated different aspects of the transportation ecosystem, including on-road to off-road traffic management, driver safety for heavy to small vehicles, and security in public transportation. In a connected vehicular environment, vehicles are capable of communicating and sharing their information. Moreover, IoT enables a vehicle to sense its internal and external environments to make certain autonomous decisions. With the help of modern-day IoT infrastructure, a vehicle owner residing in Earth's northern hemisphere can very easily track his vehicular asset remotely, even if it is in the southern hemisphere. In this chapter, we discuss the importance and applications of IoT in the vehicular systems. Figure 13.1 represents a simple architecture of a

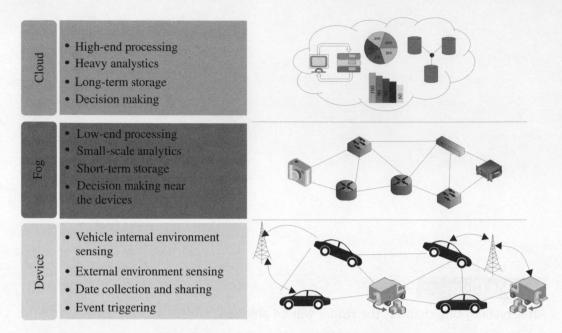

Cloud
- High-end processing
- Heavy analytics
- Long-term storage
- Decision making

Fog
- Low-end processing
- Small-scale analytics
- Short-term storage
- Decision making near the devices

Device
- Vehicle internal environment sensing
- External environment sensing
- Date collection and sharing
- Event triggering

Figure 13.1 Architecture of vehicular IoT

vehicular IoT system. The architecture of the vehicular IoT is divided into three sub-layers: device, fog, and cloud.

- Device: The device layer is the bottom-most layer, which consists of the basic infrastructure of the scenario of the connected vehicle. This layer includes the vehicles and road side units (RSU). These vehicles contain certain sensors which gather the internal information of the vehicles. On the other hand, the RSU works as a local centralized unit that manages the data from the vehicles.

- Fog: In vehicular IoT systems, fast decision making is pertinent to avoid accidents and traffic mismanagement. In such situations, fog computing plays a crucial role by providing decisions in real-time, much near to the devices. Consequently, the fog layer helps to minimize data transmission time in a vehicular IoT system.

- Cloud: Fog computing handles the data processing near the devices to take decisions instantaneously. However, for the processing of huge data, fog computing is not enough. Therefore, in such a situation, cloud computing is used. In a vehicular IoT system, cloud computing helps to handle processes that involve a huge amount of data. Further, for long-term storage, cloud computing is used as a scalable resource in vehicular IoT systems.

13.1.1   Components of vehicular IoT

Modern cars come equipped with different types of sensors and electronic components. These sensors sense the internal environment of the car and transmit the sensed data to a processor. The on-road deployed sensors sense the external environment and transmit the sensed data to the centralized processor. Thereafter, based on requirements, the processor delivers these sensed data to fog or cloud to perform necessary functions. These processes seem to be simple, but practically, several components, along with their challenges, are involved in a vehicular IoT system. Figure 13.2 depicts the components required for vehicular IoT systems.

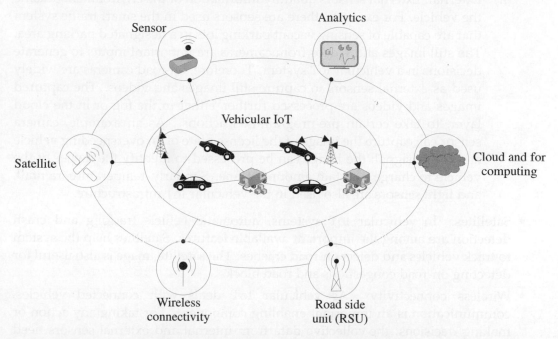

Figure 13.2 Components of vehicular IoT

Check yourself

Advanced driver-assistance systems (ADAS), Intelligent transportation system (ITS), Vehicle-to-vehicle (V2V) communication, Safety-as-a-Service (Safe-aaS) [2]

- Sensors: We have already discussed how sensors play a crucial role in an IoT-based ecosystem. Similarly, in vehicular IoT, sensors monitor different environmental conditions and help to make the system more economical, efficient, and robust. Traditionally, two types of sensors, internal and external, are used in vehicular IoT systems.

(i) Internal: These types of sensors are placed within the vehicle. The sensors are typically used to sense parameters that are directly associated with the vehicle. Along with the sensors, the vehicles are equipped with different electronic components such as processing boards and actuators. The internal sensors in a vehicle are connected with the processor board, to which they transmit the sensed data. Further, the sensed data are processed by the board to take certain predefined actions. A few examples of internal sensors are GPS, fuel gauge, ultrasonic sensors, proximity sensors, accelerometer, pressure sensors, and temperature sensors.

(ii) External: External sensors quantify information of the environment outside the vehicle. For example, there are sensors used in the smart traffic system that are capable of sensing vacant parking lots in a designated parking area. The still images and videos from cameras are important inputs to generate decisions in a vehicular IoT system. Therefore, on-road cameras are widely used as external sensors to capture still images and videos. The captured images and videos are processed further, either in the fog or in the cloud layer, to take certain pre-programmed actions. As an example, camera sensor can capture the image of the license plate of an overspeeding vehicle at a traffic signal; the image can be processed to identify the owner of the vehicle to charge a certain amount of fine. Similarly, temperature, rainfall, and light sensors are also used in the vehicular IoT infrastructure.

- Satellites: In vehicular IoT systems, automatic vehicle tracking and crash detection are among the important available features. Satellites help the system to track vehicles and detect on-road crashes. The satellite image is also useful for detecting on-road congestions and road blocks.

- Wireless connectivity: As vehicular IoT deals with connected vehicles, communication is an important enabling component. For taking any action or making decisions, the collective data from internal and external sensors need processing. For transmitting the sensed data from multiple sensors to RSU (roadside unit) and from RSUs to the cloud, connectivity plays an indispensable role. Moreover, in the vehicular IoT scenario, the high mobility of the vehicles necessitates the connectivity type to be wireless for practical and real-time data transmission. Different communication technologies, such as Wi-Fi, Bluetooth, and GSM, are common in the vehicular IoT systems.

- Road Side Unit (RSU): The RSU is a static entity that works collaboratively with internal and external sensors. Typically, the RSUs are equipped with sensors, communication units, and fog devices. Vehicular IoT systems deal with time-critical applications, which need to take decisions in real time. In such a situation, the fog devices attached to the RSUs process the sensed data and take necessary action promptly. If a vehicular system involves heavy computation, the RSU transmits the sensed data to the cloud end. Sometimes, these RSUs also work as an intermediate communication agent between two vehicles.

- Cloud and fog computing: We have already discussed the importance of fog computing and cloud in the context of IoT applications. In vehicular IoT systems, fog computing handles the light-weight processes geographically closer to the vehicles than the cloud. Consequently, for faster decision making, fog computing is used in vehicular IoT systems. However, for a heavy-weight process, fog computing may not be a suitable option. In such a situation, cloud computing is more adept for vehicular IoT systems. Cloud computing provides more scalability of resources as compared to fog computing. Therefore, the choice of the application of fog and cloud computing depends on the situation. For example, the location and extent of short on-road congestion from a certain location can be determined by fog computing with the help of sensed data. Further, the congestion information can be shared by the RSU among other on-road vehicles, thereby suggesting that they avoid the congested road. On the other hand, for determining regular on-road congestion, predictions are typically handled with the help of cloud computing. For the regular congestion prediction, the cloud end needs to process a huge amount of instantaneous data, as well as, historical data for that stretch of road spanning back a few months to years.

- Analytics: Similar to different IoT application domains, in vehicular IoT, analytics is a crucial component. Vehicular IoT systems can be made to predict different dynamic and static conditions using analytics. For example, strong data analytics is required to predict on-road traffic conditions that may occur at a location after an hour.

Points to ponder

- The sensors attached to the different parts of a vehicle, such as the battery and fuel pump, transmit the data to the cloud for analyzing the requirements for the maintenance of those parts.

- The evolution of IoT enables a user to lock, unlock, locate their cars, even from a remote location.

13.1.2 Advantages of vehicular IoT

The evolution of IoT resulted in the development of a connected vehicular environment. Moreover, the typical advantages of IoT architectures directly impact the domain of connected vehicular systems. Therefore, the advantages of IoT are inherently included in vehicular IoT environments. A few selected advantages of vehicular IoT are depicted in Figure 13.3.

(i) Easy tracking: The tracking of vehicles is an essential part of vehicular IoT. Moreover, the system must know from which location and which vehicle the system is receiving the information. In a vehicular IoT system, the tracking

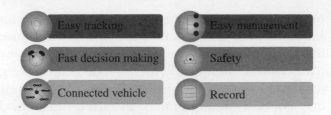

Figure 13.3 Advantages of vehicular IoT

of vehicles is straightforward; the system can collect information at a remote location.

(ii)   Fast decision making: Most of the decisions in the connected vehicle environment are time critical. Therefore, for such an application, fast and active decision making are pertinent for avoiding accidents. In the vehicular IoT environment, cloud and fog computing help to make fast decisions with the data received from the sensor-based devices.

(iii) Connected vehicles: A vehicular IoT system provides an opportunity to remain connected and share information among different vehicles.

(iv)   Easy management: Since vehicular IoT systems consist of different types of sensors, a communication unit, processing devices, and GPS, the management of the vehicle becomes easy. The connectivity among different components in a vehicular IoT enables systems to track every activity in and around the vehicle. Further, the IoT infrastructure helps in managing the huge number of users located at different geographical coordinates.

(v)   Safety: Safety is one of the most important advantages of a vehicular IoT system. With easy management of the system, both the internal and external sensors placed at different locations play an important role in providing safety to the vehicle, its occupants, as well as the people around it.

(vi)   Record: Storing different data related to the transportation system is an essential component of a vehicular IoT. The record may be of any form, such as video footage, still images, and documentation. By taking advantage of cloud and fog computing architecture, the vehicular IoT systems keep all the required records in its database.

13.1.3   Crime assistance in a smart IoT transportation system

In this section, we discuss a case study on smart safety in a vehicular IoT infrastructure. The system highlights a fog framework for intelligent public safety in vehicular environments (fog-FISVER) [1]. The primary aim of this system is to ensure smart transportation safety (STS) in public bus services. The system works through the following three steps:

(i) The vehicle is equipped with a smart surveillance system, which is capable of executing video processing and detecting criminal activity in real time.

(ii) A fog computing architecture works as the mediator between a vehicle and a police vehicle.

(iii) A mobile application is used to report the crime to a nearby police agent.

Architecture

The architecture of the fog-FISVER consists of different IoT components. Moreover, the developers utilized the advantages of the low-latency fog computing architecture for designing their system. Fog-FISVER is based on a three-tiered architecture, as shown in Figure 13.4. We will discuss each of the tiers as follows:

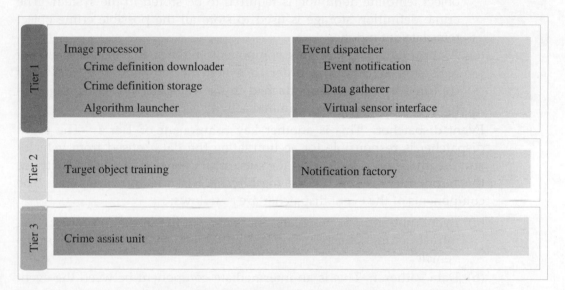

Figure 13.4 Architecture of Fog-FISVER

(i) Tier1—In-vehicle FISVER STS Fog: In this system component, a fog node is placed for detecting criminal activities. This tier accumulates the real sensed data from within the vehicle and processes it to detect possible criminal activities inside the vehicle. Further, this tier is responsible for creating crime-level metadata and transferring the required information to the next tier. For performing all the activities, Tier 1 consists of two subsystems: *Image processor* and *event dispatcher*

- Image Processor: The image processor inside Tier 1 is a potent component, which has a capability similar to the human eye for detecting criminal activities. Developers of the system used a deep-learning-based approach for enabling image processing techniques in the processor. To implement the fog computing architecture in the vehicle, a Raspberry-Pi-3 processor board is used, which is equipped with a high-quality camera. Further, this architecture uses template matching and correlation to detect the presence of dangerous articles (such as a pistol or a knife) in the sub-image of a video frame. Typically, the image processor stores a set of crime object templates in the fog-FISVER STS fog infrastructure, which is present in Tier 2 of the system. The image processor is divided into the following three parts:

(a) Crime definition downloader: This component periodically checks for the presence of new crime object template definitions in fog-FISVER STS fog infrastructure. If a new crime object template is available, it is stored locally.

(b) Crime definition storage: In order to use template matching, the crime object template definition is required to be stored in the system. The crime definition storage is used to store all the possible crime object template definitions.

(c) Algorithm launcher: This component initiates the instances of the registered algorithm in order to match the template with the video captured by the camera attached in the vehicles. If a crime object is matched with the video, criminal activity is confirmed.

- Event dispatcher: This is another key component of Tier 1. The event dispatcher is responsible for accumulating the data sensed from vehicles and the image processor. After the successful detection of criminal activity, the information is sent to the fog-FISVER STS fog infrastructure. The components of the event dispatcher are as follows:

(a) Event notifier: It transfers the data to the fog-FISVER STS fog infrastructure, after receiving it from the attached sensor nodes in the vehicle.

(b) Data gatherer: This is an intermediate component between the event notifier and the physical sensor; it helps to gather sensed data.

(c) Virtual sensor interface: Multiple sensors that sense data from different locations of the vehicle are present in the system. The virtual sensor interface helps to maintain a particular procedure to gather data. This component also cooperates to register the sensors in the system.

(ii) Tier 2—FISVER STS Fog Infrastructure: Tier 2 works on top of the fog architecture. Primarily, this tier has three responsibilities—keep updating the new object template definitions, classifying events, and finding the most suitable police vehicle to notify the event. FISVER STS fog infrastructure is divided into two sub-components:

- Target Object Training: Practically, there are different types of crime objects. The system needs to be up-to-dated regarding all crime objects. This sub-component of Tier 2 is responsible for creating, updating, and storing the crime object definition. The algorithm launcher uses these definitions in

Tier 1 for the template matching process. The template definition includes different features of the crime object such as color gradient and shape format. A new object definition is stored in the *definition database*. The database requires to be updated based on the availability of new template definitions.

- Notification Factory: This sub-component receives notification about the events in a different vehicle with the installed system. Further, this component receives and validates the events. In order to handle multiple events, it maintains a queue.

(iii) Tier 3 consists of mobile applications that are executed on the users' devices. The application helps a user, who witnesses a crime, to notify the police.

Summary

The primary aim of this chapter is to explain the details of vehicular IoT; it also provides a description of its basic architecture. This chapter also highlights the crucial components of a vehicular IoT system, which would help a learner to understand the requirement of these components. Further, the advantages of vehicular IoT is discussed. Finally, a unique case study, fog FISVER STS, is discussed, which would help a learner to visualize the application of IoT in real-world situations and the necessity of vehicular IoT solutions.

Exercises

(i)   What is the role of cloud and fog computing in vehicular IoT?

(ii)   What are the applications of IoT in transportation?

(iii) What are the advantages of vehicular IoT?

(iv)   Give an example of image processing in vehicular IoT.

(v)   What are roadside units (RSUs)?

(vi)   How can data analytics help in a vehicular IoT system?

(vii) What are the uses of a camera sensor in vehicular IoT?

(viii) How can a vehicular IoT system ensure the safety of drivers?

(ix)   Design a use case for developing an IoT-based driver sleep detection system. Please mention all types of sensors required for developing the same.

References

[1]   Neto, A. J. V., Z. Zhao, J. J. P. C. Rodrigues, H. B. Camboim, and T. Braun. 2018. "Fog-Based Crime-Assistance in Smart IoT Transportation System." *Speciical Issue on Cyber-Physical-Social Computing and Networking, IEEE Access* 6: 11101–11111.

[2]   Roy, C., A. Roy, S. Misra, and J. Maiti. 2018. "Safe-aaS: Decision Virtualization for Effecting Safety-as-a-Service." *IEEE Internet of Things Journal* 5(3): 1690–1697.

Healthcare IoT

After reading this chapter, the reader will be able to:

- Relate to the applicability of IoT in real-life scenarios
- List the salient features of healthcare IoT
- Understand and examine the basic implementation aspects of healthcare IoT
- Understand the requirements, challenges, and advantages in implementing IoT in healthcare
- Relate to the appropriate use of various IoT technologies through a real-life use case of healthcare IoT system

14.1   Introduction

Internet of Things (IoT) has resulted in the development and emergence of a variety of technologies that has had a huge impact on the medical field, especially wearable healthcare. The salient features of IoT encourage researchers and industries to develop new IoT-based technologies for healthcare. These technologies have given rise to small, power-efficient, health monitoring and diagnostic systems. Consequently, the development of numerous healthcare technologies and systems has rapidly increased over the last few years. Currently, various IoT-enabled healthcare devices are in wide use around the globe for diagnosing human diseases, monitoring human health conditions, caring/monitoring for elders, children, and even infants. Moreover, IoT-based healthcare systems and services help to increase the quality of life for common human beings; in fact, it has a promising scope of revolutionizing healthcare in developing nations. IoT-based healthcare devices provide access and knowledge about human physiological conditions through hand held devices. With this development, users can be aware of the risks in acquiring various diseases and take necessary precautions to avoid preventable diseases. The basic skeleton of an IoT-based healthcare system is very similar to the conventional IoT architectures. However, for IoT-based healthcare services, the sensors are specifically designed to measure and quantify different physiological conditions of its users/patients. A typical architecture for healthcare IoT is shown in Figure 14.1. We divide the architecture into four layers. The detailed description of these layers are as follows:

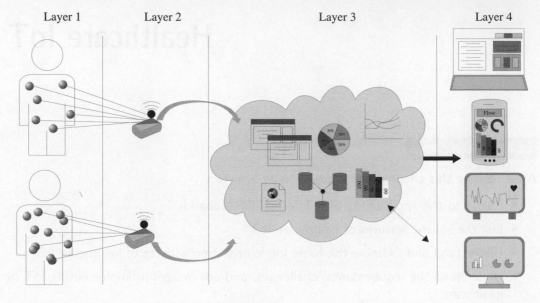

Figure 14.1 Architecture of healthcare IoT

(i)   Layer 1: We have already explained in previous chapters that sensors are one of the key enablers of IoT infrastructure. Layer 1 contains different physiological sensors that are placed on the human body. These sensors collect the values of various physiological parameters. The physiological data are analyzed to extract meaningful information.

(ii)   Layer 2: Layer 1 delivers data to Layer 2 for short-term storage and low-level processing. The devices that belong to Layer 2 are commonly known as local processing units (LPU) or centralized hubs. These units collect the sensed data from the physiological sensors attached to the body and process it based on the architecture's requirement. Further, LPUs or the centralized hubs forward the data to Layer 3.

(iii) Layer 3: This layer receives the data from Layer 2 and performs application-specific high-level analytics. Typically, this layer consists of cloud architecture or high-end servers. The data from multiple patients, which may be from the same or different locations, are accumulated in this layer. Post analysis of data, some inferences or results are provided to the application in Layer 4.

(iv) Layer 4: The end-users directly interact with Layer 4 through receiver-side applications. The modes of accessibility of these services by an end user are typically through cellphones, computers, and tablets.

Check yourself

Internet of Medical Things, Health Level 7 International (HL7), Fast Healthcare Interoperability Resources (FHIR)

14.1.1 Components of healthcare IoT

A typical IoT healthcare architecture is composed of several components that are essential to generate the whole architecture. Figure 14.2 depicts different components and their usage in an IoT healthcare system. Each of these components plays a distinct role in the smooth execution of the system as a whole. In this section, we discuss the different components for a basic healthcare IoT system.

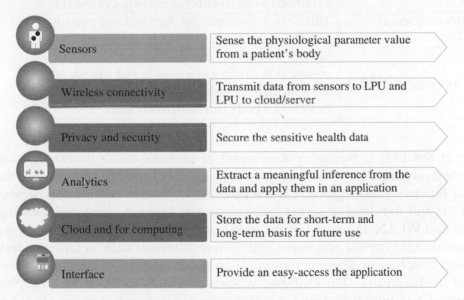

Figure 14.2 Components of healthcare IoT

(i) Sensors: We have already explained that Layer 1 mainly consists of physiological sensors that collect the physiological parameters of the patient. Few commonly used physiological sensors and their uses are depicted in Table 14.1.

(ii) Wireless Connectivity: Without proper connectivity and communication, the data sensed by the physiological sensors are of no use in an IoT-based healthcare system. Typically, the communication between the wearable sensors

Table 14.1 Commonly used healthcare sensors

Sensor	Purpose
Pulse and oxygen in blood (SpO2)	These sensors are used to measure the pulse and oxygen levels in blood.
Airflow	Also known as breathing sensor. An airflow sensor measures the change in respiratory rate.
Temperature	With change in different physiological conditions, the body temperature of a healthy adult also changes. Thus, measuring the body temperature is a routine, yet essential part of medical investigations. The temperature sensor helps to measure the body temperature.
Blood pressure	The blood pressure sensor measures the systolic, diastolic, and mean arterial pressure of the blood.
Glucometer	A glucometer measures the glucose levels in blood.
Galvanic skin response (GSR)	A GSR sensor measures the intensity of stress on a human. This sensor estimates the stress by measuring the variations in electrical characteristics of the skin.
Electrocardiogram (ECG)	This device measures the electrical and muscular activity of the heart.
Electromyogram (EMG)	EMG is a very important device that measures the health of a muscle and a nerve cell. With the help of EMG, the disruption of nerve and muscle of a body can be determined.

and the LPU is through either wired or wireless connectivity. The wireless communication between the physiological sensors and LPU occurs with the help of Bluetooth and ZigBee. On the other hand, the communication between the LPU and the cloud or server takes place with Internet connectivity such as Wi-Fi and WLAN. In Layer 4 of the healthcare IoT architecture, the healthcare data are received by the end users with different devices such as laptops, desktops, and cellphones. These communication protocols vary depending on the type of device in use. For example, when a service is received by a cellphone, it uses GSM (global system for mobile communications). On the other hand, if the same service is received on a desktop, it can be through Ethernet or Wi-Fi. Communication and connectivity in healthcare IoT is an essential component.

(iii) Privacy and Security: The privacy and security of health data is a major concern in healthcare IoT services. In a healthcare IoT architecture, several devices connect with the external world. Moreover, between LPU and the server/cloud, different networking devices work via network hops (from one networked device to another) to transmit the data. If any of these devices are compromised, it may result in the theft of health data of a patient, leading to serious security breaches and ensuing lawsuits. In order to increase the security of the healthcare data, different healthcare service providers and organizations are implementing healthcare data encryption and protection schemes [3, 4].

(iv) Analytics: For converting the raw data into information, analytics plays an important role in healthcare IoT. Several actors, such as doctors, nurses, and patients, access the healthcare information in a different customized format. This customization allows each actor in the system to access only the information pertinent to their job/role. In such a scenario, analytics plays a vital role in providing different actors in the system access to meaningful information extracted from the raw healthcare data . Analytics is also used for diagnosing a disease from the raw physiological data available [1, 2].

(v) Cloud and Fog Computing: In a healthcare IoT system, several physiological sensors are attached to a patient's body. These sensors continuously produce a huge amount of heterogeneous data. For storing these huge amounts of heterogeneous health data, efficient storage space is essential. These data are used for checking the patient's history, current health status, and future for diagnosing different diseases and the symptoms of the patient. Typically, the cloud storage space is scalable, where payment is made as per the usage of space. Consequently, to store health data in a healthcare IoT system, cloud storage space is used. Analytics on the stored data in cloud storage space is used for drawing various inferences. The major challenges in storage are security and delay in accessing the data. Therefore, cloud and fog computing play a pivotal role in the storage of these massive volumes of heterogeneous data.

(vi) Interface: The interface is the most important component for users in a healthcare IoT system. Among IoT applications, healthcare IoT is a very crucial and sensitive application. Thus, the user interface must be designed in such a way that it can depict all the required information clearly and, if necessary, reformat or represent it such that it is easy to understand. Moreover, an interface must also contain all the useful information related to the services.

Points to ponder

- As healthcare data is private, a popular US legislation—Health Insurance Portability and Accountability (HIPAA)—protects through data privacy and security provisions.

- Drones are used to deliver medicines in disaster rescue and management scenarios.

14.1.2   Advantages and risk of healthcare IoT

IoT has already started to penetrate the domain of medical science. In healthcare, IoT has become significantly popular due to its various features, which have been covered previously in this book.   Healthcare IoT helps in managing different healthcare subsystems efficiently.   Although it has many advantages, healthcare IoT has some risks too, which may be crucial in real-life applications. In this section, we discuss the different advantages and risks of healthcare IoT as depicted in Figure 14.3.

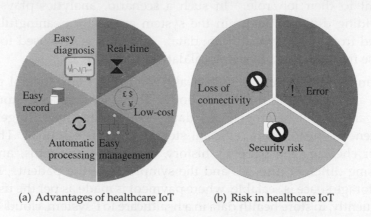

(a) Advantages of healthcare IoT   (b) Risk in healthcare IoT

Figure 14.3 Advantages and risk in healthcare IoT

Advantages of healthcare IoT
The major advantages of healthcare IoT can be listed as follows:

- Real-time: In healthcare sectors, different components, such as the condition of the patients, availability of doctors and beds in a hospital, medical facilities with their monetary charges, can vary dynamically with time. In such a dynamic scenario, one of the important characteristics of an IoT-based healthcare system is real-timeliness. A healthcare IoT system enables users, such as doctors, end users at the patient-side, and staff in a healthcare unit, to receive real-time updates about the healthcare IoT components, as mentioned earlier. Moreover, a healthcare IoT system can enable a doctor to observe a patient's health condition in real-time even from a remote location, and can suggest the type of care to be provided to the patient. On the other hand, users at the patient-end can easily take different decisions, such as where to take a patient during critical situations. Moreover, the staff in a healthcare unit are better aware of the current situation of their unit, which includes the number of patients admitted, availability of the doctors and bed, total revenue of the unit, and other such information.

- Low cost: Healthcare IoT systems facilitate users with different services at low cost. For example, an authorized user can easily find the availability of the beds in a hospital with simple Internet connectivity and a web-browser-based portal. The user need not visit the hospital physically to check the availability of beds and facilities. Moreover, multiple registered users can retrieve the same information simultaneously.

- Easy management: Healthcare IoT is an infrastructure that brings all its end users under the same umbrella to provide healthcare services. On the other hand, in such an infrastructure, the management of numerous tangible and intangible entities (such as users, medical devices, facilities, costs, and security) is a challenging task. However, healthcare IoT facilitates easy and robust management of all the entities.

- Automatic processing: A healthcare unit consists of multiple subsystems, for which manual interventions are required. For example, to register a patient with a hospital, the user may be required to enter his/her details manually. However, automatic processing features can remove such manual intervention with a fingerprint sensor/device. Healthcare IoT enables end-to-end automatic processing in different units and also consolidates the information across the whole chain: from a patient's registration to discharge.

- Easy record-keeping: When we talk about a healthcare IoT system, it includes a huge number of patients, doctors, and other staff. Different patients suffer from different types of diseases. A particular disease requires particular treatment, which requires knowledge of a patient's health history, along with other details about them. Therefore, the timely delivery of health data of the patient to the doctor is important. In such a situation, the permanent storage of the patients' health data along with their respective details is essential. Similarly, for the smooth execution of the healthcare unit, details of the staff with their daily activity in a healthcare unit are also required for storage. A healthcare unit must also track its condition and financial transactions for further development of the unit. A healthcare IoT enables the user to keep these records in a safe environment and deliver them to the authorized user as per requirement. Moreover, these recorded data are accessible from any part of the globe.

- Easy diagnosis: We have already explained that a healthcare IoT system stores the data of the patient in a secure manner. Sometimes, for diagnosing a disease, a huge chunk of prior data is required. In a healthcare IoT system, the diagnosis of the disease becomes easier with the help of certain learning mechanisms along with the availability of prior datasets.

Risk in healthcare IoT

We have already discussed the different advantages of the healthcare IoT. However, in a healthcare IoT system, there are multiple risks as well. Here, we discuss the various risks associated with a healthcare IoT system.

- Loss of connectivity: A healthcare IoT system consists of different physiological sensors that sense and transmit the sensed data to a centralized unit. Moreover, continuous data transmission from the patient is expected in a good healthcare system. Intermittent connectivity may result in data loss, which may result in a life-threatening situations for the patient. Proper and continuous connectivity is essential in a healthcare IoT system.

- Security: A healthcare IoT system contains the health data of different patients associated with the system. The healthcare system must keep the data confidential. This data should not be accessible to any unauthorized person. On the other hand, different persons and devices are associated with a healthcare IoT system. In such a system, the risk of data tampering and unauthorized access is quite high.

- Error: Data analytics helps a healthcare IoT system to predict the patients' condition and diagnosis of diseases. A huge amount of data needs to be fed into the system in order to perform accurate analytics. Moreover, the management of a huge amount of data is a crucial task in any IoT-based system. Particularly, in the healthcare system, errors in data may lead to misinterpretation of symptoms and lead to the wrong diagnosis of the patient. It is a challenging task to construct an error-free healthcare IoT architecture.

14.2 Case Studies

14.2.1 AmbuSens system

In many developing countries, patients need to be transferred from primary-care to tertiary-care hospitals for proper diagnosis and treatment. During the transit, the hospitals at both ends—the referring one as well as the referred one—do not have any information about the patient's health condition during transit. In such situations, the hospitals are unable to suggest any precautionary measures in the event of some emergency during transit. Consequently, many patients die during the transit due to lack of proper suggestive care by medical experts. To overcome these shortcomings, the Smart Wireless Applications and Networking (SWAN) laboratory at the Indian Institute of Technology Kharagpur developed a system: AmbuSens. The system was primarily funded by the Ministry of Human Resource and Development (MHRD) of the Government of India. This product system is a very crucial part of the healthcare IoT system. The primary objectives of the AmbuSens system are summarized as follows:

- Digitization and standardization of the healthcare data, which can be easily accessed by the registered hospital authorities.

- Real-time monitoring of the patients who are in transit from one hospital to another. At both hospitals, doctors can access the patients' health conditions.

- Accessibility by which multiple doctors can access the patient's health data at the same time.

- Provision of confidentiality to the health data of the patients in the cloud.
- In the AmbuSens system, wireless physiological sensor nodes are used. These sensor nodes make the system flexible and easy to use.

Architecture

The AmbuSens system is equipped with different physiological sensors along with a local hub. These sensors sense the physiological parameters from the patient's body and transmit those to a local data processing unit (LDPU). The physiological sensors and LDPU form a wireless body area network (WBAN). Further, this local hub forwards the physiological data to the cloud for storing and analyzing the health parameters. Finally, the data are accessed by different users. The detailed layered architecture of the AmbuSens system is depicted in Figure 14.4.

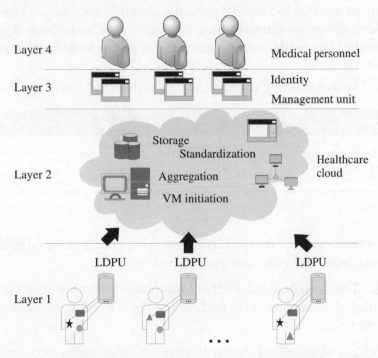

Figure 14.4 Layered architecture of AmbuSens

(i) Layer 1: This layer consists of multiple WBANs attached to a patient's body. These WBANs acquire the physiological data from the patient and transmit them to the upper layer. The physiological sensors are heterogeneous, that is, each of these sensors senses different parameters of the body. Moreover, the physiological sensors require calibration for acquiring the correct data from a patient's body. Layer 1 takes care of the calibration of the physiological sensor nodes. Further, in order to deliver the patient's physiological data from the sensor node to the LDPU, it is essential to form a proper WBAN. The formation of WBAN takes place by connecting multiple physiological sensor nodes to the LDPU so that the sensors can transmit the data to the LDPU, simultaneously.

(ii)   Layer 2: In the AmbuSens system, cloud computing has an important role. Layer 2 is responsible for handling the cloud-related functions. From Layer 1, WBANs attached to the different patients deliver data to the cloud end. The cloud is used for the long-term analysis and storage of data in the AmbuSens system. Moreover, the previous health records of the patients are stored in the cloud in order to perform patient-specific analysis. A huge volume of health data is produced by the WBANs, which are handled by the cloud with the help of big data analytics for providing real-time analysis.

(iii) Layer 3: In the AmbuSens system, the identity of the patients remains anonymous. An algorithm is designed to generate a dynamic hash value for each patient in order to keep the patient's identity anonymous. Moreover, in the AmbuSens system, at different time instants, a new hash value is generated for the patients. The entire hashing mechanism of the AmbuSens is performed in this layer.

(iv)   Layer 4: The users simply register into the system and use it as per requirement.

> **Check yourself**
>
> Smart Healthcare System [5], Caring Healthcare System [6], Criticality-aware System [7]

Hardware

In the AmbuSens system, a variety of hardware components are used such as sensors, communication units, and other computing devices.

- **Sensors**: The sensors used in the AmbuSens system are non-invasive. The description of the sensors used for forming the WBAN in the AmbuSens system are as follows:

  (i)   Optical Pulse Sensing Probe: It senses the *photoplethysmogram* (PPG) signal and transmits it to a GSR expansion module. Typically, PPG signals are sensed from the ear lobe, fingers, or other location of the human body. Further, the GSR expansion module transfers the sensed data to a device in real-time.

  (ii) Electrocardiogram (ECG) unit and sensor: The ECG module used in AmbuSens is in the form of a kit, which contains ECG electrodes, biophysical 9" leads, biophysical 18" leads, alcohol swabs, and wrist strap. Typically, the ECG sensor measures the pathway of electrical impulses through the heart to sense the heart's responses to physical exertion and other factors affecting cardiac health.

(iii) Electromyogram (EMG) sensor: This sensor is used to analyze and measure the biomechanics of the human body. Particularly, the EMG sensor is used to measure different electrical activity related to muscle contractions; it also assesses nerve conduction, and muscle response in injured tissue.

(iv) Temperature sensor: The body temperature of patients changes with the condition of the body. Therefore, a temperature sensor is included in the AmbuSens system, which can easily be placed on the body of the patient.

(v) Galvanic Skin Response (GSR) sensor: The GSR sensor is used for measuring the change in electrical characteristics of the skin.

- **Local Data Processing Unit (LDPU)**: In AmbuSens, all the sensors attached to the human body sense and transmit the sensed data to a centralized device, which is called an LDPU. An LDPU is a small processing board with limited computation capabilities. The connectivity between the sensors and the LDPU follows a single-hop star topology. The LDPU is programmed in such a way that it can receive the physiological data from multiple sensor nodes, simultaneously. Further, it transmits the data to the cloud for long-term storage and heavy processing.

- **Communication Module**: Each sensor node consists of a Bluetooth (IEEE 802.15.1 standard) module. The communication between the sensor nodes and the LDPU takes place with the help of Bluetooth, which supports a maximum communication range of 10 meters in line of sight. The LDPU delivers the data to the cloud with 3G/4G communication.

Front End

In the AmbuSens system, three actors—doctor, paramedic/nurse, and patient—are able to participate and use the services. The web interface is designed as per the requirements of the actors of the system. Each of the actors has an option to log in and access the system. The confidentiality of a patient and their physiological data is important in a healthcare system. Therefore, the system provides different scopes for data accessibility based on the category of an actor. For example, the detailed health data of a patient is accessible only to the assigned doctor. These data may not be required for the nurse; therefore, a nurse is unable to access the same set of data a doctor can access. The system provides the flexibility to a patient to log in to his/her account and download the details of his/her previous medical/treatment details. Therefore, in AmbuSens, the database is designed in an efficient way such that it can deliver the customized data to the respective actor.

Each of the users has to register with the system to avail of the service of the AmbuSens. Therefore, in this system, the registration process is also designed in a customized fashion, that is, the details of a user to be entered into the registration form is different for different actors. For example, a doctor must enter his/her registration number in the registration form.

Summary

Healthcare is a crucial application domain of IoT. A healthcare IoT ecosystem contains distinct types of components, which combine and collectively participate in providing an efficient healthcare IoT infrastructure. Considering all these aspects of healthcare IoT, this chapter provided an overview of the real healthcare IoT system. From this chapter, the reader can easily visualize the practical implementation aspects of the healthcare IoT system. The case study discussed in this chapter provided a visualization of the importance and various aspects of an IoT healthcare system from theory to practice.

Exercises

(i)   List the components of healthcare IoT.

(ii)   Why privacy and security is important for healthcare?

(iii) What is a wireless body area network (WBAN)?

(iv)   What is the difference between electrocardiogram (ECG) and electromyogram (EMG) sensors?

(v)   List the advantages of healthcare IoT.

(vi)   List the risks associated with healthcare IoT systems.

(vii) How can data analysis be used in healthcare IoT?

(viii) What is a local processing unit (LPU)?

(ix)   Discuss an idea for developing an IoT-based healthcare system, where we can include fingerprint sensor.

(x)   Why is cloud computing important for a healthcare IoT system?

References

[1]   Sheriff, C. I., T. Naqishbandi, and A. Geetha. 2015. "Healthcare Informatics and Analytics Framework." *Proceedings of International Conference on Computer Communication and Informatics (ICCCI), January 2015.*

[2]   Kumar, S. and M. Singh. 2019. "Big Data Analytics for Healthcare Industry: Impact, Applications, and Tools." *Big Data Mining and Analytics* 2(1).

[3]   Elhoseny, M., G. Ramírez-González, Osama M. Abu-Elnasr, S. A. Shawkat, N. Arunkumar, and Ahmed Farouk. 2018. "Secure Medical Data Transmission Model for IoT-Based Healthcare Systems." *IEEE Access* 6: 20596–20608.

[4]   Maria de Fuentes, J., L. Gonzalez-Manzano, A. Solanas, and F. Veseli. 2018. "Attribute-Based Credentials for Privacy-Aware Smart Health Services in IoT-Based Smart Cities." *Computer* 51(7): 44–53.

[5]   Catarinucci, L., D. de Donno, L. Mainetti, L. Palano, L. Patrono, M. L. Stefanizzi, and L. Tarricone. 2015. "An IoT-Aware Architecture for Smart Healthcare Systems." *IEEE Internet of Things Journal* 2(6): 515–526.

[6]   Laplante, P. A., M. Kassab, N. L. Laplante, and J. M. Voas. 2018. "Building Caring Healthcare Systems in the Internet of Things." *IEEE Systems Journal* 12(3): 3030–3037

[7]   Roy, A., C. Roy, S. Misra, Y. Rahulamathavan, and M. Rajarajan. 2018. "CARE: Criticality-Aware Data Transmission in CPS-Based Healthcare Systems." In *Proceedings of 2018 IEEE International Conference on Communications Workshops (ICC Workshops), Kansas City, MO*, pp. 1–6

[16] Catarinucci, L., D. De Donno, L. Mainetti, L. Palano, L. Patrono, M. L. Stefanizzi, and L. Tarricone. 2015. "An IoT-Aware Architecture for Smart Healthcare Systems." IEEE Internet of Things Journal 2(6): 515–526.

[17] Pace, P., G. Aloi, R. Gravina, G. Caliciuri, G. Fortino, and A. Liotta. 2019. "An Edge-Based Architecture to Support Efficient Applications for Healthcare Industry 4.0." IEEE Transactions on Industrial Informatics 15(1): 481–489.

[18] Ray, P. P., D. Dash, and D. De. 2019. "Edge Computing for Internet of Things: A Survey, E-Healthcare Case Study and Future Direction." Journal of Network and Computer Applications 140: 1–22.

Paradigms, Challenges, and the Future

After reading this chapter, the reader will be able to:

- Assess the various evolving aspects and paradigms of IoT
- Understand the most prominent challenges encountered during the design and development of IoT solutions
- Research upcoming and emerging domains, which find significant applicability in IoT

15.1 Introduction

Since the inception of IoT, the technology has passed through multiple stages of development and revolutionized the industrial as well as consumer sectors rapidly. However, developers face multiple challenges while developing their ideas into reality. The IoT devices are usually resource-constrained in terms of their computational capability, battery power, as well as, storage. These limitations mandate the developers to develop routines that execute complex operations on the devices with ease, making IoT applications dependent on external platforms such as cloud and fog computing. Another significant issue is the rising number of IoT devices. IoT has penetrated diverse domains and extends its scope to smart homes, vehicles, cities, utility meters, and others. Such an increase in the number of interacting devices increases the consumption of resources and causes congestion in the network. In addition to that, the devices, along with their interconnections, have a non-trivial topology, which leads to complex networks. The IoT industries, therefore, need to develop new algorithms to deal with the increasing lattices in the complex network as well as multiple access protocols to avoid delays and packet drops. The increasing number of devices also consumes a lot of power, which calls for the development of new low-power hardware and schemes that involve smart energy harvesting and its consumption.

IoT devices include both static as well as mobile nodes, depending on the user's applications. Although mobility in such environments may or may not be random, developers need to be prepared for any pattern while providing their services. Since the precise prediction of mobility is challenging, it can only be analyzed statistically. Additionally, IoT environments also have multiple access points in close proximity. Such a deployment of access points creates interference, causing a reduction in signal-to-noise-plus-interference ratio (SINR) values, which causes a decline in signal quality and packet drops. IoT service providers need to impose smart channel selection as well as schemes to overcome these interferences. Developers also need to create new models to mimic the mobility of the users and propose schemes to facilitate smooth handoffs among the access points for uninterrupted service.

15.2 Evolution of New IoT Paradigms

As mentioned earlier, since the inception of IoT, it has successfully been used by multiple industries for running their operations and also providing their services to the users. IoT has found scope in diverse operations, which has led to the origin of several paradigms based on the nature of data sources/devices/peripherals and their corresponding applications. Some of these areas and the respective IoTs are explained in this section.

15.2.1 Internet of battlefield things (IoBT)

This category is responsible for connecting soldiers with IoT. Researchers in IoBT aim to develop a suite with embedded biometric and location sensors for soldiers. Data from these sensors allows the soldiers to keep track of the troops and also share information regarding foes; it makes the whole team situationally aware. Moreover, smart analysis using machine learning algorithms opens the scope for designing superior tactics in real-time. However, IoBT also has its challenges, mostly regarding energy constraints and data rates. Soldiers need to transfer sizeable data with minimum delay, which mandates the need for optimized consumption of bandwidth and battery. Finally, IoBT systems must be robust and durable enough to withstand the rigors of sustained outdoor and battlefield use.

15.2.2 Internet of vehicles (IoV)

This category of IoT is responsible for communications among smart connected vehicles, usually through vehicular ad-hoc networks (VANETs). Smart vehicles consist of a myriad of sensors that include cameras, GPS, infrared, and others. IoV facilitates these vehicles to communicate with other vehicles, its drivers, roadside units (RSU), and other mobile and fixed infrastructures. Intuitively, IoV supports intra-vehicle, vehicle-to-vehicle (V2V), vehicle-to-infrastructure (V2I), vehicle-to-cloud (V2C), and vehicle-to-pedestrian (V2P) communication. Although IoV faces a combination of challenges related to mobility, changing states, and dynamic signal quality, it has several advantages. Developers design IoVs such that they are environmentally safe, improve road safety, enhance user convenience, as well as, increase revenue of manufacturers.

15.2.3   Internet of underwater things (IoUT)

This category of IoT aims to interconnect underwater sensors and communication infrastructure with the terrestrial Internet. The sensor nodes in IoUT consist of smart devices and are usually powered by batteries. They are also much smaller in size than normal sensors and support wireless communication based on acoustic signals. However, IoUT has a significant lacuna in its communication model as radio waves do not fare well, underwater. Although developers use sizeable antennas for the signals to penetrate through water, the signals suffer high attenuation and absorption. Underwater, the signals also suffer from high propagation delays and bit error rates. Optical signals also seem impractical underwater due to the high absorption rate. Additionally, they can only cover short distances. Researchers, however, aggregate data from the sensors situated underwater to a sink node at the surface of the water, which forwards it to the terrestrial base stations in a multi-hop manner. The sink nodes in IoUT contains both acoustic as well as radio antennas for its purpose.

15.2.4   Internet of drones (IoD)

IoT operates toward enhancing user experience while minimizing user intervention. IoD is the category concerned with the deployment and management of unmanned ariel vehicles. Service providers use IoD for various applications such as package delivery, wildlife surveillance, rescue operations, agriculture, photography, and others. However, developers need to deal with flying the drones in controlled/uncontrolled airspace and dictate navigation coordinates. For the seamless operation of IoD, developers need to fuse air traffic control networks, cellular networks, automation, and the Internet [1].

15.2.5   Internet of space (IoSpace)

This category of IoT relies on low earth orbit (LEO) satellites for providing seamless connectivity services over uneven demographic areas. However, such satellites have disadvantages concerning development and deployment cost, and loss due to failure in orbit. These satellites have the potential to reduce network latencies significantly.

Researchers have been recently working hard toward the development of small cubic satellites called CubeSats to overcome the challenges mentioned here [2]. In addition to these difficulties, satellites also present challenges related to tracking, synchronization, and handoff. We expect that technologies such as software defined networks (SDN) and network function virtualization (NFV) will play a major role in addressing these issues.

15.2.6 Internet of services (IoS)

This category is specific for manufacturers and service providers, that is, the industries. With IoS, manufacturers bring hardware and software under one umbrella. For instance, a car manufacturer builds a car with installed sensors. They later release software updates over the Internet to enhance user experience. The manufacturers may also charge for the upgrades, which generates revenue for the company. Additionally, this model also paves the way for crypto-currency as a payment method. Applications of IoS extends to factory monitoring, sensing and actuation of factory units, and generation of remote alarms in case of emergency. IoS also reaches out to smartphones that already have multiple sensors. Companies use these sensors and develop Internet-enabled apps for users.

15.2.7 Internet of people (IoP)

The Internet contains a plethora of profiles representing people and interconnected links as relations among them. The IoP interconnects these peer-to-peer networks. Researchers in social computing extensively use social graphs for representation and inferences. The IoP supporting applications facilitate direct device-to-device, people-to-people, as well as company-to-people communications. IoP further opens scope for crypto-currency as a means to transfer incentives/payments in return for services. Such structures enable smooth interaction among service providers and consumers. IoP also provides a platform for carrying out transparent and secure payments.

15.2.8 Internet of nano things (IoNT)

The interrelated systems in IoT, which usually include combinations of sensors and actuators, can be miniaturized to tiny devices with dimensions in the scale of nanometers. These devices are application-specific and occupy minimal space; they include miniaturized sensors in vehicles, as well as those responsible for monitoring the environment. Communication at the nano-scale is rendered possible in two ways: 1) electromagnetic (EM) and 2) chemotaxis communications. Electromagnetic communications at the nano-scale typically use the Terahertz band of the spectrum. However, this results in significant power issues, a limited range of communication, and severe susceptibility to interferences. Parallelly, the use of chemotaxis as a means of communication is achieved through exploiting the population dynamics of bacteria and viruses. Messages are passed in the form of chemical signatures and molecules, which are often facilitated by specifically cultured bacteria and viruses. Nano-scale IoT also finds scope in healthcare, where researchers are working actively in fighting diseases with the help of programmable bacteria/viruses/nanoparticles. However, designing such nanodevices is a non-trivial task, which the developers need to study rigorously.

15.2.9   Internet of everything (IoE)

The IoE comprises four pillars and concerns itself with the communication among them. These four pillars are people, data, processes, and things.

(i)   People: Communication among people is analogous to the IoP mentioned earlier.

(ii)   Data: Data from sensors are analyzed for inferencing and making decisions.

(iii)   Process: Information is delivered to the concerned people/machine/infrastructure.

(iv)   Things: This is analogous to the things in IoT.

The main difference between IoT and IoE is that IoT only concerns itself with the non-human aspects of technology, while IoE consists of all the other factors, which include machine-to-people (M2P) and technology-assisted peer-to-peer (P2P) interactions in addition to the features of IoT.

15.3   Challenges Associated with IoT

IoT has numerous advantages up its sleeve. However, with the advent of these technologies and heterogeneity of the nature of devices, IoT also has several challenges that researchers are trying to overcome actively. In this section, we mention a few such challenges.

15.3.1   Mobility

IoT supports unconditional M2M communication. The devices in the system, given their heterogeneity concerning configurations and usage, that is, pedestrians, vehicles, cycles, drones, robots, and others, have diverse mobility patterns. These patterns cannot be precisely predicted and need to be stochastically analyzed, which makes efforts toward seamless connectivity and quality of service tricky. Developers need to devise ways to make dynamic decisions on the handoff, synchronization, and others such issues efficiently. Tasks such as allocation of identifiers to mobile devices, handoff strategies, coverage estimation, path planning, mobility prediction, and others are some of the research domains which are directly associated with addressing this challenge.

15.3.2 Addressing

With the advent of IoT and its advantages, its adoption by the people as well as, industries are growing at an uncontrollable rate. Such an exponential increase in the number of devices exhausts the number of available IP addresses, leading to IP conflicts. In addition to that, there are very few standards or industry-recommended schemes toward addressing IoT administrators. Recently, IoT has already seen a paradigm shift from IPv4 to IPv6 addressing schemes in some industries but it is yet to be popularly adopted by the masses. Typical research challenges in this domain include addressing strategies, sub-netting strategies, and others.

15.3.3 Power

IoT devices are usually resource-constrained concerning power and computational capability. These devices need to last for a long time irrespective of their limited battery power. Such limitations call for green computing schemes for smart harvesting and consumption of power. Alternatively, it also calls for new hardware designs that consume minimum power for operation. Various upcoming research solutions focus on developing high-density batteries/cells for enabling long-term use of IoT systems. Research in this domain includes the design of low-power processors and hardware, design of low-power consuming computation techniques and algorithms, energy harvesting, alternative sources of energy, and others.

15.3.4 Heterogeneous connectivity

IoT is a vast collection of heterogeneous networks made up of long-range, as well as, short-range connectivity technologies. Some of the integrated sectors in IoT also rely on their own proprietary connectivity solutions. The proprietary nature of connectivity is commonly encountered in applications such as military, heavy industries, and others. Heterogeneity in connectivity can be significantly challenging to manage as often connectivity devices may be vendor-specific, industry-specific, or even task-specific. Some antennas may be more powerful than the others, or maybe close to one another, inducing high interferences. Coverage is also an issue in such environments. Additionally, as the devices move away from access points, they may lose contact midway, which is an open problem so far. For example, the majority of connectivity technologies are still present in industries are wired. Despite the high maintenance costs and physical space occupancy, wired solutions are considered more reliable and secure for industrial uses. Additionally, legacy connectivity technologies are still present in industries and they are majorly wired and mostly analogous. These industries and industrial systems need to be connected to the Internet. The main challenge in such situations is the amalgamation or provision of a unified solution which can handle analog as well as digital content, and that too for different vendor-specific devices and protocols. Typical research in this domain consists of work on protocol conversions, bandwidth allocation, task offloading, big-data analytics, cloud computing, and others.

15.3.5 Communication range

The wide expanse and reach of IoT have led to some of its major challenges, addressing which have given IoT some powerful new solutions. The usefulness of IoT has led to its solutions being deployed in areas with proper connectivity as well as areas, especially remote ones, where there is barely any connectivity. Both of these scenarios have their own unique set of challenges, which need to be addressed separately. For example, the deployment of low-power wireless IoT solutions in urban areas tend to frequently encounter interference and noise due to the presence of other powerful wireless solutions operating at the same frequency spectrum. In contrast, the deployment of IoT solutions in remote places such as forests and rural areas often do not have the proper network infrastructure to provide these solutions with basic Internet access. The rise of IoD in providing communication coverage to such areas through relaying of signals, enabling backhaul network access, and others is a prime example of a powerful, yet economical solution rising due to the communication demand–supply gap in IoT.

15.3.6 Security

Due to the lack of powerful and unified security standards in IoT and an increasing number of devices, IoT is vulnerable to threats from malicious attackers and bots. Although encryption seems to be a logical answer in this scenario, a significant chunk of these devices lack the storage and computational capabilities required for supporting complex mathematical operations, which come with encryption. Some manufacturers put a built-in password for security, which is temporarily helpful in a few limited scenarios. However, attackers have enough resources to crack such default passwords, which compromises the whole system. IoT is not restricted to low-power devices, and even so, these low-power devices eventually connect to remote platforms such as a server/fog/cloud. Gaining access to the network or the remote infrastructure by compromising the low-power devices is a reality in the present-day technological realm. Vulnerability to attacks such as phishing, flooding, denial of service, man-in-the-middle attacks, and others, can easily trigger a chain of anomalies, which may bring down a whole network or enterprise. Typical research in this domain includes works on hardware-level security, processor/chip-level security, physically unclonable functions (PUFs), network security, cryptography, blockchains, crypto-currency, encryption, and others.

15.3.7   Device size

Manufacturers usually design IoT devices for enhancing user experience at low cost. Further, such devices are usually small in size, equipped with unique IDs and wireless communication antennae. The low cost and size make it difficult to incorporate processing power and storage in the device. It also causes space concerns to introduce a battery. Finally, these devices end up being resource-constrained in terms of operational capability, battery power, and storage. Typical research in this aspect of IoT includes nano and microelectronics, photonics, device fabrication, and the new and upcoming paradigm of quantum computing.

15.3.8   Interoperability

IoT devices serve a myriad of applications with numerous manufacturers deploying multiple units. These devices with different purposes and different manufacturers need to interact with one another to work in harmony. With the increasing number of devices and no universal standards, researchers are working actively to enable the devices to achieve common goals automatically. Chapter 9 on interoperability covers the various aspects associated with this challenge.

15.4   Emerging Pillars of IoT

IoT is a massive paradigm with far-reaching implications across vastly interdisciplinary domains. However, some standalone paradigms, are nowadays commonly associated with IoT due to the benefits of association these technologies bring to themselves as well as to IoT. We discuss some of these emerging pillars of IoT in the subsequent subsections.

15.4.1   Big data

Manufacturers and users are deploying numerous IoT devices while serving a plethora of applications. Along with these IoT devices and applications, the rate of data generation also increases, leading to large datasets (petabytes or gigabytes). These data may be structured/unstructured; developers need to analyze these data for finding hidden patterns and generating inferences. For comprehending these inferences and decisions, developers are turning toward big data analytics. The network traffic or data is classified as big data if it satisfies specific characteristics of 1) volume, 2) variety, 3) value, 4) velocity, and 5) veracity. Big data analytics has the potential to process data from IoT devices in real time and store them using various storage schemes. Once acquired, this voluminous data can be used for studying patterns in network behavior, usage, customer experiences, mobility, connectivity issues, among many other interesting features [3].

15.4.2 Cloud/fog/edge computing

Commerical device manufacturers design IoT devices and solutions for providing affordable services to the general public. Specific use cases of IoT, such as those for industries, militaries, and other such applications, are also catered to, mostly through proprietary solutions. However, the sheer volume and variety of data that is available for further processing needs powerful resources and infrastructures. Although a significant number of IoT devices are smaller in size and resource-constrained, their massive-scale use in various applications eventually leads to a formidable amount of information to be processed and handled. Due to such limitations, these IoT devices need to depend on external platforms, particularly cloud/fog/edge computing [4] [5] schemes, to address their processing issues and generate meaningful information from the gathered data. The choice of platform is application dependent, that is, while cloud computing has unlimited resources, fog/edge computing reduces operational latencies significantly. Starting as a research paradigm, these domains have gathered worldwide acceptance and are mostly included in mainstream IoT architectures and applications.

15.4.3 5G and beyond

The launch of 3G technology facilitated robust and speedy voice, text, and data services to the users. 4G was similar to 3G but with a higher data rate, enabling its users to adopt video-based communication and making it a new normal in the communication industry. The new 5G technologies provide services with much higher speeds, as they focus on providing ubiquitous high-speed connectivity to all device types. Features such as downloading full HD movies in a matter of seconds characterize this technology. 5G technology features a fusion of high data rates with low latencies, ubiquitous coverage, and smart infrastructures to support real-time applications, enabling remote monitoring and control [6]. Researchers envision 5G as a driving force for IoT. The features of 5G have also started the race for beyond 5G technologies and paradigms. Envisioned as operating in the Terahertz band of the frequency spectrum, beyond 5G technologies are being speculated to have data rates in the tune of Gbps [7].

15.4.4 Artificial intelligence (AI)/Machine learning (ML)

Owing to the deployment of numerous IoT devices and applications, the complexity and size of data over and beyond the networks have increased significantly. The IoT data may or may not be structured and often consists of hidden patterns, which have to be derived through data processing and statistical inferences. Modern-day AI/ML-based tools have significantly powerful inferencing procedures, which can outperform almost any of the standard statistical methods. Additionally, the need for automation in a significant chunk of IoT devices and applications is another compelling reason for the rapid emergence and adoption of AI/ML with IoT [8]. AI/ML has been used for extracting information from raw data, be it from agricultural sensors, smart home sensors, or network security analyzers. AI/ML tools are mostly data-driven. However, new methods in these domains are rapidly cropping up; methods which do not always have to rely on voluminous data for generating inferences or predicting trends.

15.4.5 Cognitive communication networks

This domain is yet another upcoming pillar of IoT, which, although focused only on the communication aspects, has the potential to revolutionize the existing IoT architectures and the way data and signals are handled in a network. Cognitive communication networks or simply, cognitive networks are capable of sensing the present network's parameters, conditions, and plans. Based on the sensed information, the cognition engine can devise pathways and strategies for best achieving the end goals for a certain task. For example, an IoT node transmits data to a remote server through a fixed gateway, which has access to the Internet. However, this gateway also serves 10,000 other such IoT nodes. The situation undoubtedly raises the issue of congestion at the gateway (considering normal channel bandwidth) due to large waiting message queues. As the traditional IoT networks are designed to follow somewhat similar architectures, the load on networks is highly unbalanced and unevenly distributed. This uneven traffic load also results in long waiting times, packet drops, and noise in the data being transmitted over the network. Now consider a smart network which can sense delays and queues in a certain path and has the autonomy to choose an alternative path to facilitate data transmission between the IoT node and the remote server. This mechanism can be considered as a rudimentary example of a cognitive network.

15.4.6 Network function virtualization (NFV)

NFV is an interesting and practical concept, which proposes the virtualization of major network elements such that software virtualizes network hardware by providing the same functionalities and added features. The concept of NFV arose due to the difficulty of reconfiguring (both changing and upgrading) installed network infrastructures. Physically going to every network element for changes (updates, software patches) can be a significant challenge in terms of time, money, and human resources, especially for enterprise-grade networks consisting of tens of thousands of network elements. NFV utilizes the concept of virtualization to provide services similar to network elements through standard servers.

15.4.7 Software-defined networks (SDN)

IoT environments are highly dynamic concerning mobility and the changing states of the access points. With the implementation of fog and edge computing, the states of the service providing nodes will change; so will the users. For the seamless transition of data routes, self-organization, configuration optimization, and smart transmissions, SDN has emerged as a popular choice. SDN reduces the complexity of traditional networks by introducing a centralized control structure through the separation of control and data planes of network elements. A centralized view of the whole network ushers in the benefits of better controllability, network stability, and increased efficiencies. It is to be noted that SDN and NFV are separate paradigms. NFV simply virtualizes the network elements in a traditional network, such that the core operating procedure remains the same. In contrast, SDN introduces an entirely new way of handling network traffic by separating the control and data planes to provide a centralized architecture for the whole network.

15.4.8 Phantom networks

"Phantom networks" paradigm strives to develop intangible communication infrastructures. Being a relatively new paradigm, which is still under development, this paradigm relies on the Terahertz (THz) band for communication between aerially diffused nano-relays. The aerial nano-relays are deployed through ground-based pumps, which spray the water-suspended mixture of the nano-communication relays in the deployment area. However, unlike traditional network infrastructures, this paradigm is prone to the effects of wind, rain, humidity. Additionally, the factors of node density in an area to ensure reliable throughput and quality of service and settling time of these aerially suspended nodes play decisive roles in deciding the network lifetime and performance of the network. This paradigm is highly interdisciplinary and requires the operational knowledge of multiple domains such as nanotechnology, communication, networking, fluid dynamics, and others. Typical application areas include military communication and communication for emergency response during disaster management.

Summary

In this chapter we discuss the present challenges and upcoming paradigms associated with IoT.

Exercises

(i) What are the new evolving paradigms as a result of the use of IoT in various domains?

(ii) Discuss the salient features and differences between the Internet of vehicles and the Internet of drones.

(iii) What are the common challenges associated with the adoption of IoT in any new domain?

(iv) How is NFV different from SDN?

(v) What are the challenges associated with beyond-5G communication?

(vi) Discuss the role of AI/ML in IoT.

(vii) How is cognitive communication poised to enhance the usability of IoT networks?

(viii) What are the typical features of a data for it to be characterized as big data?

(ix) How is cloud-based storage different from regular offsite storage in IoT networks?

References

[1] Gharibi, M., R. Boutaba, and S. L. Waslander. 2016. "Internet of Drones." *IEEE Access* 4: 1148–1162.

[2] Akyildiz, I. F. and A. Kak. 2019. "The Internet of Space Things/CubeSats." *IEEE Network* 33(5): 212–218.

[3] Fahad, A., N. Alshatri, Z. Tari, A. Alamri, I. Khalil, A. Y. Zomaya, S. Foufou, and A. Bouras. 2014. "A Survey of Clustering Algorithms for Big Data: Taxonomy and Empirical Analysis." *IEEE Transactions on Emerging Topics in Computing* 2(3): 267–279

[4] Bera, S., S. Misra, and J. J. Rodrigues. 2014. "Cloud Computing Applications for Smart Grid: A Survey." *IEEE Transactions on Parallel and Distributed Systems* 26(5): 1477–1494.

[5] Sarkar, S., S. Chatterjee, and S. Misra. 2015. "Assessment of the Suitability of Fog Computing in the Context of Internet of Things." *IEEE Transactions on Cloud Computing* 6(1): 46–59.

[6] Agiwal, M., A. Roy, and N. Saxena. 2016. "Next Generation 5G Wireless Networks: A Comprehensive Survey." *IEEE Communications Surveys & Tutorials* 18(3): 1617–1655.

[7] Huq, K. M. S., S. A. Busari, J. Rodriguez, V. Frascolla, W. Bazzi, and D. C. Sicker. 2019. "Terahertz-enabled Wireless System for Beyond-5G Ultra-fast Networks: A Brief Survey." *IEEE Network* 33(4): 89–95.

[8] Poniszewska-Maranda, A., D. Kaczmarek, N. Kryvinska, and F. Xhafa. 2019. "Studying Usability of AI in the IoT Systems/Paradigm through Embedding NN Techniques into Mobile Smart Service System." *Computing*: 101(11): 1661–1685.

PART FIVE
IOT HANDS-ON

Chapter **16**

Beginning IoT Hardware Projects

Learning Outcomes

After reading this chapter, the reader will be able to:

- Understand the common hardware platforms, sensors, and actuators used in IoT
- Assess the importance of each sensor or hardware in various applications
- Understand the code structure required to operate these hardware and sensors /actuators connected to them
- Relate the IoT hardware and sensors according to the requirements of their applications

16.1 Introduction to Arduino Boards

Arduino is an open-source platform which comprises both hardware and software modules. The hardware are typically processor boards with analog and digital pins for various functions. Arduino IDE (integrated development environment) is the software that is used to communicate to the hardware and program them to perform specific functions. We cover the process right from the installation of Arduino IDE to loading codes to the processor boards.

16.1.1 Arduino vs. Raspberry Pi: Choosing a board

An Arduino board has a mounted micro controller chip to perform simple tasks. The board can execute single tasks at a time, in a repetitive manner. The main focus area of Arduino is integration of hardware to the board rather than complex computation or multitasking. For example, the Arduino board can sense the ambient temperature of an area and display it on an output console. Raspberry Pi is a more complex system, much like a minicomputer. It has an advanced processor, uses Linux-based operating

Table 16.1 Some well-known Arduino compatible processor boards and their features

Board	Microcontroller	Operating Voltage	Digital I/O Pins	Analog I/O Pins	Clock Speed
Arduino Uno	ATmega328P	5V	14	6	16MHz
Arduino Due	AT91SAM3X8E	3.3V	54	12	84MHz
Arduino Mega 2560	ATmega2560	5V	54	16	16MHz
NodeMCU	ESP8266	3.3V	11	1	80MHz

systems, and is able to perform multiple tasks with complex computation. Raspberry Pi gives the option for Wi-Fi connectivity and access point formation, something which would require external components to be connected on to a regular Arduino board. The choice of selecting Raspberry Pi or Arduino depends solely on one's application. If users want to perform rudimentary, basic level tasks then Arduino is the best option; it avoids any kind of complexities. If the intended application needs multiple tasks to be done with complex computations, Raspberry Pi is a much more suitable choice as it has pretty much everything integrated on-board. Once the users decide which board to choose, they can narrow down the options based on their requirement and the available features on the board such as operating voltage, RAM size, clock speed, storage, and additional wireless connectivity modules.

16.1.2   Arduino installation and setup

Installer files are available for different variants of the operating system on the official website of Arduino. Download the installer (Figure 16.1) and execute the file. The installer will direct you (the user) to follow the steps such as giving permissions and path for storing the files (Figure 16.2).

After successful installation of the Arduino IDE (Figure 16.3), you will see a shortcut created on your desktop if you did not uncheck the option during installation. Alternately, you can always search for the program *Arduino* in your system's search bar. An editor with toolbars on the top of the screen opens up when you open the IDE. This is the editor where you will write your Arduino sketch (Arduino code is called a sketch!). Figure 16.4 shows a blank sketch with two predefined functions (more on that later). You can find most of the settings for the Arduino IDE under the *File* → *Preferences* section, shown in Figure 16.5. There are many basic Arduino sketches available in the *Examples* section under *File* option, shown in Figure 16.6. You can go through them to get a basic idea about programming in Arduino. As you keep adding new boards and components to your IDE, new examples are automatically added. Different types of sensors and actuators can be connected to the processor board for various functionalities. The integration of these components require specific libraries to be installed in the IDE. A library generally consists of one or more C language

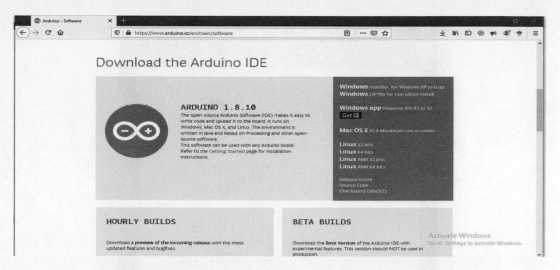

Figure 16.1 Download Arduino from the official website

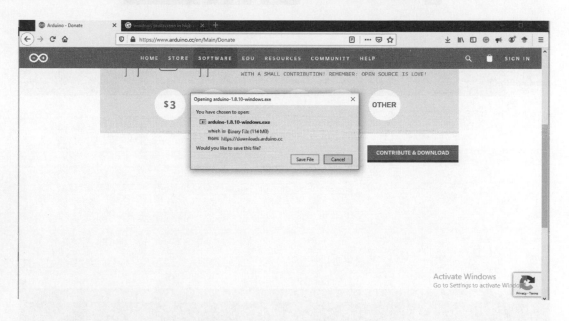

Figure 16.2 Save the Arduino installer file

declaration and definitions of sensor-specific pins and functions. You can add a library by going to *Sketch → Include Library manage libraries* and search for a particular library in the search bar, as shown in Figure 16.7. The processor board has to be connected to the system through a USB port. After connecting the board, your system assigns a port to the connected device. This port is detected by the Arduino IDE in the *port* option under *Tools* as shown in Figure 16.8. Selecting the correct port is important as it will decide to which device the sketch is uploaded. Next, the Arduino IDE has to know

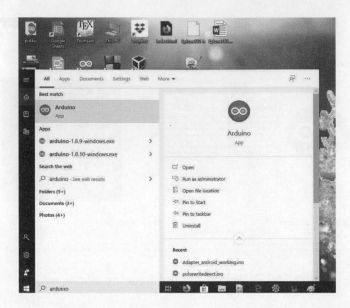

Figure 16.3 Arduino IDE after installation

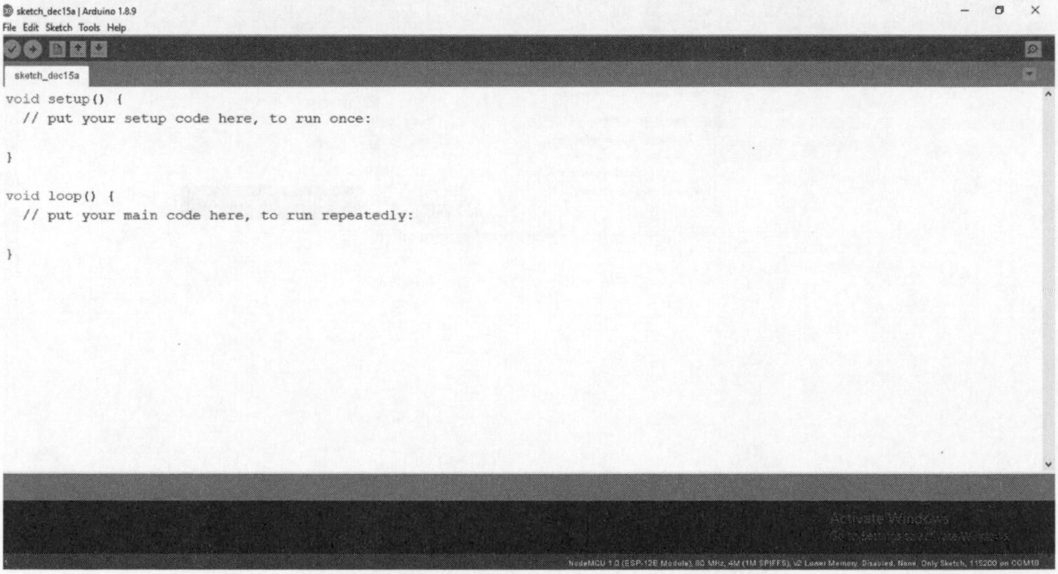

```
void setup() {
  // put your setup code here, to run once:

}

void loop() {
  // put your main code here, to run repeatedly:

}
```

Figure 16.4 An empty Arduino sketch with predefined functions

the type of board that is connected. Navigate to *Tools → Board* and you will see a list of Arduino supported boards, as shown in Figure 16.9. The basic variants of Arduino boards are already installed in the IDE. Any new board can be added to the IDE by adding it from the *Boards Manager* option placed right at the top of the existing boards list. The *Boards Manager* will take you to a screen as shown in Figure 16.10. Here you can search for a specific board or package and install it. Once you have set up your

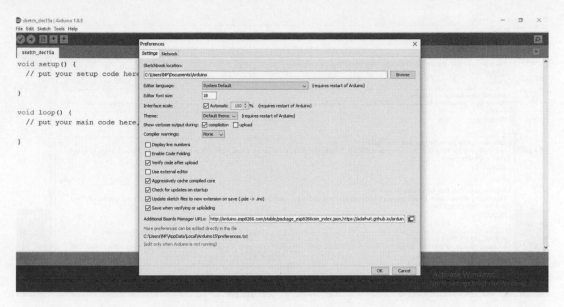

Figure 16.5 Preferences setting window

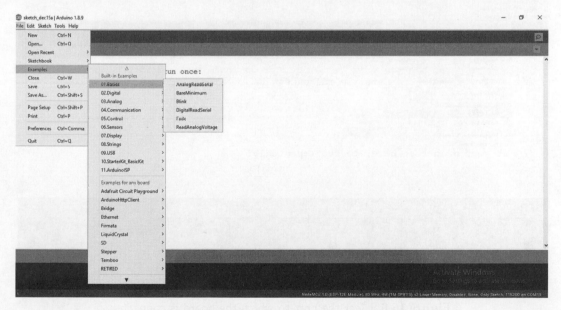

Figure 16.6 Built-in examples from installed libraries

Arduino IDE with the mentioned settings, you are ready to write your sketch and upload it to your board. Arduino gives you the option to print the output from your board to the system screen. It is called the serial console which can be accessed under the *Tools* option. Figure 16.11 shows the serial console of an Arduino IDE. Notice the second option at the bottom right corner of the console. It allows you to select the

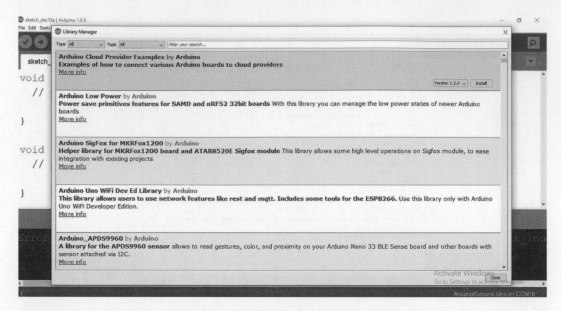

Figure 16.7 Install library from Library Manager

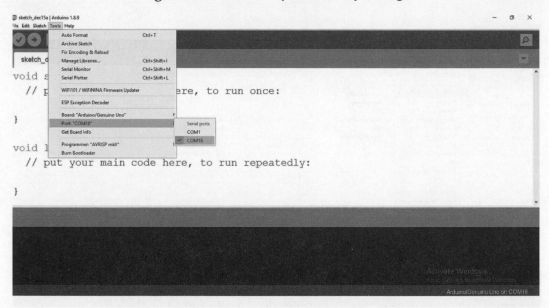

Figure 16.8 Select the port to which the board is connected

baud rate for your console. We will discuss baud rate in more detail in the subsequent sections. For now, you must make sure that the baud rate on your serial console is the same as the baud rate declared in your sketch to see the correct output.

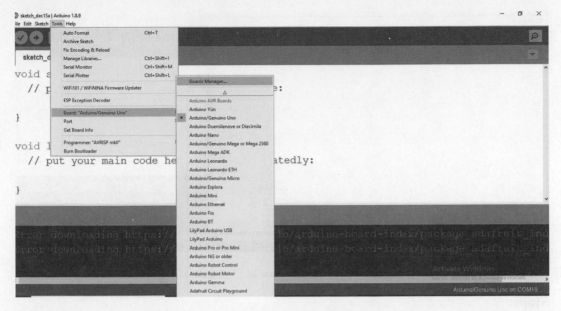

Figure 16.9 Select the board type

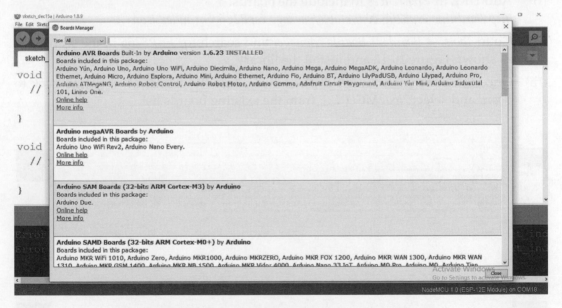

Figure 16.10 Install new boards from Boards Manager

16.1.3   Setting up Arduino IDE for NodeMCU

NodeMCU is a variant of processor boards supported by the Arduino IDE. It requires addition of external URLs in the Arduino IDE setting to let the Arduino identify the board. The step-by-step process to include NodeMCU in Arduino IDE is explained here.

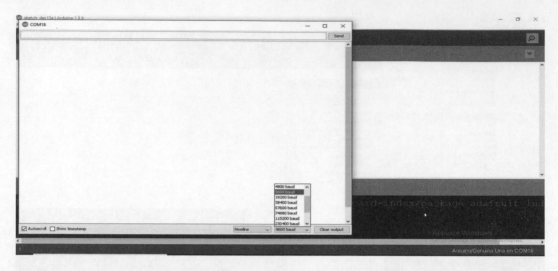

Figure 16.11 Serial console of Arduino IDE

Steps:

(i)   Add URL in *Preferences* to include the boards.
       http://arduino.esp8266.com/stable/package_esp8266com_index.json

(ii)   Go to *Tools → Board → Boards Manager* and search for *ESP8266* in the search bar
       as shown Figure 16.12 and install the library.

(iii) Once the library is successfully installed, close the *Boards Manager*, go back to
       Board and select *NodeMCU 12E* from the existing boards list.

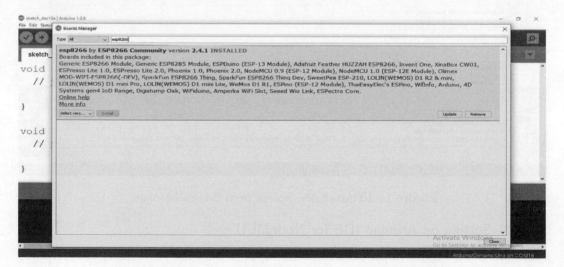

Figure 16.12 Install ESP8266 board from Boards Manager

16.2 Writing an Arduino Sketch

An Arduino program is referred to as a *sketch*. The coding language for Arduino is very similar to C and C++. A sketch follows a basic skeleton which can be altered according to the desired output. An Arduino sketch has two functions: *setup()* and *loop()*, as shown in Figure 16.4. The *setup()* function is called once during the setup phase of Arduino on powering the board. The *loop()* function, as the name suggests, runs in loop until the board is powered. The *setup()* function has the declaration of baud rate, input and output pins, and call to other functions that may be required for setting up the board. Once the sketch is ready, it is uploaded to the connected board using the upload button in the top left corner under the toolbar, as shown in Figure 16.13.

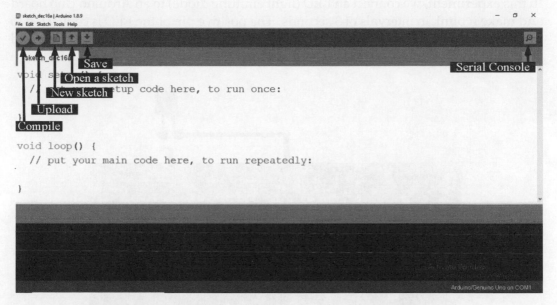

Figure 16.13 Different options available on Arduino IDE editor screen

16.3 Hands-on Experiments with Arduino

Let us see a few examples that use an Arduino Uno board to perform basic operations.

16.3.1 Printing on the serial console

This program prints a message on the serial console of the Arduino IDE. Once the code is complied and uploaded to the board, open the serial console and set the baud rate to 4800 to see the output.

```
void setup() {
Serial.begin(4800); //set the baud rate
}
void loop() {
Serial.println(''Print your statement here!''); //println prints the statement in
    new line
}
```

Output on serial console: Print your statement here!

16.3.2 LED interface with Arduino

In this experiment, we connect an LED (light emitting diode) to an Arduino Uno board and make it blink in intervals of 3 seconds. The positive pin of the LED is connected to pin 5 of the Arduino board and the negative pin of the LED is connected to the ground pin of the board (Figure 16.14).

Figure 16.14 Circuit for connecting an LED with an Arduino board

Connection description
- Connect the positive terminal of the LED to the digital pin 5 of the Arduino board through a 1kΩ resistor.
- Connect the negative terminal of the LED to the ground pin of the Arduino board.

```
int LED=5;
void setup(){
Serial.begin(4800); //set the baud rate
pinMode(LED, OUTPUT); // Declare the mode of pin
Serial.println(''LED Blink'');
}
void loop(){
digitalWrite(led, HIGH);// make the LED pin high
delay(3000); //delay of 3000 milliseconds
digitalWrite(led, LOW); //make the LED pin low
delay(3000);
}
```

Output on serial console: LED Blink!

Circuit output: Led blinks in intervals of 3 seconds.

The setting of baud rate and declaration of pin mode is done in the *setup()* function so that it is done only once when the board is powered while the output LED pin is assigned HIGH and LOW values in the *loop()* function so that the LED keeps blinking while the board is powered.

16.3.3 DHT Sensor interface with NodeMCU

A DHT (digital humidity and temperature) sensor is used to measure the humidity and temperature of a place. A separate library is provided by a vendor called Adafruit to use this sensor. You can download the DHT sensor library from the Library manager under the sketch option. Figure 16.15 shows the connection between NodeMCU and DHT sensor.

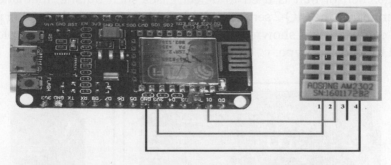

Figure 16.15 Circuit for connecting DHT with NodeMCU

Connection description
- Let the DHT pins be numbered 1, 2, 3, 4 moving from left to right.
- Connect pin 1 to 3.3 V of NodeMCU.
- Connect pin 2 to D1 of NodeMCU.
- Connect pin 4 to ground of NodeMCU.

- Leave pin 3 open.
- Power your NodeMCU using a micro USB adapter.

```
#include <DHT.h>
#define DHTPIN D1
#define DHTTYPE DHT22
DHT dht(DHTPIN, DHTTYPE);
float humidity; //Store humidity
float temp; //Stores temperature
void setup(){
Serial.begin(4800);
dht.begin();
}
void loop(){
humidity = dht.readHumidity();
temp= dht.readTemperature();
Serial.print(''Humidity: '');
Serial.print(humidity);
Serial.print('' %, Temp: '');
Serial.print(temp);
Serial.println('' C'');
delay(5000); // measure temperature every 5 second
}
```

Output on serial console: Humidity: 55.30%, Temp: 22.4 C

16.3.4 MQ-2 Gas sensor interface with NodeMCU

MQ-2 is a gas sensor that is used to detect gases such butane, methane, LPG and smoke. We interface the MQ-2 sensor with NodeMCU through the analog pin A0. The circuit with connections is shown in Figure 16.16. You can set the threshold according to the value for each gas type.

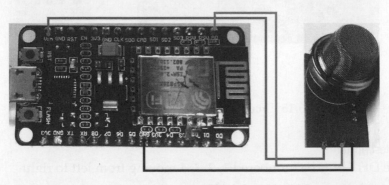

Figure 16.16 Circuit for connecting MQ-2 gas sensor with NodeMCU

Connection description

- Connect pin VCC of MQ-2 sensor to 3.3 V of NodeMCU.
- Connect pin GND of MQ-2 sensor to ground of NodeMCU.
- Connect pin A0 of MQ-2 sensor to A0 of NodeMCU.
- Power your NodeMCU using a micro USB adapter.

```
int gas = A0;
int gasThreshold = 400;

void setup() {
  pinMode(gas, INPUT);
  Serial.begin(4800);
}
void loop() {
  int gasValue = analogRead(gas); // read sensor value from the analog pin
  Serial.println(gasValue);
  if (gasValue > gasThreshold) // if sensed value exceeds the threshold
  {
    Serial.println(``Gas detetcted!'');
  }
  else
  {
    Serial.println(``No gas detected'');
  }
  delay(1000); //check value with a delay of 1 second
}
```

Serial Console Output: Gas detected!

16.3.5 Ultrasonic sensor interface with NodeMCU

An ultrasonic sensor uses ultrasound to detect the distance from objects or obstacles. We connect the ultrasonic sensor with NodeMCU (Figure 16.17) and calculate the distance of the object from the sensor.

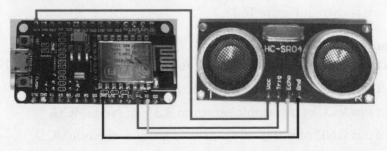

Figure 16.17 Circuit for connecting Ultrasonic sensor with NodeMCU

Connection description

- Connect pin echo of ultrasonic sensor to D1 of NodeMCU through 1kΩ resistor.
- Connect pin trig of ultrasonic sensor to D2 of NodeMCU.
- Connect pin VCC of ultrasonic sensor to 3.3 V of NodeMCU.
- Connect pin GND of ultrasonic sensor to ground of NodeMCU.
- Power your NodeMCU using a micro USB adapter.

```
int echo = D1; //connect trigger pin of ultrasonic sensor with D1 of NodeMCU
int trigger = D2; //connect echo pin of ultrasonic sensor with D2 of NodeMCU
long traveltime; // variable to store time duration
void setup(){
pinMode(trigger, OUTPUT); // declare trigger pin as output
pinMode(echo, INPUT); // declare echo pin as input
Serial.begin(9600);
}
void loop(){
digitalWrite(trigger, LOW);
delayMicroseconds(10); //delay of 10 micro seconds
digitalWrite(trigger, HIGH);
delayMicroseconds(10);
digitalWrite(trigger, LOW);
traveltime = pulseIn(echo, HIGH); //get return time of ultrasound wave
distance= duration*0.034/2; //calculate the distance
Serial.print(''Distance= '');
Serial.println(distance); // print distance on serial console
}
```

Output on serial console: Distance = 10.03

16.3.6 Obstacle detection using NodeMCU

In this setup, we use the ultrasound sensor to detect the distance of the sensor from an obstacle and trigger an alarm when the obstacle comes closer than a pre-programmed distance. We connect a buzzer through an NPN transistor with the NodeMCU. The connections are shown in Figure 16.18.

Connection description

- Connect pin echo of ultrasonic sensor to D2 of NodeMCU through 1kΩ resistor.
- Connect pin trig of ultrasonic sensor to D3 of NodeMCU.
- Connect pin VCC of ultrasonic sensor to 3.3 V of NodeMCU.
- Connect pin GND of ultrasonic sensor to ground of NodeMCU.
- Connect positive terminal of the buzzer to Vin of NodeMCU.
- Connect the negative terminal of the buzzer to collector pin of BC337 transistor.

- Connect the base pin of BC337 to D1 through 1 kΩ.
- Connect the emitter terminal of BC337 to ground of NodeMCU.
- Power your NodeMCU using a micro USB adapter.

Output on serial console: Obstacle detected! **Circuit output:** Buzzer sounds for 5 seconds when obstacle is detected.

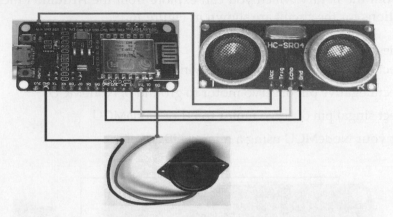

Figure 16.18 Circuit for alarm on obstacle detection

```
int buzzer = D1; //buzzer pin
int echo = D2; //connect trigger pin of ultrasonic sensor with D1 of NodeMCU
int trigger = D3; //connect echo pin of ultrasonic sensor with D2 of NodeMCU
long traveltime; // variable to store time duration
void setup(){
  pinMode(buzzer, OUTPUT); // declare buzzer pin as output
  pinMode(trigger, OUTPUT); // declare trigger pin as output
  pinMode(echo, INPUT); // declare echo pin as input
  Serial.begin(9600);
}
void loop(){
  digitalWrite(trigger, LOW);
  delayMicroseconds(10); //delay of 10 micro seconds
  digitalWrite(trigger, HIGH);
  delayMicroseconds(10);
  digitalWrite(trigger, LOW);
  traveltime = pulseIn(echo, HIGH); //get return time of ultrasound wave
  distance= duration*0.034/2; //calculate the distance
  if (distance <= 10){
    Serial.println(Obstacle detected!); // print distance on serial console
    digitalWrite(buzzer, HIGH);
    delay(5000);
    digitalWrite(buzzer, LOW);
    }
}
```

16.3.7   Servo motor interface with NodeMCU

In this experiment, we make a circuit to control a servo motor using NodeMCU. The servo motor moves in 0° to 180° angles. The Arduino IDE provides a built-in library *Servo* for controlling the servo motor. We include this library in our sketch and call library functions to control the movement of our servo motor. There are other functions from the library which you can explore from the Arduino official website. The connections for the circuit are shown in Figure 16.19.

Connection description

- Connect positive pin servo motor to Vin of NodeMCU.
- Connect negative pin of servo motor to ground of NodeMCU.
- Connect singal pin of servo motor to D1 of NodeMCU.
- Power your NodeMCU using a micro USB adapter.

Figure 16.19 Circuit for connecting servo motor to NodeMCU

```
#include <Servo.h> //include servo library
int servo = D2; //declare server pin at D2
Servo servoObject; //create a Servo type object
void setup() {
  servoObject.attach(servo); //connect servo object to the servo pin
}
void loop(){
  servoObject.write(0); //move motor to 0 Degree angular position
  delay(1000); //wait in the position for 1 second
  servoObject.write(90); //move motor to 90 Degree angular position
  delay(1000); //wait in the position for 1 second
  servoObject.write(180); //move motor to 180 Degree angular position
  delay(1000); //wait in the position for 1 second
}
```

Circuit output: Servo motor's shaft rotates in sequence as per the input value.

16.3.8 Relay interface with NodeMCU

Relay is an actuator that uses electromagnetic effect to act as a switch. It can close and open a circuit when electricity is passed through it. In this experiment, we use a relay to control an LED. Relay can be used to control the power supply to any other connected device. The connections for the circuit are shown in Figure 16.20.

Connection description

- Connect positive pin relay to Vin of NodeMCU.
- Connect negative pin of relay to ground of NodeMCU.
- Connect signal pin of relay to D1 of NodeMCU.
- Connect common terminal of relay to ground of NodeMCU.
- Connect normally open terminal to negative terminal of LED.
- Connect the positive terminal of the LED to D2 of NodeMCU
- Power your NodeMCU using a micro USB adapter.

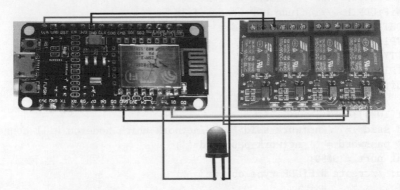

Figure 16.20 Circuit for connecting relay to NodeMCU

```
int relayPin= D1 //relay signal connected at D1
int LED = D2 //LED connected at D2
void setup() {
  pinMode(relayPin, OUTPUT);
  pinMode(led, OUTPUT);
  digitalWrite(led, HIGH);
}
void loop(){
  digitalWrite(relayPin, HIGH);
  delay(5000);
  digitalWrite(relayPin, LOW);
  delay(5000);
}
```

Circuit output: LED blinks every 5 seconds.

16.3.9 Data transmission between NodeMCU and remote server

NodeMCU uses an ESP8266 chip to enable wireless communication. It can act as a client, server, or access point. We perform a simple experiment to send data from NodeMCU to a remote server in the same network through a UDP (user datagram protocol) socket program. The circuit is similar to the DHT integration circuit in Figure 16.15.

Connection description

- Let the DHT pins be numbered 1, 2, 3, 4 moving from left to right.
- Connect pin 1 to 3.3 V of NodeMCU.
- Connect pin 2 to D1 of NodeMCU.
- Connect pin 4 to ground of NodeMCU.
- Leave pin 3 open.
- Power your NodeMCU using a micro USB adapter.

```
#include <ESP8266WiFi.h> //include wifi header file
#include <WiFiUDP.h> //include UDP header file
#include <DHT.h>;  //include DHT header file
#define DHTPIN D1
#define DHTTYPE DHT22
DHT dht(DHTPIN, DHTTYPE);
float humidity; //Store humidity
float temp; //Stores temperature
char packet[1024];
const char* ssid  = ''<network_ssid>''; //network which nodemcu will connect
const char* password = ''<network_password>'';
unsigned int port = 9000;
WiFiUDP Udp; //create WiFiUDP type object
void setup(){
Serial.begin(4800);
dht.begin();
WiFi.begin(ssid, password);
Udp.begin(port);
while (WiFi.status() != WL_CONNECTED) {
  delay(500);
  Serial.print(''Connecting to WiFi'');
  }
}
void loop(){
  Udp.begin(port);
  Udp.beginPacket(''<YOUR_IP>'', port);
  temp= String(dht.readTemperature());
  temp.toCharArray(packet, 1024);
  Udp.write(packet); //send data
  Udp.endPacket();
  delay(5000); // send data every 5 seconds
}
```

```
#Python UDP Server Code
import socket
UDP_IP = ''<YOUR IP>''
UDP_PORT = 9000
sock = socket.socket(socket.AF_INET,socket.SOCK_DGRAM) # UDP socket
sock.bind((UDP_IP, UDP_PORT))
while True:
  received_data,addr= sock.recvfrom(1024) # 1024 buffer size
  print received_data
```

16.3.10   Pulse sensor interface with NodeMCU

We integrated temperature and humidity sensor, gas sensor, ultrasonic sensor and others. Let us now sense our own pulse using a basic biomedical sensor. In this experiment, we will integrate a pulse sensor with NodeMCU. Until now we have been using the serial console to see the output but here we will see the plot of our pulse data. The circuit connection is shown in Figure 16.21.

Connection description

- Connect positive pin of pulse sensor to 3.3 V of NodeMCU.
- Connect negative pin of pulse sensor to ground of NodeMCU.
- Connect signal pin of pulse sensor to A0 of NodeMCU.
- Power your NodeMCU using a micro USB adapter.

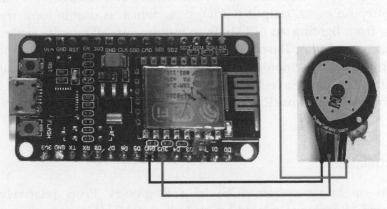

Figure 16.21 Circuit for connecting a pulse sensor to NodeMCU

```
int pulse = A0;
void setup() {
  pinMode(pulse, INPUT);
  Serial.begin(4800);
}
void loop() {
  int pulseValue = analogRead(pulse);
  Serial.println(pulseValue);
}
```

Output on serial console: <pulse amplitude>
Output on serial plotter: Navigate to *Tools* → *Serial Plotter* to see the plot.
Note that the plot is visible only when the number values are printed without any text.
You can make further changes to the code to calculate the pulse rate.

16.4   Introduction to Raspberry Pi Boards

The Raspberry Pi is a single-board personalized computer developed by the Raspberry Pi Foundation (a charitable organization) in the UK in the year 2009. The main motive of the developers was to promote the study of computer science and its exposure to computers among people of all ages at a low cost. The Raspberry Pi is comparable to the size of a credit card and equipped with an HDMI (high-definition multimedia interface) port for connecting with regular monitors/TVs. It attaches to a standard keyboard and mouse through USB ports. It also has an Ethernet port, is Wi-Fi enabled, and also has a Bluetooth. The Raspberry Pi also features 40 general-purpose input/output (GPIO) pins, which is usable for a wide range of applications, from lighting an LED bulb to building an HD media center to creating a gaming machine. Today, Raspberry Pis are being used extensively by students and researchers for developing new systems all around the world. More details regarding its hardware and scope are available at *www.raspberrypi.org*. The Raspberry Pi also has a smaller version by the name Raspberry Pi Zero and Raspberry Pi Zero-W with minor variations.

16.4.1   Installation

Installing an operating system (OS) in a Raspberry Pi is relatively easy and straightforward. Raspbian, created by by Debian GNU/Linux, is the most popular OS distribution for Raspberry Pis. Raspbian offers both headless (command line) access to the Raspberry Pi and also a GUI (graphical user interface), which supports all the features of a typical PC. New versions of Raspbian are available for download at www.raspberrypi.org. However, the installation needs a *disk imager* software (available for both Windows and Linux) for burning the downloaded *.img* (Raspbian) file into a microSD card (at least 8 GB). This microSD card goes into the slot in the

Raspberry Pi. The final step is to connect a monitor via the HDMI port, keyboard and mouse through the USB ports, and power it through a micro USB port. The Raspberry Pi boots up, and a sequence of lines appears on the monitor. After some time, the desktop appears, which is very similar to that of a Linux desktop. The default user name is *pi*, and the password is *raspberry*. For security, we strongly recommend changing the password immediately.

Apart from the installation method mentioned here, some developers also use NOOBS for installing Raspbian. NOOBS is an OS installation manager, which contains Raspbian, along with a few other options. The NOOBS also needs to be burned into the microSD card; it guides the user through the installation process. There are also a few SD cards available in the market that have NOOBS preloaded. It is advisable to download these versions straight from the official website, as shown in Figure 16.22.

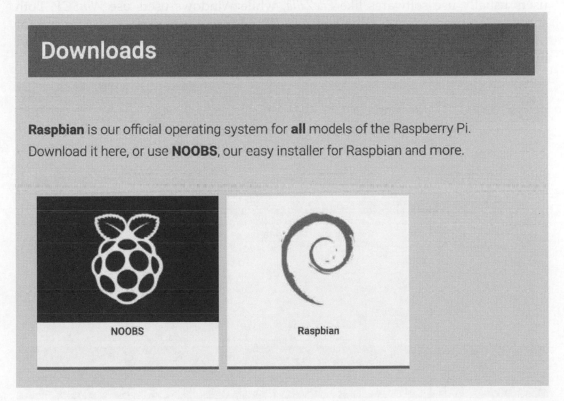

Figure 16.22 Official download page

16.4.2   Remotely accessing the Raspberry Pi

The Raspberry Pi does not mandate the use of a plugged package of monitor, keyboard, and mouse. Raspbian has the provision of accessing its command line remotely through Secure Socket Shell (SSH). However, it needs to have a wired/wireless connection to a network with a valid IP address. In addition to the network connection, the users need to enable SSH in the Raspberry Pi. For enabling the SSH, type *sudo raspi-config* in the terminal and navigate to *Interfacing Options* using the arrow keys, as shown in Figure 16.23. Under *Interfacing Options*, navigate to *SSH* and choose Yes for enabling SSH, as shown in Figure 16.24. Once enabled, the Raspberry Pi is ready for remote access. For Linux users, it is possible to access Raspberry Pi directly through the terminal. The command for doing so is *ssh user_name@IP_address*, followed by a prompt that queries for the *password*. Note that the terminal does not display any character while entering the password. On the other hand, Windows users need to install an SSH software (preferably Smartty or Putty) and enter the *hostname* (IP address), *user name*, and *password* for accessing the Raspberry Pi's terminal.

Raspbian also allows file transfers from a remote system over the network. Linux users usually use softwares like *FileZilla*, while Windows users use *WinSCP*. Both software prompts users for the *hostname* (IP address), *user name*, and *password* for initiating file transfer. The file transfer appears over the same port as SSH, i.e., port number 22.

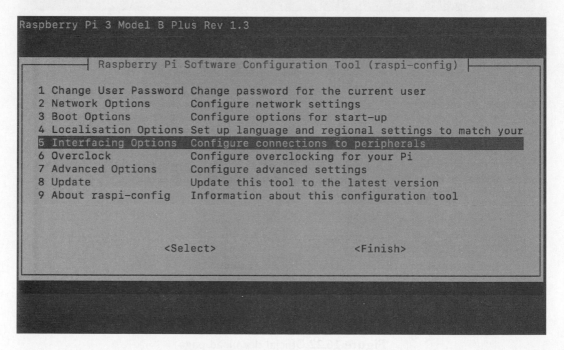

Figure 16.23 Enable SSH (Step 1)

16.4.3   Introduction to Python basics

Python is one of the most popular high-level object-oriented programming languages. It has a myriad of modules and packages that significantly increases the productivity of developers. Analogous to most Linux distributions, Raspbian has Python versions

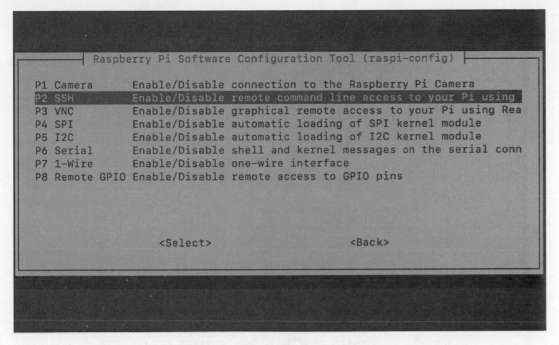

Figure 16.24 Enable SSH (Step 2)

2.x and 3.x preinstalled, which enables users to dive into programming directly after installation. The users may run programs directly on the terminal by opening the Python console or through *.py* scripts. Usually, developers write their programs on Python IDEs running on remote systems and transfer the files to the Raspberry Pi for execution. On the other hand, a few prefer writing the codes directly using editors available in Raspbian (vim, nano, etc.).

16.4.4 Accessing GPIO pins

As mentioned earlier, Raspberry Pi has 40 GPIO pins. It has two 5 V and two 3.3 V pins, and a number of ground (0 V) pins, all of which are unconfigurable. The rest of the pins are configurable, where the pins read 3.3 V as high and 0 V as low. In addition, the GPIO pins also support pulse-width modulation (PWM), serial peripheral interface (SPI), inter-integrated circuit (I2C), and serial communication. These GPIO pins are easily configurable through Python to other programming languages. Note that GPIO pins need to be programmed and connected very carefully as wrong configurations may damage the board.

Before diving into programming, it is necessary to know the pin numbers. GPIO pins are labeled in two formats: Broadcom (BCM) and Board (BOARD). The BOARD format labels the pins according to the actual number on the board. On the other hand, the BCM format labels the pins according to the way the chip onboard the Raspberry Pi addresses them. Figure 16.25 shows the GPIO pins labeled in BCM and

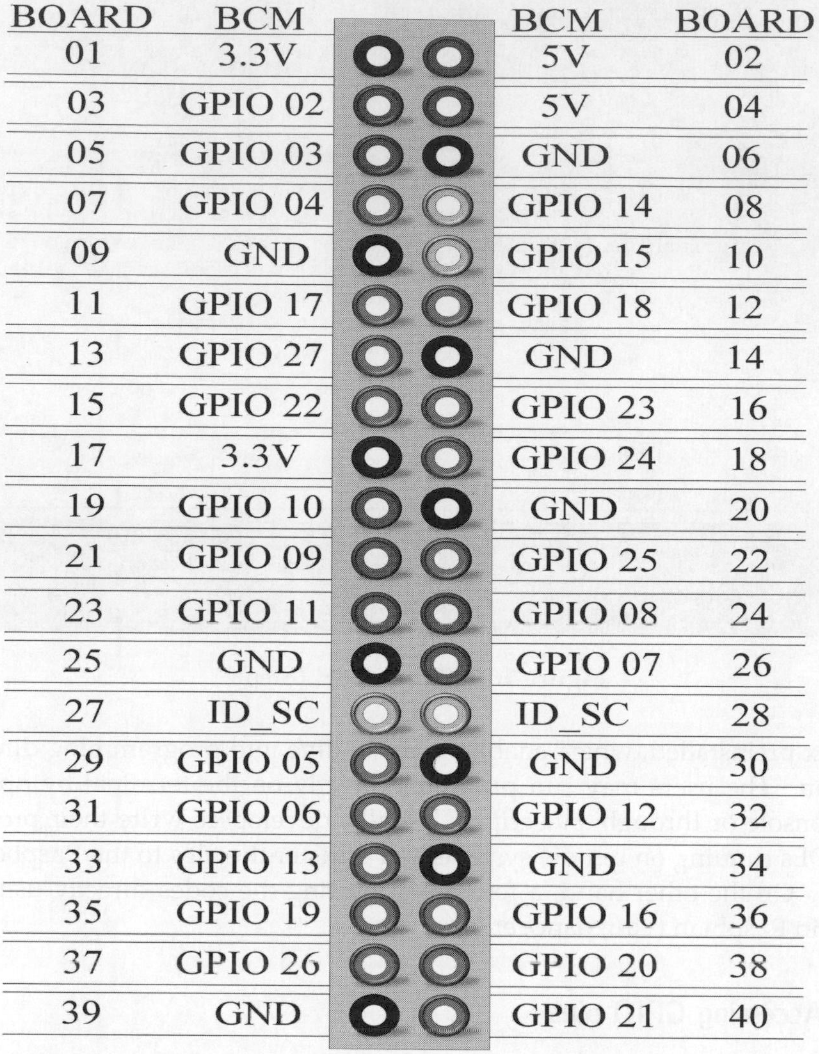

BOARD	BCM			BCM	BOARD
01	3.3V	⬤	⬤	5V	02
03	GPIO 02	⬤	⬤	5V	04
05	GPIO 03	⬤	⬤	GND	06
07	GPIO 04	⬤	◯	GPIO 14	08
09	GND	⬤	◯	GPIO 15	10
11	GPIO 17	⬤	⬤	GPIO 18	12
13	GPIO 27	⬤	⬤	GND	14
15	GPIO 22	⬤	⬤	GPIO 23	16
17	3.3V	⬤	⬤	GPIO 24	18
19	GPIO 10	⬤	⬤	GND	20
21	GPIO 09	⬤	⬤	GPIO 25	22
23	GPIO 11	⬤	⬤	GPIO 08	24
25	GND	⬤	⬤	GPIO 07	26
27	ID_SC	◯	◯	ID_SC	28
29	GPIO 05	⬤	⬤	GND	30
31	GPIO 06	⬤	⬤	GPIO 12	32
33	GPIO 13	⬤	⬤	GND	34
35	GPIO 19	⬤	⬤	GPIO 16	36
37	GPIO 26	⬤	⬤	GPIO 20	38
39	GND	⬤	⬤	GPIO 21	40

Figure 16.25 GPIO pin numbering in BCM and BOARD modes

BOARD modes. The following lines in Python allows access to the GPIO pins: import RPi.GPIO as GPIO

GPIO.setmode(GPIO.BCM) The first line imports the necessary library functions. In case the GPIO library is not preinstalled, running *sudo pip install –upgrade RPi.GPIO* or *sudo apt-get install python-rpi.gpio python3-rpi.gpio* on the terminal installs the required packages. The GPIO.setmode() function sets the mode for addressing the pins. In the code mentioned earlier, we have set it to BCM, which may be altered to BOARD based on user requirement. In the next section, we illustrate GPIO and Python programming using a few experiments.

16.4.5   Configuring WiFi on Raspberry Pi

Connecting the Raspberry Pi through the GUI is straightforward. However, it gets relatively tricky while connecting through the command line. The user must know the SSID (service set identifier) and password for connecting to a WiFi router. The Raspberry Pi will not connect to the router in case of any error. For automatically connecting to the router, the details needs to be written into the *wpa_supplicant.conf* file in */etc/wpa_supplicant/wpa_supplicant.conf*. We accomplish this using nano and enter the following:

network={

ssid="*Router SSID*"

psk="*Router Password*"

}

Restart the Raspberry Pi for the changes to take effect and observe the Raspberry Pi connect to the router automatically.

16.5   Hands-on Experiments with Raspberry Pi

16.5.1   Printing on console/terminal

This is a very basic experiment in any programming language. We will use the objective of printing *Hello World* in the console to show how to run Python programs on the Raspberry Pi. The easiest way is to do it on the Python console. Type *python* on the terminal to open the console. On the next line, type *print("Hello World")* and press enter. This prints the characters written between the double quotes in the console. Typing *exit()* exits from the Python console and returns to the terminal.

A few developers prefer writing the same program as a script, where they type in the program using editors (such as nano) and execute it from the terminal. Let us write the *Hello World* script. Open a file using nano with the file extension as *.py* by typing *sudo nano print_text.py*. Type *print("Hello World")* on it. Save and exit to return to the terminal. The file is ready for execution. Type *python print_text.py* on the terminal and see the result on the screen. We use these steps to run the python programs in all subsequent experiments. We will mostly focus on assembling the hardware in each of the experiments from now on.

16.5.2   LED interface

In this experiment, we develop a program for making an LED bulb blink using Raspberry Pi and Python. We need jumper wires, an LED bulb, and a resistor for implementation. The resistor regulates the current flowing into the bulb, which avoids damage. Thus, though the resistor is optional, we highly recommend using it. We connect one end of the bulb to GPIO 18 (BCM mode) using the jumper wires and the other to the ground, as shown in Figure 16.26. Now open a file as *pi_led_blinker.py* using nano and type the following.

Connection description
- Connect the positive terminal of the LED to GPIO 18 through a 1kΩ resistor.
- Connect the negative terminal of the LED to the ground pin of Raspberry Pi.
- Power the board using a USB adapter.

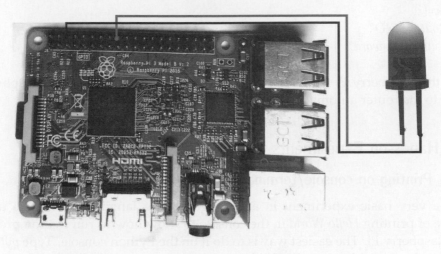

Figure 16.26 Circuit for connecting LED with Raspberry Pi

```
import RPi.GPIO as GPIO # Import the GPIO library
import time # Import the time module to enable sleep

GPIO.setwarnings(False) # Ignore warnings for simplicity
GPIO.setmode(GPIO.BCM) # Use BCM pin numbering
GPIO.setup(18, GPIO.OUT, initial=GPIO.LOW) # Setting GPIO pin 18 as an output pin
    with initial value low or LED off

while True: # Run forever
  GPIO.output(18, GPIO.HIGH) # Turn LED on
  time.sleep(1) # LED on for 1 second
  GPIO.output(18, GPIO.LOW) # Turn LED off
  time.sleep(1) # LED off for 1 second
```

Save the file as explained earlier and execute the program by typing *python pi_led_blinker.py* in the terminal. The LED bulb should be blinking now. In case it does not blink, turn the LED around and change the polarity. It should be blinking now. Note that indentations are very important in Python. Any mistake in indentation will lead to errors.

Circuit output: LED blinks every 1 second.

16.5.3   PiCamera interface

Raspberry Pi has a camera module port for connecting a camera board (also developed by the Raspberry Pi Foundation). The module is a high quality 8 megapixel camera, capable of capturing images and videos in HD. It attaches to the Raspberry Pi through its ribbon cable onto the camera module port. For enabling the camera, type *sudo raspi-config* in the terminal and navigate to *Interfacing Options*, as in Figure 16.23. Then, as shown in Figure 16.27, find the option that says *Camera* and enable it. Connect the PiCam to the board as shown in Figure 16.28. Users may control the Raspberry Pi camera or PiCam through the command line or through Python programming.

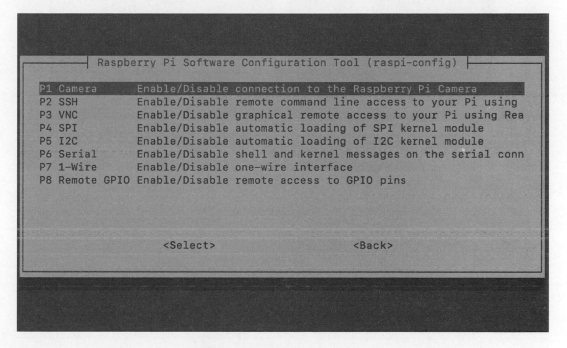

Figure 16.27 Enable camera

Connection description
- Connect the ribbon cable of the PiCam to the dedicated slot in the Raspberry Pi.
- Power the board using a USB adapter.

For capturing still images through the command line and saving it in the Desktop, type *raspistill -o Desktop/image_name.jpg* in the terminal. The terminal opens a prompt for a 5 second preview and then captures the image. Similarly, for recording videos, type *raspivid -o Desktop/video_name.h264* in the terminal. For capturing images through Python programming, execute the following program.

(a) A close view of the connected PiCam   (b) Complete setup of PiCam with Raspberry Pi

Figure 16.28 Connecting PiCam with Raspberry Pi

```
from picamera import PiCamera #Import the camera library
import time

camera = PiCamera() #Create camera object

camera.start_preview() #Start camera preview
time.sleep(5) #Continue preview for 5 seconds
camera.capture('Desktop/image_name.jpg') #Capture image and save
camera.stop_preview() #Exit
```

Output: Captures an image and saves it with a preview window of 5 seconds. Similarly for recording videos, execute the following program.

```
from picamera import PiCamera #Import the camera library
import time

camera = PiCamera() #Create camera object

camera.start_preview() #Start camera preview
camera.start_recording('Desktop/video_name.h264') #Record video and save
time.sleep(5) #Continue recording for 5 seconds
camera.stop_recording() #Stop recording video
camera.stop_preview() #Exit
```

Output: Starts camera preview for 5 seconds, records and saves video.

16.5.4　DHT Sensor interface

The DHT (digital humidity and temperature) sensor is a low-cost digital module for recording temperature and humidity. In this experiment, we demonstrate how to read the data from a DHT sensor using Raspberry Pi. Before diving into programming, we need to install the packages from Github. In case the git module is not installed in the Raspberry Pi, install it using *sudo apt-get install git-core*. Then download the DHT library using *git clone https://github.com/adafruit/Adafruit_Python_DHT.git*, install as *sudo apt-get install build-essential python-dev* from the Adafruit_Python_DHT directory, and finally type *sudo python setup.py install*.

Connection description

- Connect the terminals of DHT 1, 2, and 4 to 3.3 V, GPIO 4 and ground of Raspberry Pi.
- Power the board using a USB adapter.

The DHT sensor has 3 pins: *signal, VCC,* and *ground* (left to right). We connect the signal pin to GPIO pin 4, and the rest to the designated pins, as shown in Figure 16.29. For reading the data from the DHT sensor and printing it in the console, write and execute the following code.

Figure 16.29 Connecting DHT sensor with Raspberry Pi

```
import Adafruit_DHT #import the DHT library

while True: #Run forever
    h, t = Adafruit_DHT.read_retry(11, 4) #The first argument is for the version of
        the DHT sensor (11 in our case) and the second argument is for the GPIO pin
        number
    print('Temperature: {0:0.1f} C Humidity: {1:0.1f} %'.format(t, h)) #Print data
        from the sensor
```

Output: Temperature: 23.66 °C Humidity: 45%

16.5.5   Client–server socket programming

Socket programming is very essential for connecting two devices over the network. To test this, consider two devices (one or both Raspberry Pis). One device needs to act as a *server*, which keeps listening for new connections, and the other needs to act as a *client*, which connects to the *server*. Make sure that the devices are connected to the same network and save the IP addresses for entering into the programs. The following code is for the device acting as the *server*.

```
import socket #Import the socket library

s = socket.socket() #Create socket object
port = 5000 #Port number for the socket (change according to need)
s.bind((''IP Address of Server'', port)) #bind server to IP address and port
s.listen(5) #start listening for connections. Supports 5 connections simultaneously

while True: #Loop forever
  c, addr = s.accept() #Accept connection from clients
  print("Client connected!!")
  c.send(''Connection Established'') #Send string to client
  c.close() #Close connection
```

The following code is for the device acting as the *client*.

```
import socket #Import the socket library

s = socket.socket() #Create socket object

port = 5000 #Port number should be same as server

s.connect((''IP Address of Server'', port))
print (s.recv(1024)) #Print data received from the server
s.close() #Close connection
```

Server Output: Client connected!!
Client Output: Connection established

16.5.6   Serially reading data from Arduino

Arduino and Raspberry Pi are the most popularly used single-board processors in the world. While Arduino provides a simple yet powerful platform for connecting peripherals, Raspberry Pi provides high operational capability as well as network connectivity. Keeping in mind the features of both the boards, combining them to work as a single board is beneficial. We now demonstrate how to connect the two boards over serial connection. Toward this, connect the Arduino's USB cable to one of the USB ports in the Raspberry Pi, as shown in Figure 16.30. Type *lsusb* in the terminal to check for successful connection. Next, we need the port in which the Arduino is connected to the Raspberry Pi. We achieve this by typing *ls /dev/tty** in the terminal, as shown in Figure 16.31. Note the port and enter it in the following code.

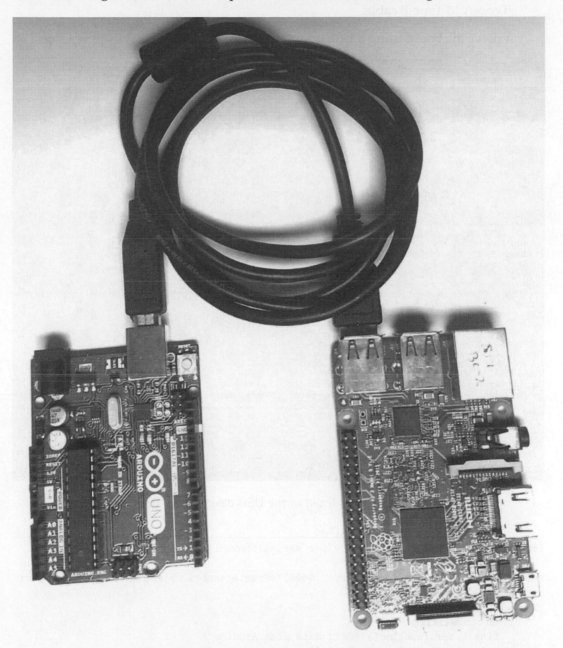

Figure 16.30 Connecting Arduino with Raspberry Pi

Connection description

- Connect the Arduino to any of the USB ports of Raspberry Pi through a USP Type-A to Type-B cable.

- Power the board using a USB adapter.

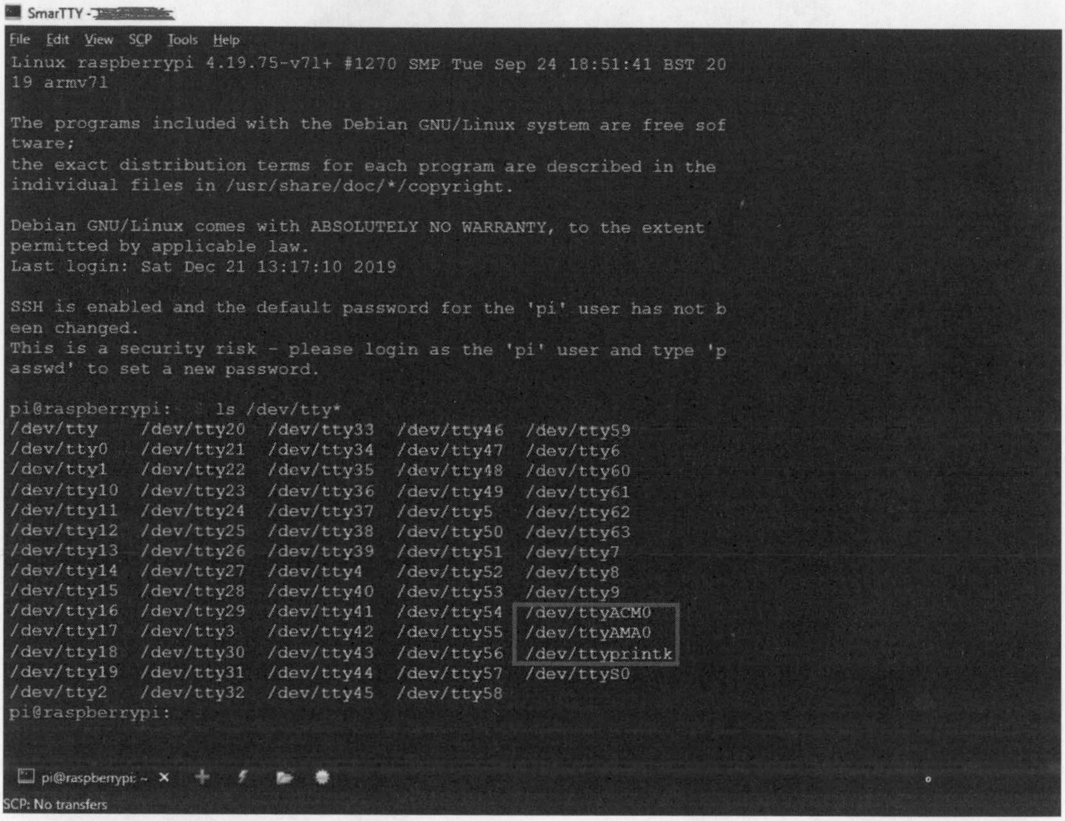

Figure 16.31 Arduino listed in the USB device list of Raspberry pi

```
import serial #Import the library for serial communication

ser = serial.Serial('Arduino port', 9600) #Create serial object with Arduino's port
    name and 9600 baud rate
while True: #Loop forever
    if(ser.in_waiting >0):
        line = ser.readline() #Read data from Arduino
        print(line) #Print received data
        ser.write(b'3') #Send 3 to Arduino
```

On the other end, i.e., in the Arduino, burn the following sketch using the Arduino IDE for establishing connection.

```
int r = 1;
void setup(){
  Serial.begin(9600);
}
void loop(){
 Serial.println(''Hello'') //Send Hello to Raspberry Pi
  if(Serial.available()){     //Connection from Raspberry Pi to Arduino
    r = r * (Serial.read() - '0'); //Convert char value to integer
    Serial.println(r); //Print data from Raspberry Pi
  }
}
```

Raspberry Pi Output: Hello
Arduino output: 3

Summary

This chapter covered the basics of sensing and actuation circuits, which were enabled using various processing platforms and hardware. This chapter initiates the reader on how to get started with some of the basic Arduino and Raspberry Pi-based boards, and how to integrate external sensors and actuators with them. The circuit diagram for the integration of various sensors/actuators, their sample codes, and outputs are provided for a few necessary sensors and boards. We hope that the readers of this book will be able to connect what they learned in the previous chapters with the basic mechanisms outlined in this chapter. The readers could also try implementing various communication and connectivity technologies using the basic background we have provided in this chapter.

Exercises

(i)   How are Arduino boards different from Raspberry Pi boards?

(ii) What are the common Arduino boards available in the market?

(iii) How are analog sensors connected to Arduino boards?

(iv) What is a sketch in the context of Arduino?

(v)   Design a program to infinitely blink a sequence of 4 LEDs connected to an Arduino board, one after the other with a delay of 500 ms.

(vi) Design a program to sense temperature using a temperature sensor connected to an Arduino board. If the temperature exceeds 40 °C, a servo motor is actuated for 10 seconds. The actuated motor rotates between 0 and 180 degrees.

(vii) Design a program to sense obstacles using an ultrasonic proximity sensor connected to a NodeMCU board. Upon detection of an obstacle within a range of fewer than 5 cms, the board transmits a warning message to a remote server-based client.

(viii) Design a program to capture an image through the RPi camera. Once the image is clicked, an LED connected to the RPi board blinks twice, otherwise, the LED is in the OFF state.

(ix) Design a program to capture an image through the RPi camera and simultaneously record the temperature and humidity of the environment using a DHT connected to the same Raspberry Pi board. The data from the camera and the DHT sensor are transmitted to a remote server through a socket.

IoT Analytics

After reading this chapter, the reader will be able to:

- Describe the common analytical tools and machine learning algorithms used with IoT data
- Assess the importance and applicability of each algorithm
- Understand the operating principle of each of these analytical methods
- Assess the performance of various analytical and learning algorithms and methods through the use of various performance metrics
- Relate to the uses of various learning algorithms through examples

17.1 Introduction

In previous chapters, we learned that sensors are an intrinsic part of IoT. These sensors collect data from the environment and serve different IoT-based applications. The raw data from a sensor require processing to draw inferences. However, an IoT-based system generates data with complex structures; therefore, conventional data processing on these data is not sufficient. Sophisticated data analytics are necessary to identify hidden patterns. In this chapter, we discuss a few traditional data analytics tools that are popular in the context of IoT applications. These tools include k-means, decision tree (DT), random forest (RF), k-nearest neighbor (KNN), and density-based spatial clustering of applications with noise (DBSCAN) algorithms. Before discussing these algorithms, let us understand some of the basics related to machine learning (ML).

17.1.1 Machine learning

The term "machine learning" was coined by Arthur Lee Samuel, in 1959. He defined machine learning as a "field of study that gives computers the ability to learn without being explicitly programmed".

ML is a powerful tool that allows a computer to learn from past experiences and its mistakes and improve itself without user intervention. Typically, researchers envision IoT-based systems to be autonomous and self-adaptive, which enhances services and user experience. To this end, different ML models play a crucial role in designing intelligent systems in IoT by leveraging the massive amount of generated data and increasing the accuracy in their operations. The main components of ML are statistics, mathematics, and computer science for drawing inferences, constructing ML models, and implementation, respectively.

> **Points to ponder**
>
> - ML is an important tool, which is used by different social networking websites such as *facebook* and *twitter*.
> - Autonomous vehicles use ML to determine their paths and speeds.

17.1.2 Advantages of ML

Applications fueled by ML open a plethora of opportunities in IoT-based systems, from triggering actuators to identifying chronic diseases from images of an eye. ML also enables a system to identify changes and to take intelligent actions that relatively imitates that of a human. As ML demonstrates a myriad of advantages, its popularity in IoT applications is increasing rapidly. In this section, we discuss the different advantages of ML, as depicted in Figure 17.1

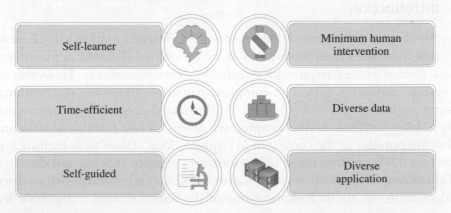

Figure 17.1 Advantages of ML

(i) **Self-learner:** An ML-empowered system is capable of learning from its prior and run-time experiences, which helps in improving its performance continuously. For example, an ML-assisted weather monitoring system predicts the weather report of the next seven days with high accuracy from data collected in the last six months. The system offers even better accuracy when it analyzes weather data that extends back to three more months.

(ii) **Time-efficient:** ML tools are capable of producing faster results as compared to human interpretation. For example, the weather monitoring system generates a weather prediction report for the upcoming seven days, using data that goes back to 6–9 months. A manual analysis of such sizeable data for predicting the weather is difficult and time-consuming. Moreover, the manual process of data analysis also affects accuracy. In such a situation, ML is beneficial in predicting the weather with less delay and accuracy as compared to humans.

(iii) **Self-guided:** An ML tool uses a huge amount of data for producing its results. These tools have the capability of analyzing the huge amount of data for identifying trends autonomously. As an example, when we search for a particular item on an online e-commerce website, an ML tool analyzes our search trends. As a result, it shows a range of products similar to the original item that we searched for initially.

(iv) **Minimum Human Interaction Required:** In an ML algorithm, the human does not need to participate in every step of its execution. The ML algorithm trains itself automatically, based on available data inputs. For instance, let us consider a healthcare system that predicts diseases. In traditional systems, humans need to determine the disease by analyzing different symptoms using standard "if–else" observations. However, the ML algorithm determines the same disease, based on the health data available in the system and matching the same with the symptoms of the patient.

(v) **Diverse Data Handling:** Typically, IoT systems consist of different sensors and produce diverse and multi-dimensional data, which are easily analyzed by ML algorithms. For example, consider the profit of an industry in a financial year. Profits in such industries depend on the attendance of laborers, consumption of raw materials, and performance of heavy machineries. The attendance of laborers is associated with an RFID (radio frequency identification)-based system. On the other hand, industrial sensors help in the detection of machiney failures, and a scanner helps in tracking the consumption of raw materials. ML algorithms use these diverse and multi-dimensional data to determine the profit of the industry in the financial year.

(vi) **Diverse Applications:** ML is flexible and can be applied to different application domains such as healthcare, industry, smart traffic, smart home, and many others. Two similar ML algorithms may serve two different applications.

17.1.3   Challenges in ML

An ML algorithm utilizes a model and its corresponding input data to produce an output. A few major challenges in ML are listed as follows:

(i)   **Data Description:** The data acquired from different sensors are required to be informative and meaningful. Description of data is a challenging part of ML.

(ii) **Amount of Data:** In order to provide an accurate output, a model must have sufficient amount of data. The availability of a huge amount of data is a challenge in ML.

(iii) **Erroneous Data:** A dataset may contain noisy or erroneous data. On the other hand, the learning of a model is heavily dependent on the quality of data. Since erroneous data misleads the ML model, its identification is crucial.

(iv) **Selection of Model:** We have already discussed the use of ML algorithms in different applications. Multiple models may be suitable for serving a particular purpose. However, one model may perform better than others. In such cases, the proper selection of the model is pertinent for ML.

(v)   **Quality of Model:** After the selection of a model, it is difficult to determine the quality of the selected model. However, the quality of the model is essential in an ML-based system.

17.1.4   Types of ML

Typically, ML algorithms consist of four categories: (i) Supervised (ii) Unsupervised (iii) Semi-supervised (iv) Reinforcement Learning (Figure 17.2). In this section, we briefly explore different categories of ML. Before discussing further, we determine the meaning of labeled- and unlabeled-data. As the name suggests, labeled data contain certain meaningful tags, known as labels. Typically, the labels correspond to the characteristics or properties of the objects. For example, in a dataset containing the images of two birds, a particular sample is tagged as a crow or a pigeon. On the other hand, the unlabeled dataset does not have any tags associated with them. For example, a dataset containing the images of a bird without mentioning its name.

(i)   **Supervised Learning:** This type of learning supervises or directs a machine to learn certain activities using labeled datasets. The labeled data are used as a supervisor to make the machine understand the relation of the labels with the properties of the corresponding input data. Consider an example of a student who tries to learn to solve equations using a set of labeled formulas. The labels

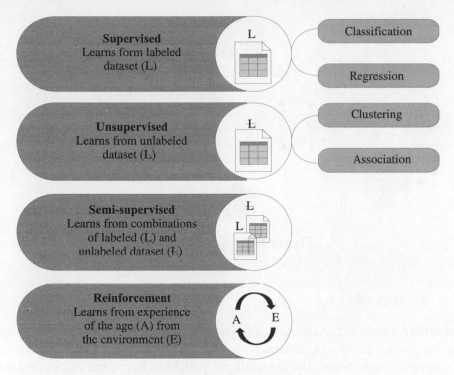

Figure 17.2 Types of ML

indicate the formulae necessary for solving an equation. The student learns to solve the equation using suitable formulae from the set. In the case of a new equation, the student tries to identify the set of formulae necessary for solving it. Similarly, ML algorithms train themselves for selecting efficient formulae for solving equations. The selection of these formulae depends primarily on the nature of the equations to be solved. Supervised ML algorithms are popular in solving classification and regression problems. Typically, the classification deals with predictive models that are capable of approximating a mapping function from input data to categorical output. On the other hand, regression provides the mapping function from input data to numerical output. There are different classification algorithms in ML. However, in this chapter, we discuss three popular classification algorithms: (i) k-nearest neighbor (KNN), (ii) decision tree (DT), and (iii) random forest (RF).

We use regression to estimate the relationship among a set of dependent variables with independent variables, as shown in Figure 17.3. The dependent variables are the primary factors that we want to predict. However, these dependent variables are affected by the independent variables. Let x and y be the independent and dependent variables, respectively. Mathematically, a simple regression model is represented as:

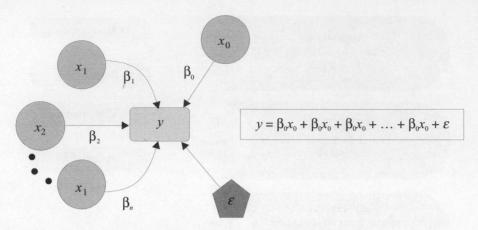

$$y = \beta_0 x_0 + \beta_0 x_0 + \beta_0 x_0 + \dots + \beta_0 x_0 + \varepsilon$$

Figure 17.3 Regression model

$$y = \beta_0 x_0 + \beta x + \epsilon \tag{17.1}$$

where β represents the amount of impact of variable x on y and ϵ denotes an error. In the given equation, x_0 creates β_0 impact on y, which indicates that the value of y can never be 0. Similarly, for multiple variables, say n, the regression model is represented as:

$$y = \sum_{i=0}^{n} \beta_i x_i + \epsilon \tag{17.2}$$

> **Check yourself**
>
> Support vector machine, Multilayer perceptron, Deep neural network, Convolutional neural network, Recurrent neural network

(ii) **Unsupervised Learning:** Unsupervised learning algorithms use unlabeled datasets to find scientific trends. Let us consider an example of the student similar to that described in the case of supervised learning, and illustrate how it differs in case of unsupervised learning. As already mentioned, unsupervised learning does not use any labels in its operations. Instead, the ML algorithms in this category try to identify the nature and properties of the input equation and the nature of the formulae responsible for solving it. Unsupervised learning algorithms try to create different clusters based on the features of the formulae and relate it with the input equations. Unsupervised learning is usually applied to solve two types of problems: clustering and association. Clustering divides the data into multiple groups. In contrast, association discovers the relationship or association among the data in a dataset.

(iii) **Semi-Supervised Learning:** Semi-supervised learning belongs to a category between supervised and unsupervised learning. Algorithms under this category use a combination of both labeled and unlabeled datasets for training. Labeled data are typically expensive and are relatively difficult to label correctly. Unlabeled data is less expensive than labeled data. Therefore, semi-supervised learning includes both labeled and unlabeled dataset to design the learning model. Traditionally, semi-supervised learning uses mostly unlabeled data, which makes it efficient to use, and capable of overcoming samples with missing labels.

(iv) **Reinforcement Learning:** Reinforcement learning establishes a pattern with the help of its experiences by interacting with the environment. Consequently, the agent performs a crucial role in reinforcement learning models. It aims to achieve a particular goal in an uncertain environment. Typically, the model starts with an initial state of a problem, for which different solutions are available. Based on the output, the model receives either a reward or a penalty from the environment. The output and reward act as inputs for proceeding to the next state. Thus, reinforcement learning models continue learning iteratively from their experiences while inducing correctness to the output.

17.2   Selected Algorithms in ML

17.2.1   *k*-nearest neighbor (KNN)

KNN learning falls under the category of supervised learning algorithms. It is also known as lazy or instance-based learning. Typically, KNN learning is one of the simplest algorithms, as it needs no explicit training model. In KNN, the selection of a suitable *k* is essential to split the dataset into *k* clusters. Table 17.1 summarizes the advantages and disadvantages of KNN.

Table 17.1 Advantages and disadvantages of KNN

Advantages	Disadvantages
*Easily implementable	*Noise sensitive
*Easy handling of missing data	*Needs a large memory space
*Does not learn during the period of training; therefore, is faster compared to other ML algorithms	*Does not work well for large and high dimensional dataset

Steps for KNN: KNN follows five simple steps:

(i) Select a value of k, where k indicates the number of nearest sample points necessary for the application.

(ii) Compute the distance between the given data points to all the available training data points.

(iii) Sort the distances in ascending order.

(iv) Consider the kth nearest distances and identify the corresponding class of data points.

(v) Identify the majority of the class of those k nearest neighbor data points as the class of unknown data points.

Selection of k: In the KNN algorithm, k indicates the number of nearest neighbors necessary for predicting the outputs in a particular model. An appropriate selection of k is crucial to avoid anomalies. Such anomalies usually lead to wrong outputs as well as overfitting. The selection of a small value of k may give rise to a high variance in the output of the class level of the new dataset, which affects stability. However, a high value of k may bias the output in the class level of the new dataset.

Distance Computation: Typically, the KNN algorithm uses Euclidean, Manhattan, and Minkowski distance formulae for computing the distance between the unknown data point and the given sample data points. Let there be two points A and B; let a_i and b_i indicate their coordinates. Then, we compute the distances as follows:

- Euclidean distance $\mathcal{E}_d$ among different data points is computed as:

$$\mathcal{E}_d(A, B) = \sqrt{\sum_{i=1}^{x} (a_i - b_i)^2} \tag{17.3}$$

- Manhattan distance $\mathcal{M}_d$ among different data points is computed as:

$$\mathcal{M}_d = \sum_{i=1}^{x} |(a_i - b_i)| \tag{17.4}$$

- Minkowski distance M_d among different data points is computed as:

$$M_d = \left(\sum_{i=1}^{x} |(a_i - b_i)|^y \right)^{\frac{1}{y}} \tag{17.5}$$

Example: Let us consider a dataset consisting of two classes of objects: class M (solid squares), and class N (solid circles). We apply KNN for splitting these data into their corresponding classes. Consider an unknown data point (question mark in Figure

17.4) that needs identification of the cluster to which it belongs. Initially, we set $k = 3$, which indicates that we need 3 nearest neighbors for the classification process. The data points with distance d_1, d_3, and d_4 are the nearest neighbors to the unknown data point. As two of them are within class M, the unknown data point belongs to this cluster. Similarly, if we set $k = 5$, the neighbors located at distances d_1, d_2, d_3, d_4, and d_5 are nearest to the unknown data point; as three of them are within class M, the unknown data point is classified as an object of class M.

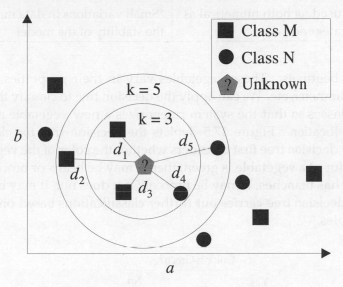

Figure 17.4 Example of KNN algorithm

17.2.2 Decision tree

The decision tree is a simple and powerful tool in ML that uses trees for determining its course of action. A tree may contain one or more branches, where each of these branches caters to a specific decision. It consists of two types of nodes: decision and leaf nodes. A decision node directs to a choice in the system, whereas a leaf node results in the final classification. The advantages and disadvantages of decision trees are listed in Table 17.2.

Steps for Decision Tree:

(i) List all the possible decisions to be made for a given problem, based on the attributes.

(ii) Determine all the possible events that may occur after making a decision.

(iii) Expand the tree, considering steps (i) and (ii), until it reaches the leaf representing the final class.

Example: Let us consider a bucket holding different types of vegetables. For simplicity, let us assume that the bucket contains broccoli, beans, tomatoes, red chilies,

Table 17.2 Advantages and disadvantages of decision tree

Advantages	Disadvantages
*Easy to visualize	*May cause over fitting if the data contain noise
*Data preparation can be done with less effort	*Proper selection of attributes of data is very essential
*Can be used for both numerical as well as categorical data	*Small variations in data may affect the stability of the model

eggplants, and beetroots. These vegetables vary in their properties, such as color, shape, length, diameter, etc. We can apply the decision tree to classify these vegetables into different classes so that the system categorizes a new vegetable and places it in the designated location. Figure 17.5 depicts the decision tree for classifying these vegetables. The decision tree first considers whether the color of the vegetable is green or not. If the color of a vegetable is green, then it may be beans or broccoli. Moreover, if the vegetable has branches, it may be broccoli; if it does not, it may be beans. In the same way, the decision tree carries out further classifications based on the properties of these vegetables.

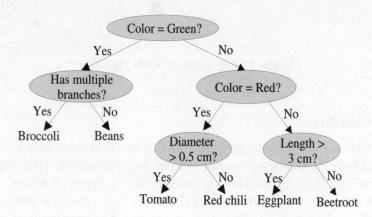

Figure 17.5 Example of a decision tree algorithm

17.2.3 Random forest

Random forest is another popular algorithm in ML, which is similar to the decision tree. This algorithm uses multiple decision trees to reduce the possibility of overfitting. The decision trees are created during the training phase; the outcome from the majority of the decision trees becomes the final decision of a random forest. Table 17.3 presents the advantages and disadvantages of the random forest algorithm.

Table 17.3 Advantages and disadvantages of random forest

Advantages	Disadvantages
*Can be used for both the classification and regression as it is based on the decision tree	*May increase the complexity as multiple decision trees need to be made.
*Provides better accuracy than decision tree in the presence of large dataset.	*Interpretation is difficult due to the presence of multiple decision trees

The random forest algorithm follows similar steps as the decision tree while creating the trees, considering multiple parameters (properties) from a given dataset. Further, it uses outputs from these trees to make final decisions.

Example: Figure 17.6 depicts an example of a random forest, where we consider the same example of vegetable classification as for the decision tree algorithm. We observe that in Figures 17.6(a) and 17.6(b), two different decision trees are constructed to classify the vegetables. The random forest generates its final decision based on the outcomes of the two different decision trees.

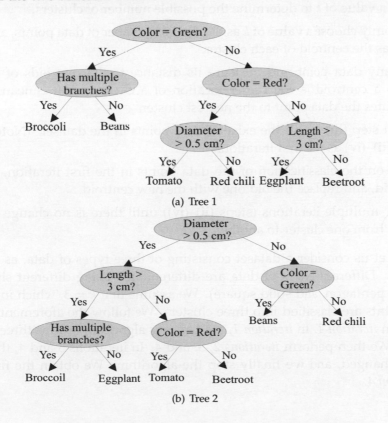

(a) Tree 1

(b) Tree 2

Figure 17.6 Example of random forest algorithm

17.2.4   *k*-means clustering

k-means clustering is an unsupervised ML algorithm used for clustering. This algorithm follows an iterative process for dividing the dataset into *k* distinct clusters. Typically, *k*-means clustering uses the Euclidean distance between a point and the cluster's centroid for its operations. Table 17.4 presents the advantages and disadvantages of *k*-means clustering.

Table 17.4 Advantages and disadvantages of *k*-means clustering

Advantages	Disadvantages
*Compact cluster as compared to that by the hierarchical clustering algorithm.	*Initial selection of the random point may affect the final results.
*Provides a faster computation, with a small value of *k*	*Difficult to decide the value of *k*

Steps for *k*-means Clustering:

(i)   Select a value of *k* to determine the possible number of clusters.

(ii)   Randomly choose a value of *k* as the distinct number of data points, and consider those as the centroid of each cluster.

(iii) Take any data point and measure its distance to the centroids of all clusters, where a centroid is the center location of a cluster. The *k*-means algorithm associates the data point to the nearest cluster.

(iv)   Repeat step (iii) for all the existing data points in the dataset. (Note: Consider steps (ii)–(iv) as the first iteration.)

(v)   Based on the classification of the data points in the first iteration, find a new centroid, and replace the old one with the new centroid.

(vi)   Go for multiple iterations (steps (ii)–(iv)) until there is no change in the data points from one cluster to another.

Example: Let us consider a dataset consisting of three types of data, as depicted in Figure 17.7. Different types of data are differentiated using different shapes (solid circle, solid pentagon, and solid square). We assume that $k = 3$, which indicates that the data points are classified into three clusters. We follow the aforementioned steps and perform *iteration 1*. In *iteration 1*, the *k*-means algorithm returns three clusters c_1, c_2, and c_3. We then perform *iterations* 2, 3, and 4. In iterations 3 and 4, the centroids remain unchanged, and we finally stop the algorithm. We obtain the final clusters from *iteration* 4.

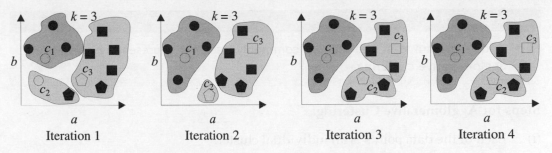

Figure 17.7 Example of k-means clustering

17.2.5 Agglomerative clustering

Agglomerative clustering falls under the category of hierarchical clustering algorithms. This clustering algorithm uses a bottom-up approach in the form of a tree. As the name suggests, it merges similar clusters to form bigger ones and finally stops by forming a single cluster. Agglomerative clustering algorithm are of three types, which are as follows:

- Single linkage: Considers the minimum distance between any two data points from two different clusters.

- Complete linkage: Considers the farthest distance between any two data points from two different clusters.

- Average linkage: Considers the average distance among all pairs of the data points.

Table 17.5 presents the advantages and disadvantages of agglomerative clustering.

Table 17.5 Advantages and disadvantages of agglomerative clustering

Advantages	Disadvantages
*All advantages of hierarchical clustering.	*All disadvantages of traditional hierarchical clustering
*Simple dendograms provide sufficient information about the clusters	*If a data point is assigned to a cluster in an initial phase, it is difficult to assign the data point in another cluster in the later phase.
*Simple and easy to implement.	

> Points to ponder
>
> - A dendrogram is a tree that is commonly used for representing clusters in agglomerative clustering.

Steps for Agglomerative Custering:

(i)   Each of the data points is an individual cluster.

(ii)   Considering all the points, form the distance matrix by computing the distance among them.

(iii)   Merge two closest clusters based on the existing distance matrix and update accordingly.

(iv)   Continue step (iii) until a single cluster forms.

Example: Let us consider a few data points, p_1, p_2, p_3, p_4, and p_5 as depicted in Figure 17.8. Each of the data points consists of coordinates X and Y, as shown in Figure 17.8(a). In order to create the clusters using these data points, we apply the agglomerative clustering algorithm. We use the single linkage to make the distance matrix table, as depicted in Figure 17.8(b). In this matrix table, we observe that p_1 is the closest to p_5. We marked the minimum distance 0.1, and consider that the data points p_1 and p_5 are in the same cluster. We depict the cluster formation in Figure 17.8(a) along with the formation of a dendrogram. Considering p_1 and p_2 to be in the same cluster, we create the next distance matrix table and observe that p_4 is the nearest data point of the cluster formed by the data points, p_1 and p_5. In the same way, we form the second cluster considering the data points p_1, p_5, and p_4. We continue the process of creating the cluster and dendrogram, considering the distance matrix table, until all the data points are in the same cluster.

17.2.6   Density-based spatial clustering of applications with noise (DBSCAN) clustering

DBSCAN is a clustering algorithm that works based on the density of the data points. This clustering algorithm belongs to the category of unsupervised learning. Typically, DBSCAN focuses on the data points that are close to a given data point and generates a cluster based on the density of the region.

The DBSCAN consists of two crucial components: (a) epsilon and (b) minimum points. Epsilon is typically represented by *EPS* and ϵ, which decides the considerable radius of a given data point. The minimum point is represented as *MinPts*, which represents the considerable minimum number of data points in a dense region.

In DBSCAN, the data points consists of three categories:

- Core points: These points consist of the number of neighbor data points more than or equal to the value of *MinPts*, within a specified ϵ.

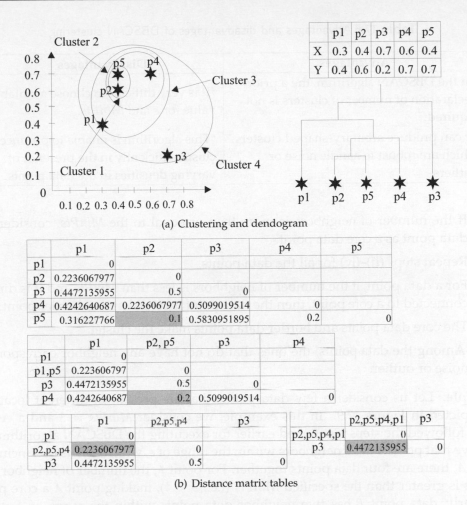

(a) Clustering and dendogram

	p1	p2	p3	p4	p5
X	0.3	0.4	0.7	0.6	0.4
Y	0.4	0.6	0.2	0.7	0.7

	p1	p2	p3	p4	p5
p1	0				
p2	0.2236067977	0			
p3	0.4472135955	0.5	0		
p4	0.4242640687	0.2236067977	0.5099019514	0	
p5	0.316227766	0.1	0.5830951895	0.2	0

	p1	p2, p5	p3	p4
p1	0			
p1,p5	0.223606797	0		
p3	0.4472135955	0.5	0	
p4	0.4242640687	0.2	0.5099019514	0

	p1	p2,p5,p4	p3
p1	0		
p2,p5,p4	0.2236067977	0	
p3	0.4472135955	0.5	0

	p2,p5,p4,p1	p3
p2,p5,p4,p1	0	
p3	0.4472135955	0

(b) Distance matrix tables

Figure 17.8 Agglomerative learning

- Border points: Border points are those points that directly connects to a core point, but have the number of neighbor data points less than *MinPts*.

- Noise/outlier points: The points that are neither core nor border points are known as noise/outlier points.

The advantages and disadvantages of DBSCAN are shown in Table 17.6.
Steps for DBSCAN clustering:

(i)   Assign the values of ϵ and *MinPts*.

(ii) Consider a random data point and check the distance, d, between the data point and its neighbors.

(iii) Count the number of neighbors of the data point that are located at a distance below a pre-defined value ϵ, i.e., $d < \epsilon$.

Table 17.6 Advantages and disadvantages of DBSCAN clustering

Advantages	Disadvantages
*In the DBSCAN algorithm, the a priori declaration of number of clusters is not required.	*It is very difficult to choose a suitable value for ϵ and *MinPts*.
*It can produce arbitrary-shaped clusters, which are robust to handle noise or outliers.	*This algorithm is unable to produce a cluster efficiently in the presence of varying densities in the data points.

(iv) If the number of neighbors is more than or equal to the *MinPts*, consider that data point as a core data point.

(v) Repeat steps (ii)–(iv) for all the data points.

(vi) For a data point, if the number of neighbors is less than *MinPts* and it is directly connected to a core point, then the data point is considered as border point.

(vii) The core data points and border data points make the cluster.

(viii) Among the data points, the ones that do not have any neighbor correspond to noise or outlier.

Example: Let us consider a few data points that are present in different locations, as depicted in Figure 17.9. In this example, we consider *MinPts* = 3 and a certain ϵ. We followed the steps mentioned earlier for executing the DBSCAN algorithm. We observe that point *A* has 3 neighbors within the range of ϵ, which means that including point *A*, there are four data points together. For point *A*, the number of neighbor data points is greater than the specified *MinPts* (i.e., 3 < 4), making point *A* a core point. Similarly, data point *B* has two neighbor data points within the range of ϵ, which indicates that there are three data points, including point *B*. In this case, *MinPts* = 3, making data point *B* a core point too. Datapoint *C* connects to the core point *B*, and has less number of neighbor data points than *MinPts* within the range of ϵ, making point *C* a border point. However, point *D* does not have any neighbor within the distance of ϵ, making point *D* a noise/outlier.

Check yourself

Other clustering algorithms and their steps for execution

17.3 Performance Metrics for Evaluating ML Algorithms

In this chapter, we discussed different algorithms in ML. These ML algorithms may provide different levels of accuracy depending on their design and the dataset.

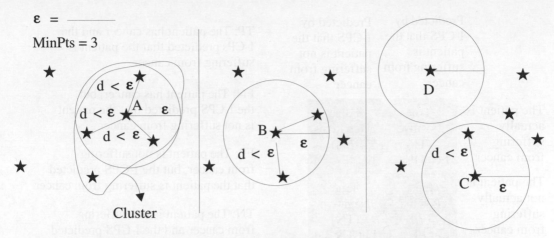

For point **A**, *MinPts* > 3 and for point **B**, *MinPts* = 3. Therefore, Points **A** and **B** are core points. For point **C**, *MinPts* < 3 and connected to core point, thus point **C** is a **border** point. Point **D** has no neighbor with ε, therefore, it is **noise/outlier**.

Figure 17.9 Example of DBSCAN clustering

Therefore, after developing an ML model, it is essential to evaluate its performance. Such evaluations help in improving the accuracy level. In this section, we will discuss different performance metrics to evaluate an ML model.

> Points to ponder
>
> - ML algorithms use part of the dataset for training the model; this part of the dataset is known as the training dataset. The ML algorithm uses the remaining part of the dataset as a test dataset for testing the trained model.
> - Artificial intelligence is primarily used for decision making, whereas machine learning is used for learning new things from a dataset.

The most common performance metrics for classification are (i) simple accuracy, (ii) precision, (iii) recall, and (iv) F-beta. Before discussing these metrics in detail, we must understand the *confusion matrix*. Let us consider the example of an IoT-based cancer prediction system (I-CPS), which uses a certain ML model. Let 10 patients visit the hospital, who possibly suffer from cancer. A group of doctors in the hospital use I-CPS to predict the status of cancer in those 10 patients based on the data from different physiological sensors. Now the prediction of I-CPS may either be correct or incorrect, depending on the sensors' data. Let us draw a confusion matrix based on the situation mentioned here. Figure 17.10 depicts the confusion matrix for the given situation.

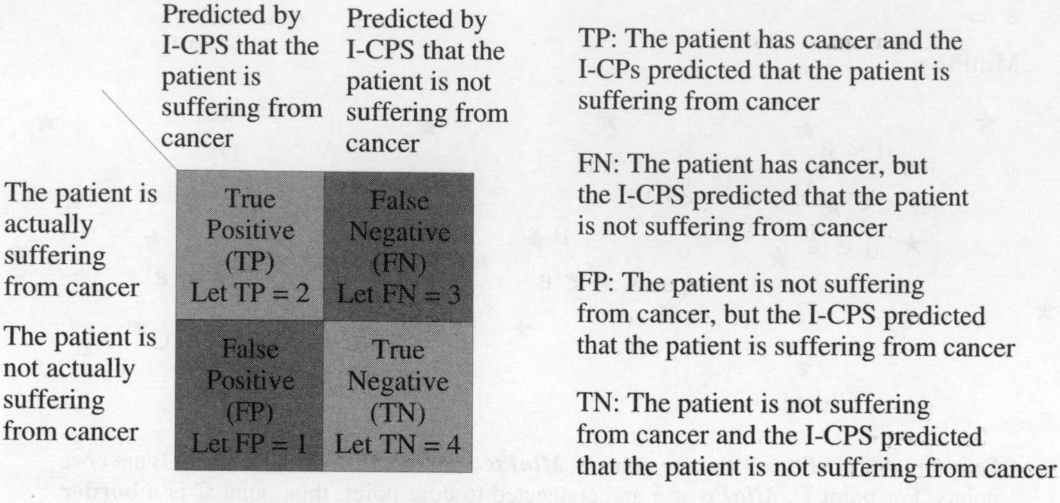

	Predicted by I-CPS that the patient is suffering from cancer	Predicted by I-CPS that the patient is not suffering from cancer
The patient is actually suffering from cancer	True Positive (TP) Let TP = 2	False Negative (FN) Let FN = 3
The patient is not actually suffering from cancer	False Positive (FP) Let FP = 1	True Negative (TN) Let TN = 4

TP: The patient has cancer and the I-CPs predicted that the patient is suffering from cancer

FN: The patient has cancer, but the I-CPS predicted that the patient is not suffering from cancer

FP: The patient is not suffering from cancer, but the I-CPS predicted that the patient is suffering from cancer

TN: The patient is not suffering from cancer and the I-CPS predicted that the patient is not suffering from cancer

Figure 17.10 Confusion matrix

(i) *Simple accuracy:* This is the ratio of the number of predicted samples divided by the total number of predicted samples. Further, we extend the simple accuracy in terms of percentage. We denote simple accuracy as $\mathcal{A}$. Therefore,

$$\mathcal{A} = \frac{Number\ of\ samples\ that\ are\ predicted\ correctly}{Total\ number\ of\ predicted\ samples} \tag{17.6}$$

Using the confusion matrix, simple accuracy is:

$$\mathcal{A} = \frac{TP + TN}{TP + TN + FN + FP} \tag{17.7}$$

According to Figure 17.10, $TP = 2$, $TN = 6$, $FP = 1$, and $FN = 1$. We derive the simple accuracy of the model used in the I-CPS as:

$$\mathcal{A} = \frac{2 + 4}{2 + 4 + 3 + 1} = 0.6 \tag{17.8}$$

Further, we convert the value of $\mathcal{A}$ into percentage as $0.6 \times 100\% = 60\%$.

(ii) *Precision:* Precision ($\mathcal{P}$) is the ratio of all the true positives (TP) divided by the summation of true positive (TP) and false positive (FP) outcomes of the model. In the particular example of I-CPS:

$$\mathcal{P} = \frac{TP}{TP + FP} = \frac{2}{2 + 3} = \frac{2}{3} = 0.67 \tag{17.9}$$

Similar to $\mathcal{A}$, we convert the value of $\mathcal{P}$ into percentage as $0.67 \times 100\% = 67\%$.

(iii) *Recall:* Recall ($\mathcal{R}$) is the ratio of true positive (TP) divided by the summation of true positive (TP) and false negative (FN).

$$R = \frac{TP}{TP + FN} = \frac{2}{2 + 4} = \frac{2}{6} = 0.33 \tag{17.10}$$

We convert the value of $\mathcal{R}$ into percentage as $0.33 \times 100\% = 33\%$

(iv) F_β: This performance metric is used when the performance of a model is difficult to measure using precision ($\mathcal{P}$) and recall ($\mathcal{R}$). Mathematically, we derive F_β as:

$$F_\beta = \frac{1}{\frac{\beta}{\mathcal{P}} + \frac{(1-\beta)}{\mathcal{R}}} \tag{17.11}$$

where β is a factor that gives importance to precision and recall. A greater value of precision provides greater importance to precision.

For the I-CPS, let $\beta = 0.3$. Then, we compute the F_β as:

$$F_\beta = \frac{1}{\frac{0.3}{67} + \frac{(0.7)}{33}} = 38.92\% \tag{17.12}$$

So far, we discussed the performance metrics for classification. Similarly, for regression, different performance metrics exist to measure the error of the data points with the regression line. In model fitting, typically, the datasets are divided into two parts: (a) training data and (b) test data. The training phase of the model is oblivious to the test dataset. In such cases, the consideration of *error* is essential. The error value is the difference between the observed and estimated values in a model. In this chapter, we discuss a few common metrics of regression: (i) Sum of square error (SSE); (ii) mean absolute error (MAE) (iii); root mean square error (RMSE); and (iv) mean square error (MSE)

(i) *Sum of Squares Error (SSE):* This gives the difference between the actual data points and the mean of all the data points. Mathematically, SSE is:

$$SSE = \sum_{i=1}^{n} (y_i - \bar{y})^2 \tag{17.13}$$

(ii) *Mean Absolute Error (MAE):* MAE uses the absolute difference between the actual and predicted data points. Mathematically,

$$MAE = \frac{1}{n} \sum_{i=1}^{n} |y_i - \bar{y}| \tag{17.14}$$

(iii) *Root Mean Square Error (RMSE):* This error computes the total standard deviation of all the errors in a model. Mathematically,

$$\text{RMSE} = \sqrt{\frac{1}{n} \sum_{i=1}^{n} (y_i - \bar{y})^2} \qquad (17.15)$$

(iv) *Mean Square Error (MSE):* This error is similar to SSE and MAE. However, MSE is derived as the square of the difference between the absolute and predicted values in a model. Mathematically,

$$\text{MSE} = \frac{1}{n} \sum_{i=1}^{n} (y_i - \bar{y})^2 \qquad (17.16)$$

Summary

In this chapter, we discussed the common analytical tools used in IoT. This discussion will help a learner understand the advantages and challenges of ML in IoT. The chapter provides a basic idea and an introduction to different types of ML algorithms with examples. Finally, we discussed different performance metrics for evaluating the ML algorithms.

Exercises

(i) What is machine learning (ML)? Why do we use ML?

(ii) What are the major challenges in ML?

(iii) What are the types of ML?

(iv) Compare supervised and unsupervised learning based on basic definition, type of data used, and types of problems handled.

(v) List the differences between *k*-means and KNN.

(vi) What are the basic performance metrics used for ML?

Conceptual Questions

Introduction

This portion of the book spans the contents of the entire book. It is designed to provide readers with an idea about the various concepts required for utilizing IoT in real life. We present, here, various conceptual questions, which one comes across in real life; they can be answered using the concepts, protocols, and methodologies covered in the previous chapters of this book. This portion also provides an informal guideline for teachers and instructors to formulate questions relating to IoT and its technologies.

The questions provided in here are significantly different from the ones covered at the end of each chapter. They require the reader to delve deeper into the concepts and fundamentals of the covered technologies, which will require some effort on the reader's part. At times, the readers will have to consult other books, online resources, and materials. This exercise will enable them to explore, beyond the introductory level, the various concepts covered in this book.

Questions

Q1 What are the roles of end-users and the service provider in cloud computing?

Q2 What is a hypervisor?

Q3 Let there be a task of 480 TB from a host machine that needs to be processed in the cloud. The service provider has three data centers (DC), DC_1, DC_2, and DC_3, which are capable of processing the task. The following table describes the capacity of different data centers: To which of the DCs do you think it best to allocate the task? Justify your answer.

Q4 Let there be four fog nodes F_1, F_2, F_3, and F_4, connected with several IoT devices. The data generated from the IoT devices can be processed either in single or multiple fog nodes. The fog nodes, F_1, F_2, F_3, and F_4, can accommodate up to 20 GB, 90 GB, 28 GB, and 46 GB respectively. Further, these fog nodes, F_1, F_2, F_3,

Table CQ.1 Data Center Configurations

Parameters	DC_1	DC_1	DC_2
Available storage (TB)	1043	560	1095
Processing speed (Gbps)	875	950	865
Distance from the host (km)	380	101	356

and F_4 take 12 μs; 18 μs; 28 μs and 22 μs to complete 10 GB of task. A cloud architecture is also available to execute different tasks; it can process 20 GB of task in 8 μs. An IoT device requires to process 246 GB of data. The time taken to transmit data packets to the cloud is longer than the time taken to transmit data packets to fog nodes. However, the device is associated with a critical application. Assume that the device can split the data into multiple segments and the segments can be assigned among the fog nodes and the cloud for processing. For the given scenario, compute the total time required to process 246 GB of data.

Q5 The power required to transmit a bit in Zigbee is 180 μW/bit. An implementation has 200 Zigbee slave nodes (where each node is attached to four sensors), each of which periodically transmits data to a master node at intervals of 120 seconds. The packet from each node consists of a 64-bit node identifier and a 64-bit destination identifier besides the 8-bit sensor values and 8-bit sensor identifier from each sensor. How much power will be consumed by the implementation (except at the master node) for 1 hour of operation.

Q6 The power required to transmit a bit in Zigbee is 180 μW/bit. An implementation has 20 Zigbee slave nodes (where each node is attached with two sensors), each of which periodically transmits data to a master node at intervals of 2 seconds. Consider that the packet from each node consists of a 64-bit node identifier and a 64-bit destination identifier besides the 8-bit sensor values and 8-bit sensor identifier from each sensor. If the slave node transmit rate is changed to 1 packet per 10 seconds, calculate the savings in energy for 1 hour of operation.

Q7 A Zigbee-based processor generates 1 million symbols per hour. This processor uses a 915 MHz band with a data rate of 250 kbps, using binary phase shift keying (BPSK) with an allocation of 4 bits per symbol. Calculate:

- Baud rate
- Time taken to transfer 1 million symbols using the calculated baud rate.

Q8 Each transmission slot in Bluetooth is 0.625 μs long. A voice data of type HV3 (HTML viewer 3) takes one slot every 3.75 ms (6th slot) for transmission and consists of 30 bytes of data with no error correction. Calculate the time taken to transmit 300 bytes of voice data over this configuration.

Q9 Each transmission slot in Bluetooth is 0.625 μs long. A voice data of type HV2 takes one slot every 2.5 ms (4th slot) for transmission and consists of 20 bytes of data with 10 bytes of forward error correction code. Calculate the time taken to transmit 1.5 kB of voice data over this configuration.

Q10 Each transmission slot in Bluetooth is 0.625 μs long. A voice data of type HV1 takes one slot every 1.25 ms (2nd slot) for transmission and consists of 10 bytes of data with 20 bytes of forward error correction code. Calculate the time taken to transmit 1.2 kB of voice data over this configuration.

Q11 An active NFC (near-field communication) device has a carrier amplitude (A) of 2 mV and a modulated ASK (amplitude-shift keying) wave of peak amplitude (k) 0.2 mV. The device has a data rate of 212 kbps, and employs Manchester encoding. Calculate the % modulation index of this active NFC device's signal.

Q12 A passive NFC device has a 10% ASK, IEEE 802.30 Manchester encoded data rate of 106 kbps. It receives an 8-bit data in the form 10101100. Draw the digital waveform.

Q13 Thread is based on the IEEE802.15.4 2006 specifications. It transmits 127 byte packets with a data rate of 250 kbps over the 2.4 GHz band. Out of the 127 bytes, the payload occupies a maximum of 63 bytes. A Thread router alternatively transmits sensor values of 9 bytes and 40 bytes per second to a border router. Considering ideal transmission and full throughput, what is the total volume of data transmitted through this Thread network in an hour?

Q14 Considering the scenario mentioned in Q13, is the Thread's data-carrying capacity exceeded or underutilized? In either case, by how much?

Q15 Thread is based on the IEEE802.15.4 2006 specifications. It transmits 127 byte packets with a data rate of 250 kbps over the 2.4 GHz band. Out of the 127 bytes, the payload occupies a maximum of 63 bytes. Considering 20 end-devices connected to the Thread network, where each device sends 90 byte packets after every 0.5 seconds to the thread router, calculate the incoming data volume at the border router in an hour.

Q16 Considering the scenario mentioned in Q15, what is the data volume in the network up to the border router in an hour?

Q17 Considering the scenario mentioned in Q15, how many additional devices of the same configuration can be supported in the same network for the same duration?

Q18 Differentiate between the pros and cons of FHSS (frequency-hopping spread spectrum) and DSSS (direct-sequence spread spectrum). What are the advantages of using FHSS in ISA100.11A?

Q19 A certain industrial application requires the use of a wireless solution for networking sensors deployed in a factory floor. The major requirement of the application is to ensure interference-free communication. Out of the two available options, WirelessHART and ISA100.11A, which one is more suitable for this scenario and why?

Q20 A certain industrial application requires the use of a wireless solution for networking sensors deployed on a factory floor. The major requirement of the application is to ensure that all of its 5,694,967,200 devices are uniquely addressable without using any subnets. Which protocol out of WiFi, Bluetooth, ISA100.11A, and WirelessHART are suitable for this application? Justify.

Q21 For the given signal sequence, draw the O-QPSK (quadrature phase shift keying) waveform: 1100101011.

Q22 How does the concept of contention window help in avoiding collisions in the IEEE802.15.4 protocol?

Q23 How is GFSK (Gaussian frequency shift keying) operationally different from BFSK?

Q24 While designing a communication protocol for a resource-constrained device, which of the two modulating schemes, GFSK and BFSK, will be more adept for the scenario in terms of processing involved?

Q25 While designing a communication protocol, the following are the requirements:

- It should be able to send a large number of signals simultaneously.
- It should not depend on sender–receiver synchronization

Which multiplexing technique would be suitable for such a scenario?

Q26 How do inductive coupling and backscatter coupling define the range of RFID (radio-frequency identification) communication?

Q27 A smart–home installation requires high-bandwidth applications for monitoring intruders visually. The installation requires the use of multiple wired and wireless cameras integrated into a central hub within the home. Which network connectivity solution would be suitable for such an installation? Justify.

Q28 A media streaming IoT device requires wireless connectivity between an audio player and four speakers. The requirements of this device setup emphasize audio quality and resilience to interference, all the while maintaining low requirements of power. Suggest a connectivity protocol for such a setup and justify your answer.

Q29 A smart home installation requires multiple sensors to be connected to a home-based hub. The minimum connection distance between the devices (if in a mesh configuration) or between the device and the hub (if following a direct connection) has to be 60 m. Out of Zigbee or Z-Wave, which one is suitable for this setup? Why?

Q30 An IoT-based sensor network installation in a smart home requires the integration of 40,000+ devices wirelessly. The aspect of low power consumption is also crucial in such a scenario, along with the requirement for inbuilt encryption of the data being transmitted over the network. Which wireless connectivity solution is suitable for this installation? Why?

Q31 A smart warehouse installation requires an IoT-based system for cost-effective and low-power asset tracking and location management on its operating floor. The distance between the tracer and the items to be tracked should be in the range of 1 m to 10 m. Which wireless connectivity solution is best suited for this task? Justify.

Q32 Can RFID systems be used for tracking assets spread over an area of 800×800 m^2? If yes, which type of RFID antenna and arrangement is best suited for this task?

Q33 An IoT-based smart payment system requires enhanced security measures, all the while being contactless. Out of RFID and NFC, which solution best fits this application?

Q34 An IoT-network implementation requires connectivity solutions that can communicate over distances greater than 800 m. The installation needs to be low power and independent of cellular connectivity requirements and should not incur operator costs (cellular or otherwise). Considering low bandwidth requirements, which protocol best suits such requirements?

Q35 An IoT-based network installation requires extensive communication ranges. However, you cannot install your own base station, and your area of operations is limited to where the stations are set up. The data rates of the connectivity solution is not a restrictive factor. However, data encryption over the channel is a must. Out of Sigfox and NB-IoT, which connectivity solution best meets the requirements of such an installation?

Q36 According to MQTT (message queuing telemetry transport) protocol specification available from IBM:

- MQTT connection uses approximately 86 bytes of data (CONN and CONNACK).

- In a "send and reply" scenario (i.e., an API call from the device to the cloud): There will be a PUBLISH from the device to the cloud that will use 15 bytes of overhead (PUBLISH). There will also be a PUBLISH from the cloud to the device with the reply that uses another 15 bytes of overhead (PUBLISH).

- In a "server topic" scenario of pushing data, there will be a PUBLISH from the device to the cloud that will use 15 bytes of overhead (PUBLISH). If the QoS (quality of service) is set to 1, there will be a PUBACK from the cloud to the device with 2 bytes of overhead (PUBACK).

- MQTT defines a "heartbeat interval," and at this interval, the client will send a PINGREQ packet to the server. The server will, in turn, send a PINGRSP. Each of these packets is 2 bytes of overhead. The interval for the heartbeats is configurable and affects the delay after a network outage that the client and server detect as a communication error. If you set a heartbeat interval of 24 hours, you will only send one heartbeat per day, but in the event of a network outage, the client and server will not recognize the outage until the heartbeat message is sent. This may or may not be acceptable based on your application design.

Using the aforementioned information, answer the following:

A. Estimate the daily device to cloud data overhead (from the device) of MQTT only, if 100 properties are transmitted per hour. Consider the heartbeat to be set at 2 per day. Convert the final estimate in kB.

B. Estimate the daily cloud to device data overhead for MQTT if 10 properties are transmitted per hour. The heartbeat is set to one per hour. Convert the final estimate in kB.

C. Estimate the monthly power consumption (30 days) if in a scenario, 500 properties are transmitted per hour (device to cloud), with a heartbeat of 1 per hour. Each byte generated consumes 20 μW of power.

Q37 An IoT application requires the use of connectivity solutions which can talk to other IEEE802.15.4-based WPANs (wireless personal area network) as well as to IP-based networks. Which technology can be adopted for such an application?

Q38 An IoT-based application requires extended device usage over very long periods of time. The network should be able to talk to IEEE802.15.4 WPANs. Out of 6LoWPAN and Zigbee, which one is more suitable for such a scenario?

Q39 Out of RPL and LOADng, which one is designed to have low overheads and better scalability in terms of dense networks?

Q40 How is mDNS (multicast domain name system) functionally different from the traditional DNS mechanism?

Q41 Which networking mechanism is used in zero configuration networks for resolving domain names to IP addresses?

Q42 An IoT network needs resolution of domain names to IP addresses. However, it should follow a flat file structure, should work with both link-local as well as global IP addresses, and have a UDP (user datagram protocol) packet size greater than 512 bytes. Which network mechanism is suited for such a scenario?

Q43 An IoT network requires communication strategies that provides improved congestion control and forward error correction. Additionally, the factors of faster client–server handshaking and connection multiplexing are a must for the network. Which communication protocol is best suited for such a scenario and why?

Q44 The main limitation of traditional IoT networks is that it is based on an IP-based Internet architecture, which is highly host oriented. What can be a more robust alternative for IoT scenarios? Justify.

Q45 How can a TCP (transmission control protocol)/IP stack be included on a 8/16-bit microcontroller? Please note that these microcontrollers are highly constrained in terms of processing and space requirements.

Q46 What are the salient features of nanoUDP that make it different from traditional UDP?

Q47 What are the salient features of nanoTCP that make it different from traditional TCP?

Q48 An IoT network requires a communication mechanism that is capable of synchronous as well as asynchronous communication. The mechanism should support both request–response, as well as publish–subscribe models. Which protocol would be best suited for such a mechanism? Justify

Q49 An IoT network requires a communication mechanism that is capable of asynchronous communication and follows a publish–subscribe model. The minimum level of application reliability is 3, and security is not an issue. Out of the two, CoAP and MQTT, which protocol is best suited for such a mechanism?

Q50 An IoT network requires a communication mechanism that is capable of asynchronous communication, and follows a publish–subscribe model. The minimum level of application reliability is 2, and security is not mandatory. However, the suitability of the chosen protocol for LLN nodes (in the tune of thousands) should be excellent. Out of the two, CoAP and MQTT, which protocol is best suited for such a mechanism?

Q51 An IoT web-APi requires support for WebSocket addressing and WebSocket security; it follows an XML-based structure. Which protocol is best suited for such a scenario?

Q52 Which web-API can support a logical tree format, deviating from the traditional bitstream format?

Q53 What are the salient features of the REST (representational state transfer) architectural style?

Q54 An IoT-based communication protocol is to be selected for an application which should provide reliable queuing, topic-based messaging, follow a publish–subscribe model, have flexible routing, transactions and security. Which protocol is most suited for such an application? Justify.

Q55 An IoT-based communication protocol is to be selected for an application which should provide topic-based messaging, and follow a publish–subscribe model. However, the protocol should work with resource-constrained and low-bandwidth networks. Which protocol is best suited for such an application? Justify.

Q56 An IM application requires a persistent connection over TCP while communicating with clients (on smartphones) and a server. Which protocol is suitable for such an application, considering that XML messages might need parsing and should use open-source databases?

Q57 Suppose you are a multiplayer online game developer. You have the following requirements for your game:

(a)   Peer to peer implementation of "bots" or "agents"
(b)   Unrestricted "bot" to "bot" communication
(c)   File transfers that may be characterized as big files
(d)   No roster management

You have the choice of using either of the two following protocols: XMPP and AMQP. Which one would you choose? Justify your selection with the pros and cons of each of these protocols, and show the superiority of your chosen protocol for the designated task.

Q58 The design of an online application requires handling long-lived XML stream with a single XML parser instance, which means there is no need to instantiate a new XML parser every time while processing a message. Which protocol is best suited for this task?

Q59 An online application requires the ability to handle up to 5000 users at any given time. The client should not be javascript-dependent, and can work on both android as well as iphone clients. The data is typically in the form of binary streams. Which protocol is suited for such an application?

Q60 Discuss the pros and cons of choosing EPC or uCode for tagging objects.

Q61 Differentiate between URIs and URLs. Is there any difference between the two based on performance?

Q62 A city-wide IoT deployment requires device management, which should include the features of provisioning, device configuration, software upgrades, and fault management. The devices are constrained with limited bandwidth, memory, storage, and processing. Which protocol is suitable for achieving this task? Discuss

Q63 How is the Web Thing model different from the Physical Web?

Q64 Construct a decision tree to classify all possible triangles and quadrilaterals. Consider trapeziums, squares, rectangles, rhombuses, parallelograms, equilateral triangles, isosceles triangles, and scalene triangles.

Q65 A mobile phone company produces 50 mobile phones in 12 hours. In the company, there is a testing system, which checks whether a mobile phone is correctly produced or not, i.e., if there is any fault in a produced mobile phone from the company. The testing system is made up of an ML algorithm. The performances of the testing system are listed as follows:

(a)   8 mobile phones were produced not faulty, but the testing system detected them as faulty.

(b) 32 mobile phones were not faulty and the testing system detected them as not faulty.

(c) 3 mobile phone were faulty and the testing system detected them as faulty

(d) 5 mobile phone were faulty but the testing system detected them mobile phones as not faulty.

Based on the aforementioned statements, construct the confusion matrix and find:

(i) Accuracy of the testing system

(ii) Precision of the testing system

(iii) Recall of the testing system

(iv) $F\beta$ of the testing system

Q66 There is a set of vehicles, which includes three classes: car, truck, and bi-cycle. The ith car, jth truck, and kth bicycle are denoted by C_i, T_j, and B_k. Figure CQ.1 represents different coordinates of the vehicles. An unknown vehicle, p, is given to you with a coordinate $5, 5.5$. Using KNN, determine the class of point p. Assume $k = 5$.

Vehicles	C_1	C_2	C_3	C_4	C_5	T_1	T_2	T_3	T_4	T_5	T_6	B_1	B_2	B_3	B_4	B_5
x-axis	2.5	3.5	4.5	4.5	3	2.5	6.5	6.5	7.5	3	4.5	6	6.5	7.5	5	5.5
y-axis	2	3.5	4.5	2	6.5	4.5	5.5	2.5	3.5	2	7	7	4.5	6.5	3.5	5.5

Figure CQ.1 Coordinates of the vehicles

Q67 Figure CQ.2 represents the coordinates of 10 data points. If the value of k is 3, construct the clusters using k-means algorithm. Assume $k = 3$.

Points	P_1	P_2	P_3	P_4	P_5	P_6	P_7	P_8	P_9	P_{10}
x-axis	2	3.5	1.5	3	1	2	2	4	4.5	3
y-axis	5	5	3	3	2	2	1	4.5	4	2

Figure CQ.2 Coordinates of the points

Q68 Considering the data points depicted in Figure CQ.3, apply agglomerative clustering algorithm to construct the dendogram. Create the entire distance matrix table.

Q69 There is an IoT network consisting of a few IoT devices, fog nodes, and cloud. The available bandwidth to transmit a frame from the IoT to fog devices is 20

Points	P_1	P_2	P_3	P_4	P_5	P_6
x-axis	2	3.5	3.6	4.2	1.1	3.5
y-axis	3.2	1	2.1	2.7	5.1	3.9

Figure CQ.3 Coordinates of the points

Mbps, which is able to pass 18000 frames in two minutes. Each of these frames can carry a maximum of 16000 bits. Find the throughput of the transmitting channel.

Q70 In a network, there are m number of IoT devices and n number of fog devices. Two devices d_1 and d_2 produce data of 32 MB and 39 MB per minute, respectively. As d_1 and d_2 have limited computation capabilities, these devices select fog nodes to process their data. Devices d_1 and d_2 choose f_1 and f_2 to transmit the data through different medium. The propagation speed of these medium are 1.4×10^4 m/s and 1.45×10^4 m/s, respectively. It takes 60 ms for a data packet to travel from d_1 to f_1 and 50 ms to travel from d_2 to f_2. Find the distance between d_1 and f_1 and d_2 and f_2.

Q71 Two fog nodes are situated at a distance of 20 m; the time required to transmit a data packet from one fog node to another is 5 ms. Find the propagation speed.

Q72 Let different IoT devices be wirelessly connected with a fog node. As multiple IoT devices exist in the network, a random access mechanism needs to be implemented. In such a scenario, which of the random access mechanism do you prefer to choose over carrier sense multiple access/collision dection (CSMA/CD) and carrier sense multiple access/ collision avoidence (CSMA/CA). Justify your answer.

Q73 A fog device has a main memory and a small cache memory. The effective access time to fetch data from the fog device is 15% greater than the cache access time. If the main memory and cache access time are 300 ms and 200 ms, respectively, compute the cache hit ratio percentage of the fog device.

Q74 As per the mechanism of an IoT network, every IoT device receives an acknowledgment from the fog nodes after the successful delivery of a message. Further, an IoT device waits 3 ms to receive the acknowledgment and if an acknowledgment is not received by the IoT device within 3 ms, it re-transmits the same message to the fog node. Now, considering the aforementioned scenario, an IoT device wants to transmit two consecutive messages to a fog node. The IoT device starts to transmit the first message, which is dropped during the transmission due to collision with the other packet. Consequently, the IoT device does not receive any acknowledgment from the fog node and re-transmits the same message. In the second attempt, the message is successfully delivered to the fog node and this time, the acknowledgment is dropped. Therefore, the IoT device waits for 3 ms and re-transmits the same message, but the fog device discards the message and sends the acknowledgment to the IoT device. This time, the IoT device receives acknowledgment after 2 ms. Now, the IoT device transmits the second message and it is successfully delivered to the fog node. The acknowledgment of the second message is received by the IoT device after 3 ms. Draw the time sequence diagram of the message transmission between the IoT device and the fog node.

Q75 A set of IoT devices is deployed over a region to monitor different environmental parameters such as temperature, rainfall, humidity etc. Each of these IoT devices has a certain communication range; these devices are battery-powered (you can consider this as the energy source). As the IoT devices have limited computation capabilities, these devices transmit the sensed data to the fog node for processing. However, a few IoT devices are directly connected to the fog node. Therefore, an IoT device, which is not directly connected with the fog node, uses intermediate IoT devices to deliver its sensed data packet. Let there be an IoT device, d_1 that wants to transmit a data packet to the fog node, but d_1 is not directly connected to the fog node. Therefore, d_1 has to select the best suitable IoT device, as intermediate hop, among its neighbor IoT devices to deliver the packet to the fog node. Please note that the neighbor IoT devices of d_1 are those that are within their communication range. Design a mechanism that helps d_1 to select the most suitable neighbor IoT device to forward the sensed data.

Q76 There is a cloud service provider (CSP), who provides SaaS, PaaS, and IaaS. Mr. X wants to avail SaaS by using two software "Xango Photo Editor (XPE)" and "Xango Video Editor (XVE)". The price of XPE is \$0.75/day, \$20/30 days, and \$200/year. For XVE, a user needs to pay \$1/day, \$28/30 days, and \$300/year. If Mr. X uses XPE for 18 days and XVE for 23 days, initially, and then subscribes XPE for 2 months and XVE for 1 month, how much does Mr. X have to pay for the services?

Q77 A sensor-cloud platform consists of 3 sensor owners: S_1, S_2, and S_3. The capability of a sensor owner is determined by the following three factors:

(a) Duration (D) of the sensor owner for which he/she is in the business.

(b) The number of different types (T) of sensor nodes, the sensor owner can provide to the SCSP

(c) The total number of sensor nodes (N), the sensor owner is able to lease to the SCSP

The table in Figure CQ.4 shows the values of D, T, and N for different sensor owners. Based on these values, which of sensor owners should the SCSP choose and why? Derive the mathematical formula to select the sensor owner. As per

Sensor	D	T	N
S_1	3	14	21
S_2	2	18	9
S_3	7	6	30

Figure CQ.4 Values of D, T, and N

policy of the platform, a sensor owner charges two types of rent: fixed and variable charges. The fixed charges depend on the capability of the sensor owner. The variable cost of the selected sensor owner, S_i, depends on the duration of the usage of sensor nodes owned by S_i. If sensor owner S_i lends three sensor nodes for x, y, and z hours with a rent of a, b, and c units per hour, then compute the total charged price by S_i.

Q78 The engineers of a bridge under construction decide that for the long-term safety and structural integrity of the bridge, it needs various IoT-based wireless sensor nodes embedded into the construction material itself. The only significant challenge to this solution is the lack of a power source to continuously power these nodes. Suggest a suitable protocol/technology which can enable this solution. The sensor nodes are expected to be functional forever; sleep-scheduling of these nodes is not a viable solution.

Q79 A smart building application requires the deployment of a network that can communicate over powerlines as well as RF. In case one mode of communication is jammed or disabled, the other may take over. Suggest an appropriate technology for this network deployment. Highlight the features of the chosen technology to justify your answer.

Q80 How much time would a 120 kB data (control/command/data) take to be transmitted over an X-10 network? Consider that no retransmissions occur.

Q81 Draw the encoded waveform for the sequence 11001010 prior to transmission over an Insteon network.

Q82 Two Insteon devices are separated by a distance of 121 m. Considering no obstruction in between these devices, calculate the time taken to transmit 2 kB of data.

Q83 A building network requires the integration of more than 50,000 devices across the whole building. The main requirements of the proposed network include 16-bit addressing and IP-network integration. Suggest a suitable technology for this task. Justify.

Q84 Can Arduino boards be programmed in C?

Q85 Can Arduino boards be programmed in Python?

Q86 How is the Arduino serial port different from its digital I/O ports?

Q87 Why is the baud rate important while integrating Arduino boards over serial communication links?

Q88 Design an Arduino program that uses a temperature and humidity sensor to trigger a servo motor after every 10 seconds, if the threshold for temperature crosses 40 °C and humidity is less than 30%. What is PWM (pulse width modulation) in Arduino?

Q89 What is I2C communication?

Q90 Differentiate between board and BCM modes of Raspberry Pi.

Q91 How can Raspberry Pi be made to read analog signals?

Q92 How can a sensor requiring 5 V for its operation be integrated with Raspberry Pi GPIO pins, which work on 3.3 V?

Q93 Which Arduino board is suitable for building a smart speaker that can detect voice commands and inform the user through pre-recorded speech sounds after actuating the necessary devices connected to it? Justify.

Q93 How does a piezoelectric sensor work?

Q94 A silicon-based photocell has a current output of 15 **mA/cm**^{2} and an efficiency of 18%. Calculate the total current generated for a series connection of such photocells for a total area of 100 m^2. Consider that the photocell array is under full illumination and no external losses occur.

Q95 Classify the type of error if a temperature sensor reads 102 °C for an actual temperature value of 102.1 °C.

Q96 A pressure sensor **A** reads 55.3 psi for an actual pressure reading of 55.2 psi. After a while, the same sensor reads 54 psi for the same pressure. Identify the category of error in this sensor.

Q97 What is the working principle of a micro-electromechanical systems (MEMS)-based accelerometer?

Q98 Which one would be a better choice: a digital temperature sensor or an analog one? Justify.

Q99 What is quantization error, and how is it avoided?

Q100 What are the salient characteristics of the CPS paradigm, which makes it different from M2M (machine to machine) or WSN (wireless sensor network)?